NANO AND MICRO ENGINEERED MEMBRANE TECHNOLOGY

Membrane Science and Technology Series

Membrane Science and Technology Series, 10

NANO AND MICRO ENGINEERED MEMBRANE TECHNOLOGY

C.J.M. van Rijn

Aquamarijn, Micro Filtration B.V.
Beatrixlaan 2, 7255 DB Hengelo, The Netherlands

2004

ELSEVIER

Amsterdam - Boston - Heidelberg - London - New York - Oxford
Paris - San Diego - San Francisco - Singapore - Sydney - Tokyo

ELSEVIER B.V.
Sara Burgerhartstraat 25
P.O.Box 211, 1000 AE
Amsterdam, The Netherlands

ELSEVIER Inc.
525 B Street, Suite 1900
San Diego, CA 92101-4495
USA

ELSEVIER Ltd
The Boulevard, Langford Lane
Kidlington, Oxford OX5 1GB
UK

ELSEVIER Ltd
84 Theobalds Road
London WC1X 8RR
UK

First edition 2004

Library of Congress Cataloging in Publication Data
A catalog record is available from the Library of Congress.

British Library Cataloguing in Publication Data
A catalogue record is available from the British Library.

ISBN: 0-444-51489-9
ISSN: 0927-5193

⊗ The paper used in this publication meets the requirements of ANSI/NISO Z39.48-1992 (Permanence of Paper).
Printed in Hungary.

Preface

This book is about membrane science and nano-micro engineering techniques, merging into one new exciting technology referred to as 'Nano and Micro Engineered Membrane Technology'. New tools are discussed in order to design and fabricate advanced nano-micro engineered membranes, looking very much like the model membrane structures as originally depicted in many textbooks on membrane technology, but which up to now were not easy to realise. Applications cover many fields of science, such as nano-micro filtration, gas separation, optics and photonics, catalysis, microbiology, controlled drug delivery, nano-patterning, micro contact printing, MEMS, atomisation, cross-flow emulsification etc.

The formation of micro and nano pores in a thin but strong membrane structure and its potential applications is the key element in this book. A brief overview of classical methods to produce filtration membranes with tortuous pore structures (pore size 1 micron - 1 nm) is presented in chapter 1 and in the subsequent chapter 2 an overview is presented of conventional micro perforation methods for obtaining well defined pore sizes, typical larger than 5 micron with use of laser drilling, electroforming, precision etching etc. With micro engineering techniques (chapter 3), originating from the semiconductor industry, it is relatively easy to downscale well defined pore sizes to form e.g. micron and submicron pores (5 micron-100 nm) using photolithographic methods, with e.g. contact masks and wafer steppers. Also described in detail in this chapter is the formation of sub-100 nm pores with use of a laser-interference lithographic technique. This method may be considered low-cost and is applicable for large surfaces.

In chapter 4 some elementary fluid mechanics related to fluid flow in conducts and single and multiple orifices is presented covering analytical methods as well as computational fluid dynamics. Some new practical rules of thumb are derived for fluid flow in circular or slit shaped orifices in micro sieves. Much effort has been put in strength and maximum pressure load analysis (chapter 5) of perforated and unperforated membranes. New analytical expressions were obtained that were verified by a number of computer simulations and many experiments.

A separate chapter (chapter 6) has been devoted to the pioneering work of manufacturing polymeric perforated membranes because of its potential future economical impact. Phase separation micro moulding is introduced as a new method, not only to produce perforated membranes, but also to enhance the effective membrane surface area of classical flat sheet and capillary filtration membranes through micro structuring the membrane surface with high aspect ratio structures.

Large scale microfiltration applications on e.g. skim milk and lager beer are presented in chapter 7 with use of micro engineered membranes, whereas in chapter 8 a micro scale Lab-on-a-Chip microfiltration/fractionation demonstrator is discussed.

Nanotechnology and nano engineered membranes is the fascinating topic of chapter 9, with typical examples as nanopatterning, nanophotonics and nanomembrane technology.
Nanotechnology is likely to require an approach to fabrication different from that of microtechnology. Whereas microscale structures are typically formed by top-down techniques such as patterning, deposition, and etching, the practical formation of structures at nanoscale

dimensions will probably involve an additional component, i.e. bottom-up selfassembly, a process whereby structures are built up from atomic or molecular-scale units into larger and increasingly complex structures, analogous to biological systems. As our capabilities expand, some combination of top-down (lithographic) and bottom-up (including self-assembly) techniques probably will be employed for the efficient manufacturing of future membrane structures and systems.

This book closes with novel pioneering applications on atomisation (chapter 10) for deep pulmonary inhale and cross flow emulsification (chapter 11) for the manufacturing of e.g. functional foods and nano/micro emulsions.

The near future of this new cross-road technology will be on the validation of new and presented applications and on the quest for finding alternative methods to produce nano and micro engineered membranes.

George M. Whitesides
Mallinckrodt Professor of Chemistry
Harvard University
Cambridge, USA

Author's Note

Being a chemical physicist and having learned something about semiconductor technology at Philips I decided in 1993 to start a research company, Aquamarijn, on filtration membranes made with micro machining technology for leukocyte filtration. The first micro engineered membranes were manufactured on 4 January 1994 in the MESA Clean Room of the University of Twente and were made after four months harsh micro machining labour by me having not many practical Clean Room experience. The membrane (two inch diameter) was called a microsieve and had pores of 5 micron in diameter. Of course the membranes, if you know the trick, can also be made in much less time, it is one of the most simple micro engineered products, in comparison to e.g. air bag sensors, flow sensors, micro motors etc..

New things never stay new, so the technology had to be developed in a quick pace mainly by Wietze Nijdam (since 1995) and myself with enthusiastic support of the micro-mechanics research group of Jan Fluitman and Miko Elwenspoek in the mid nineties.

New technology is only successful if at a given moment others want to take over and want to do research themselves or bring developed products to the market. An example of this is a large medical company that learned from the Aquamarijn technology and also applied for patents (1998) on leukocyte filtration applications. The original patent application of 1995 by Aquamarijn on e.g. polymeric embossed microsieves and laser interference lithography for the manufacturing of nanosieves is as well general as specific and nearly all of the later research have shown that the ideas were feasible. A number of articles were published, but the PR profile of Aquamarijn was kept relatively low. In the late nineties luckily other scientific institutes embraced the pioneering work and started also fundamental research on microsieves (University of California, since 1998), hydrogen separation membranes supported by microsieve (Massachusetts Institute of Technology, since 1999) and biomedical nanosieves with 20-50 nanometer sized pores (University of Illinois at Chicago, since 2000). After the achievement of nanosieve structures (Nanotechnology 1998) several nano-oriented research groups from the University of Twente asked for our technological support on topics as nano stencilling and photonic crystals, because the use of a nano engineered membrane may sometimes be a prerequisite in their nano applications.

Funding is never easy and it took much effort and creativity to find suitable sources. Aquamarijn owes a lot of financial gratitude towards the Dutch STW organisation and the SENTER and NOVEM subsidy programs from the Dutch government. Without their loyalty and support the Dutch Research BV's, in particular Aquamarijn, would have a very hard time. Luckily it seems that the investment climate in micro/nano engineering in the Netherlands is changing in a positive way by more co-financing of venture capitalists of start-up companies with a fair back-up from the universities. Also the Dutch government of economic affairs (EZ) seems to gain more sympathy for micro and nano engineering technology and more incubating support will surely be welcomed by start-up companies with a mean grown-up time of at least five years.

I would like to thank all those who contributed to the ongoing adventure of nano and micro engineered membrane technology:

Miko Elwenspoek and Kees Eijkel (University of Twente) whom supported us technically and morally from the start in 1994 at the MESA$^+$ research institute.
Onno Raspe and Leo van der Stappen for their pioneering work on lager beer filtration at the 'Grolsche Bierbrouwerij' since 1996.
Cock Lodder and Gert Veldhuis (UT) for the initial collaboration in 1998 on laser interference lithography.
Janneke Kromkamp, Tjeerd Jongsma, Rolf Bos, Karel Horn, Marco Spits and Tjerk Gorter including the total Friesland Coberco Dairy Foods crew (The D-Force team) for their ongoing work on milk filtration and fractionating applications.
Rob Sillen, who enthusiastically supplied Aquamarijn with numerous qualified students of the Utrecht Micro Engineering Competence Centre, Hoge School van Utrecht.
Remko Boom, you will never leave him without a new idea, and his advanced 'food oriented' research group at the University of Wageningen.
Jo Janssen from Unilever Research Laboratories for the implementation of the microsieve technology in an ongoing large international research project on Emulsification processes (THAMES, Towards Highly Advanced Membrane Emulsification Systems).
Matthias Wessling, Geert Henk Koops, Jonathan Barsema, Laura Vogelaar, Miriam Girones and Lydia Versteeg from the Membrane Technology Group (UT) for the further research on microsieves and the just started pioneering work on phase separation micro and nano moulding technology, with a product potential that goes far beyond the scope of this book.
Jos Keurentjes and Marius Vorstman (University of Eindhoven) for the implementation of nano engineered membrane technology for hydrogen separation with palladium membranes.
Henri Jansen, Niels Tas, Dave Blank, Johannes Burger, Erwin Berenschot, Meint de Boer, Stefan Sánchez and Bert Otter (UT) for their scientific, technological and pictorial support.
Huub van Vossen, Johnny Sanderink, Gerrit Boom, Arie Kooy, Stan Krüger, Gerard Roelofs including other members of the MESA$^+$ pit-stop team.
Stein Kuiper who as an excellent craftsman performed the PhD job he was asked to do according to the STW research proposal on microsieves.
The DIMES institute at Delft, ASML at Veldhoven and Philips research laboratories at Eindhoven for giving adequate support, also in the challenging cases.
Alfred Colenbrander for editing and Kristan Goeting for making many tedious corrections in the manuscript.

I would like to thank all those fine researchers for their contributions to this book, with special thanks to: Tjeerd Jongsma, Rolf Bos, Remko Boom, Anneke Abrahamse, Ruud van der Sman, Gerben Brans, Albert van der Padt, Jurriaan Huskens, Marius Kölbel, Jürgen Brugger, Janneke Kromkamp, Karin Schroen, Marcel den Exter, Freek Kapteijn, Herman van Bekkum, Matthias Wessling, Geert Henk Koops, Stein Kuiper, Laura Vogelaar, Kobus Kuipers, Niek van Hulst, Tong Du Hien, Frank Gielens, Ben Mentink, Kristan Goeting, Jeroen Wissink, and last but not least Wietze Nijdam being also one out of a few ideal engineers.

The research described in this book was initiated and carried out by Aquamarijn Research, Friesland Coberco Dairy Foods Research Centre at Deventer, Grolsche Bierbrouwerij at Groenlo, MESA$^+$ Research Institute, Food Process Group of University of Wageningen, Membrane Technology Group of University of Twente, Onstream MST, Utrecht Micro Engineering Competence Centre, Chemical Process Group of University of Eindhoven, Medspray and Unilever Research Vlaardingen.

All readers are kindly asked to address their remarks, comments and questions to the author (ceesvanrijn@nanomi.com). Information: www.nanomi.com & www.microsieve.com

Dr. Cees J.M. van Rijn, August 2003.

CONTENTS

Chapter 1

Overview Membrane Technology

1. INTRODUCTION

Membranes play an essential role in nature but also in todays modern industrial society. Biological membranes have an essential function in the metabolism of all living species. Through exchange processes a single cell can be supplied with nutrients and it can excrete waste products through its outer cell-membrane. Moreover these exchange processes enable communication between cells, by exchange of e.g. hormones, which made the development of complex life forms possible.

 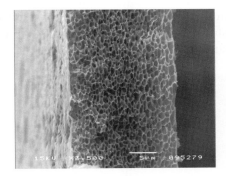

Fig. 1 SEM photographs: Left: a micro porous skeleton of a diatomaceous earth fossil. Diatom cells are enclosed by cell walls made largely of silica (SiO_2). This glassy shell, or frustule, consists of two tightly fitting halves often resembling a flat, round, or elongate box. The minute perforations allow dissolved gases and nutrients to enter and exit. Right: cross-section of a micro porous synthetic polyethersulfon membrane obtained with a phase inversion process. The membrane may have at least one outer skin layer with a mean pore size smaller than that from the inner part. (Courtesy Aquamarijn Research/Membrane Technology Group, University of Twente).

Considering that the first membrane experiments dated from the 18th century, using membranes (from Latin *membrana*, a thin sheet) of biological origin [1], industrial membrane separation with synthetic membranes has been strongly developed since the introduction of the asymmetric polymeric membrane [2] in the early sixties.

Most of the synthetic membranes produced using a phase inversion process have an inner sponge-like structure, which contributes to an additional fluid resistance and, a fortiori, a reduced operational product flux at a given transmembrane pressure. Also, the retention of particles, proteins and other coagulates is not only determined by the pores present in the skin layer, but also by the sponge support structure itself. Moreover in a few applications, such as clarification of beverages, cleaning agents must be removed completely after cleaning to prevent contamination of the beverage. Various attempts have been made to downscale sieve type filters, which do not have the afore mentioned shortcomings, such as the track-etched membranes and the Anopore membrane (see Fig. 14 and Fig. 15). However these latter membrane types lack the possibility of achieving a high porosity and/or adjusting freely the pore size.

Using nano and micro techniques originating from semiconductor technology it is possible to fabricate thin membranes with perfectly uniform pores [3]. The membrane or microsieve, was made using flat substrates, thin-layer deposition techniques, photolithography and high-resolution etching methods. Microsieves and nanosieves (with pore size < 100 nm) may be used for numerous filtration applications, but they can also be employed as a scaffold to study particles, microorganisms, lipid bilayers that are retained or adhered to the surface of the micro/nanosieve.

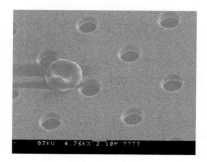 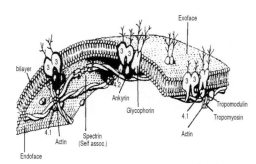

Fig. 2 Left: SEM photograph of a human blood cell retained on a microsieve membrane (Courtesy Aquamarijn Research). Right: Drawing of the lipid bilayer (cell membrane) of a human red blood cell, showing the single-pass transmembrane protein Glycophorin (Courtesy Microvascular Research [4]).

The screening of a wide variety of molecules through natural and artificial constructed membrane channels (porins) in lipid cell membrane layers is an increasing object of interest. A prerequisite for this is the construction of adequate micro and nano biotools fabricated using novel nano and micro engineering techniques.

Current biotechnological separation applications (e.g. food, beverage, bioreactors and pharmaceutical industry) are fast developing applications of artificial membranes. Also there are new membrane applications, such as affinity membranes (e.g. protein purification), biosensors (e.g. glucose sensors), biohybrid organs (e.g. artificial livers) and artificial tissue structures (e.g. artificial skin).

2. MEMBRANE PROCESSES[*]

In the chemical process industry one often encounters the problem of separating a mixture in its components. Membranes can in principle carry out most of the separation processes and can complement or form an alternative for chemical processes like distillation, extraction, fractionation, adsorption etc. Advantages of membrane filtration are among others: a low energy consumption, the separation can be carried out continuously, up-scaling is relatively simple and membrane technology can be used for almost any kind of separation.

Table 1

Some membrane separation processes arranged according to the mechanism of separation.

separation mechanism	membrane separation process
size exclusion (filtration)	nano filtration, ultra filtration, micro filtration
solubility/diffusivity	reverse osmosis, gas separation, pervaporation, liquid membranes
charge	electro dialysis

The success of using membranes is closely related to the intrinsic properties of the membrane. Interfacial interactions between membrane surface, surrounding environment and solutes govern membrane performance to a great extent. These interactions have considerable impact on transport characteristics,

---

[*] This section has been co-written with W. Nijdam, Aquamarijn Research.

selectivity, fouling propensity, bio- and hemo-compatibility of the membrane. Especially in biotechnological and medical applications, where highly adsorptive solutes such as proteins are present, and adsorption of proteins often results in fouling of the membrane, which lead to considerable losses in flux, selectivity and performance [5].

2.1   FILTRATION THROUGH SIZE EXCLUSION

Table 2

Filtration Processes with their properties and application.

FILTRATION PROCESS	PORE SIZE	SEPARATION CAPABILLITY	PRESS. (bar)	APPLICATION
Molecular Sieving	0.3-1 nm	Branched molecules versus unbranched molecules	0 - 3	Gas separation with zeolite molecular sieves
Nanofiltration	1 - 10 nm	Permeation of low molecular weight (200 – 20.000 daltons) substances	5 - 25	Purification of sugar from acids, salts from dyes, water treatment
Ultrafiltration	5-100 nm	Retention of viruses, bacteria's, dissolved substances with molecular weight between 10.000 – 500.000 daltons	0.5 - 5	Dairy industry, beverage industry, pharmaceutical industry, separation of water from crude oil, separation of fruit and vegetable extracts, waste water treatment
Microfiltration	50 nm-5µm	Retention of bacteria's, Colloids, protozoa (*Cryptosporidium, Legionella*)	0.5 - 3	Prefiltration in water treatment, sterile filtration, dye industry, beverage clarification, screening of bacteria

Molecular Sieves

In 1756 Cronstedt discovered certsin type of natural porous minerals, which when heated produced steam. He called them therefore Zeolites from the Greek for Boiling Stone (zeo, from zein, to boil, lithos - stone). In 1949, Milton working for Union Carbide produced the world's first man-made zeolites, the most important being Linde A and X. Linde A has become one the most widely used zeolites. Zeolites are highly crystalline alumino-silicate frameworks comprising $[SiO_4]^{4-}$ and $[AlO_4]^{5-}$ tetrahedral units. T atoms (Si/Al) are joined by oxygen bridges. An overall negative surface charge of the alumina-silicate framework requires counter ions e.g. Na^+, K^+ and Ca^{2+}.

Zeolites are made synthetically by a crystallization process using a template (e.g. Tetra Propyl Ammonium) that by self-assembly together with the silicon, aluminium and oxygen atoms forms a skeleton. Afterwards the template is removed (e.g. by heating, i.e. calcination) leaving small subnanosized channels along the crystal planes of the zeolite structure. The framework structure may contain cages and cavities, connected with channels, which may allow small molecules to enter. Typical channel sizes are roughly between 3 and 10 Å in diameter. In all, over 130 different framework structures are now known.

Zeolites have the ability to act as catalysts for chemical reactions which take place within the internal cavities. An important class of reactions is that the class is catalysed by hydrogen-exchanged zeolites, whose framework-bound protons acts as very strong acids. This is exploited in many organic reactions, including crude oil cracking, isomerisation and fuel synthesis. Zeolites can also serve as oxidation or reduction catalysts, often after metals have been introduced into the framework. Examples are the use of titanium ZSM-5 in the production of caprolactam, and copper zeolites in NO_x decomposition.

 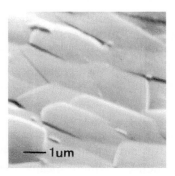

Fig. 3, Left: Unordered zeolite type A crystals, Right: top view of MFI type ordered zeolite (molecular sieve) crystals having orthogonal nanopores of 0.55 nm on a silicon substrate (Courtesy TOCK, TU Delft).

Top-layers of zeolites have also been deposited on many microporous support materials, e.g. alumina, to study their gas selective properties or to separate branched monomers (e.g. alkenes) from unbranched ones.

Nanofiltration

Nanofiltration is a relatively young description for filtration processes using membranes with a pore size ranging from 1 to 10 nm. This term has been introduced to indicate a specific domain of membrane technology in between ultrafiltration and reverse osmosis. One of the first researchers using the term 'nanofiltration' was Eriksson in 1988 [6]. But already some years earlier there was a company called FilmTec started to use this term for their NF50 membrane which was supposed to be a very open reverse osmosis membrane or a very dense ultrafiltration membrane [7].

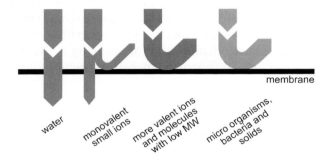

Fig. 4 Schematic drawing of separation processes that occur in nanofiltration membranes.

History of nanofiltration dates back further in time. As early as 1970 Cadotte showed that reverse osmosis membranes, a nanofiltration membrane with slightly smaller pore size, could be made out of polyethyleneimine and toluenediisocyanate, forming a polyurea [8]. This membrane, the NS 100, was different in several aspects from the known reverse osmosis membranes at that time. It was the first reverse osmosis membrane not made by a phase inversion process using cellulose acetate or polyamide, which in addition showed a high salt retention.

All ceramic and some polymeric nanofiltration membrane are considered to be nanoporous, and most of the polymeric nanofiltration membranes are regarded as 'dense', which means that no fixed pores are present in the membrane, but they contain a fine network structure that can be charged or uncharged. In several transport models for nanofiltration the pore size is used as parameter to describe the morphology of the membrane [9,10]. However, this concept of pores may be considered as hypothetical for several polymeric nanofiltration membranes, Bowen and Mukthar [11] reported that the determination of an effective pore size by transport models should not mean that those pores really exist in nanofiltration membranes. The mass transport properties is the same for ions passing through the polymer network of a specific membrane as for ions passing through pores with a given diameter in a membrane.

In general nanofiltration membranes are used to separate relatively small organic compounds and (multivalent) ions from a solvent. However in most of the applications with aqueous solutions, such as wastewater, potable water and process water, the transport mechanisms are not fully understood. In the case of separation of organic solvents new developments in the field of polymeric membranes are continuously showing up, where the solvent stability of these polymeric membranes is frequently studied and improved. Polymeric membranes may suffer from swelling when wrongly chosen.

Typical applications of nanofiltration membranes are the separation of salts from dye solutions or the separation of acids from sugar solutions for the extraction of the purest products in highly concentrated form.

Ultrafiltration and Microfiltration

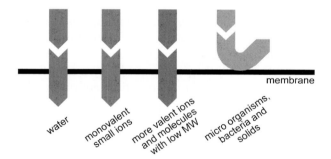

Fig. 5, Schematic drawing of separation processes that occur in ultra- and microfiltration membranes.

Ultrafiltration membranes are porous membranes (pore range 5 to 50 nm), which have a distinct and permanent open porous membrane structure through which transport occurs [12,13]. The separation of micro-organisms, bacteria and solids from the liquid is mainly based on size and is pressure driven. The term ultrafiltration has been introduced to discriminate the process with a more narrow retention behaviour from microfiltration. The membranes, in most cases polymeric, can be used both in dead-end mode and cross-flow mode. In dead-end mode all liquid is forced to pass the membrane, whereas in cross-flow mode, a tangential flow across the membrane is used to minimize fouling, thereby keeping the filtration process going [14].

Fig. 6, SEM micrographs of synthetic membrane filters. Left: Organic phase separated membrane filter, Right: Ceramic sintered alumina filter (Courtesy Aquamarijn Research).

For the retention of larger particles microporous membranes (with pore size from 50 nm to 5 μm) can be used. As an example, a feed containing macromolecules (e.g. proteins) is filtered through a membrane that contains small pores. The particles to be separated (the macromolecules) are retained from flowing through the pores, but the solution itself can flow through the pores. The reason that the particles cannot freely enter the pores is their geometric size, although interaction with the pores could play a significant role. In this case the properties of the membrane are also dependent on the structure of the membrane, together with the intrinsic properties of the membrane material, like the type of chemical end-groups at the surface. Separation of larger particles (e.g. micro-organisms) usually proceeds according to the same mechanism.

2.2   SEPARATION BY SOLUTION-DIFFUSION PROCESS

To achieve separation on a molecular scale, a relatively dense membrane is required. At the feed side, the components are being dissolved in the membrane and transported by diffusion through the membrane with a driving force acting inside the membrane. At the permeate side the components leave the membrane. Separation results from differences in the solubility of the components into the membrane material and differences in velocities of diffusion through the membrane. The driving force is solely activated by properties of the membrane material, like chemical affinity, and not of porosity (the geometry) of the membrane. Because the transport is a result of the solubility and the diffusivity, the separation process is called the 'solution-diffusion' mechanism.

The three most important separation processes will be discussed further; reverse osmosis, gas separation and pervaporation.

Reverse osmosis

In the reverse osmosis process, the processing liquid is being transported through the membrane under high pressure [15]. The membrane retains most of the ions and larger solids in the liquid and the process is often used to produce high quality water. Polymeric reverse osmosis membranes have no macroscopic pore structure, but consist of a polymer network in which solutes can be dissolved.

Like the membranes used for dialysis purposes, the membrane for reverse osmosis usually consists of a material that swells due to the processing liquid (e.g. water). This solvent-swollen network may be considered as a porous system comparable to dialysis membranes, although the structure of the latter is looser than that of reverse osmosis membranes. It is difficult to distinguish single pores, as these are not fixed in place or in time due to the flexibility of the polymer chains. The distinction between dense and porous membranes is not very discrete. During the separation process a considerable part of the membrane structure is filled with the feed. Membranes for dialysis purposes can easily take up 50 % of its volume of water. This volume may be called the pore volume, although the membrane matrix does not contain any distinct pores.

Gas Separation

Gas separation (for example separation of acid gases from natural gas or hydrogen in production processes) with polymeric membranes has become a widely used process over the last 3 decades. However, the fact that rubber balloons filled with gas loose their gas due to diffusion through the balloon wall, was already known for more than 150 years [16,17]. The concept of gas separation with membranes has been introduced by Monsanto [18]. Their membranes were able to recover hydrogen gas from the ammonia production.

For the gas separation process using membranes it is difficult to achieve a good selectivity. The separation is being performed under harsh conditions (high pressure, high temperature) and also the membrane stability may suffer under presence of the process gas [19,20]. For instance, CO_2 is a plasticizer (softener) for the polymer material, hereby weakening the membrane structure and thus decreasing the selectivity [21].

The preferred membrane for gas separation consists of a thin ($< 1\mu m$) top layer, supported by a rigid support with negligible transport resistance [22]. The thin top layer, i.e. the perm selective layer, determines the overall gas flux.

Fig. 7 SEM micrograph, cross-section of an asymmetric polymeric gasseparation membrane with a gas selective dense skin layer (Courtesy Aquamarijn Research/Membrane Technology Group, University of Twente).

Membranes can be classified on the basis of their morphology or the separation process [18]. Current gas separation membranes are either thin dense films, integrally skinned asymmetric membranes or composites mainly prepared from glassy polymers [7]. Asymmetric membranes have a dense top layer and a porous substructure and are formed by a phase inversion process [19]. Composites have a dense top layer and a porous substructure. The top layer is created in a separate step, for example by coating. In both cases, the perm selective top layer should be as thin as possible to achieve a high flux. The substructure should have good mechanical strength with negligible gas transport resistance. Thin polymeric films by themselves are too weak to withstand the high differential gas pressures required in gas separation operations. Membranes with a support layer are therefore the most widely applied. The advantage of a composite membrane is that the top layer and the support can be optimised separately.

Pervaporation

Pervaporation is a membrane separation process that can be used for the separation of liquid-liquid mixtures with an azeotropic composition or mixtures of components with a relatively small difference in volatility (e.g. dehydration of ethanol, acetic acid, removal of ethanol from a fermentation product). The transport occurs according to the solution-diffusion model through the non-porous membrane [23]. The diffusion through the membrane is in general being enhanced by a lower partial pressure on the permeate side of the membrane (vacuum or a pure inert gas). Because the process uses a low energy input it is a promising alternative for conventional techniques like distillation. The mixture flows through modules, where it passes along the membranes. The component to be removed is vaporized through the membranes, and collected at very low pressure in a vacuum vessel. It is then condensed and purged. A vacuum pump extracts the noncondensables and maintains the required vacuum. This vaporization cools down the processed mixture, which must be heated to maintain the highest flux through membranes. Main applications are: wine and beer dealcoholisation, removal of organic solvents from aqueous streams, aroma recovery/up-concentration and wastewater purification.

2.3   SEPARATION BY CHARGE

In separation processes the charge of a molecule may influence its transport properties through a medium, or a charged molecule (ion) may selectively be exchanged for another charged molecule. Ion exchangers are solid water-insoluble "high-molecular" substances that can exchange ions bound to other free-in-solution ions of the same charge. Normally ion exchangers are resins composed of cross-linking polymers that possess electrically active functional groups. Every group gives an exchange either of a very basic (anion exchanger) ion or of a very acidic (cation exchanger) ion. Two key factors determine the effectiveness of an exchanger: its affinity for the ion in question and the number of active sites available.

Ion-exchange groups may also be incorporated in membrane materials. About forty years ago ion-exchange membranes and the electrodialysis process were introduced which incorporates such membranes. Electrodialysis is a process in which solutions are desalted or concentrated electrically. The key to the process is a semi-permeable barrier that allows passage of either positively charged ions (cations) or negatively charged ions (anions) while excluding passage of ions of the opposite charge. These semi-permeable barriers are now commonly known as ion-exchange, ion-selective or electrodialysis membranes.

A special type of the ion exchange membrane is the bipolar membrane. It contains both the anionic membrane and the cationic membrane, but one of the membranes is only permeable for anions and the other only for cations. So there is no transport through the membrane possible. At the surface where the two type of membranes meet, water is split into OH^- and H^+ and these ions are transported out of the membrane. This process is forming no gasses. Electro dialysis with bipolar membranes (ED-BPM) can be used to replace electrolysis of water, but with a wider variety of applications. Of course care should be taken that the two streams are separated from each other to prevent recombination into water [5,24,25]

Using a combination of an anionic and a cationic membrane, water can be split in H^+-ions and OH^--ions. Combining this with liquids containing salts, direct acids and alkaline solutions can be made. In industrial applications this gives the opportunity for direct control of the pH of a process stream (juice industry and fermentation reactors) [26,27,28].

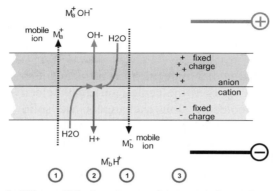

Fig. 8, Electrodialysis process that occur in and around a bipolar membrane. In an externally applied electrical field (3) water is split in the membrane and the water ions are being transported (2) out of the membrane. If the membrane is permeable for ions (1), positively charged ions will move through the membrane towards the OH^--ions. Negatively charged ions will move towards the H^+-ions.

The replacement of electrolysis by membrane electro dialysis, seemed to be very promising already two decades ago, but it was found that it is difficult to meet the specifications on a large scale. The available membranes didn't have the ideal membrane requirements like a very high selectivity (the bipolar membrane layers are also permeable to salt co-ions), the membrane permeability should be very high (yet, the membrane layers add extra membrane resistance), and a good long term stability (the membranes are being attacked by a high concentration of alkaline, being produced in electro dialysis).

3. MEMBRANE STRUCTURES*

The structure of a membrane is vital for the performance of the membrane. The membranes are either symmetrical, where the properties of the membrane do not change throughout the cross section of the membrane or they are asymmetrical (also known as composite). In the latter case the membrane is composed of a thin selective layer and a strong support layer giving mechanical

---

* This section has been co-written with W. Nijdam, Aquamarijn Research.

strength. With some techniques it is possible to create both membrane layer and support in one single preparation step.

Asymmetric membranes are in general superior compared to symmetric membranes because the flux determining top layer can be very thin. A schematic representation of symmetric and asymmetric membranes is given in Fig. 9. Asymmetric membranes offer great possibilities in optimising the membrane separation properties by varying the preparation parameters of especially the thin top layer. Also the ongoing development in the field of polymers throughout the decades resulted in the use of new polymers, like polysulfone (1965), polyether-ether-ketone (1980) and polyetherimide (1982).

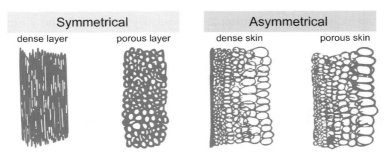

Fig. 9 Schematic drawing of different membrane types. The membranes can be made both as fibers (circular) and as flat sheet.

3.1   CERAMIC MEMBRANES

Most ceramic membranes have an asymmetrical membrane structure with either a dense or a porous skin layer. The rough porous support is made of sintered ceramic particles (alumina (Al_2O_3), titania (TiO_2), and zirconia (ZrO_2)), in which the pores are subsequently reduced in size (in three to five deposition steps) before the top layer is formed. Typically the final layer of the support has a pore size between 1 and 5 µm. The preferred shape of ceramic membranes is a rod, because flat discs have shown to be too brittle.

Building the ceramic rod starts with making the support by injection molding the inner core with the largest particle size. As long as the rod is not yet sintered (i.e. green phase), the paste of ceramic materials can be shaped freely. The rod is being prebaked (fired) and subsequently a finer particle coating is being applied by dip coating. This process of dip coating and firing is repeated several times until the final support coating is applied.

In the last step the thin top layer is being applied, often again by several steps of dip coating giving the membrane its separation properties. The whole stack is sintered for final fixation of all ceramic particles.

A typical α-alumina tube has a pore size of to 110 - 180 nm and is used as a microfiltration or an ultrafiltration membrane. For a finer membrane a γ-alumina layer can be applied on top of the α-alumina layer (see Fig. 10). By means of dip coating of a sol-gel, e.g. from a Boehmite sol (γ-AlOOH) [29] a thin layer can be formed on top of the α-alumina. This layer is subsequently dried and heated up to 700°C. By adding aluminum isopropoxide (Al(iso-OHC$_3$H$_7$)$_3$ in distilled water, a Boehmite sol can already be formed at 80°C. Every dip coating and subsequent controlled drying step forms a layer with a thickness of around 0.5 μm [30]. During drying the alumina solidifies, yielding a porous shell of alumina. The shell is then hardened (calcined) by heating for several hours at high temperatures. The duration and temperature determine the final pore size, which is for a γ-alumina layer typically between 3 and 7 nm. The grain size also influences the mechanical and physical properties of the ceramics: with reduced grain size, the membranes are stronger, less brittle and have higher temperature resistance [31,32].

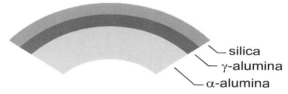

silica
γ-alumina
α-alumina

Fig. 10 Schematic view of the different layers in a ceramic rod.

For certain applications it is preferred to have an even finer pore size, for instance in molecular sieving (separating gasses, e.g. H$_2$/CH$_4$, H$_2$O/CH$_4$). In a CVD process (e.g. tetraethoxysilane (TEOS) and methyltriethoxysilane) ultra thin silica can be applied, that reduces the pore size even more. By careful adjusting the gas flows through the rod and process temperatures, voids in the γ-alumina layer can be blocked, because the silica prefers to grow inside the voids. This is beneficial for the separation behaviour of the silica layer, as the selectivity of the membrane increases. The silica layer can also be applied using a multi layer sol-gel deposition method [33,34] instead of an intermediate γ-alumina layer.

The sintering process can be applied to pure ceramic material and also to polymer/ceramic mixtures. Sintered polymeric/ceramic membranes usually are macro-porous membranes (pore size >>50 nm), but have after sintering the advantageous properties of ceramic membranes.

Advantageous properties of ceramic membranes:

Ceramics do not absorb water and do not swell. Swelling is a common problem with many materials, because when the membrane material absorbs water, the pore size increases, resulting in less retention and change in selectivity.

Ceramics are thermally stable. The membranes allow processes to be run at high temperatures. This makes the filtration for viscous fluids like oils easier, because their viscosity decreases.

Ceramics are wear resistant (physically hard). This allows harsh cross flow conditions beneficial for the removal of particles or cake layer, without damaging the membrane.

Ceramic materials, especially alumina, titania, zirconia, are chemically very resistant. Most chemicals can be filtered or used during cleaning without attacking the membrane material.

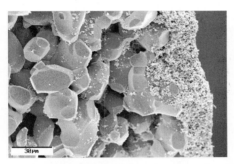

Fig. 11 SEM Photograph of a sintered alumina ceramic membrane with a pore size of 800 nm (Courtesy University of Karlsruhe).

3.2 POLYMERIC MEMBRANES

A significant research effort has been made to develop polymeric membranes in the past decades in a large variety of polymeric materials. Also a wide range of fabrication processes is available for polymeric membranes and will be discussed below.

Template leaching
By mixing two different components and moulding this mixture into a certain heterogeneous structure, after which one of the components is removed by selective leaching. This process is used in the preparation of porous glass membranes.

Stretching of a polymeric film
Dense polymeric films or foils can be stretched thereby generating voids in the film. It is a technique that is commonly used for e.g. semi permeable membranes (also used for outdoor breathable clothing) with stretching a

hydrophobic foil (for example PTFE). Best results are obtained with crystalline polymers.

Phase Inversion

A common technique to prepare asymmetric membranes is phase inversion. Whilst the process is suitable to produce the whole spectrum of membranes from microfiltration to gasfiltration, the membranes can be coated afterwards to obtain other separation characteristics (by coating or interfacial polymerisation).

Phase inversion (originally introduced by Cestings) is a very versatile technique. In one single process step various types of membranes can be made consisting of a porous support with or without a skin layer, where the skin layer can be dense or porous with a well defined pore size.

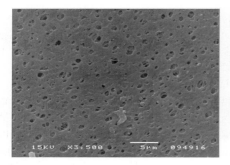 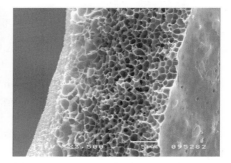

Fig. 12, SEM Photographs of liquid induced phase separated polymeric membranes, left: topview skin layer, right: cross-section showing a relatively dense skin layer on retentate side and an open skin layer at permeate side (Courtesy Aquamarijn Research/Membrane Technology Group, University of Twente).

Phase separation is the process of changing a one-phase casting solution into at least two separate phases. In all phase separation processes for membrane manufacturing, this casting solution is split into at least two phases: a solid material-rich phase that forms the porous/non-porous structure of the membrane and a material-poor phase that will be removed from the membrane. Phase separation is generally a thermodynamically driven process. After casting the solution onto a flat glass plate (or rotating cylinder in a continuous process) the formation of the pores starts by changing the composition of the solution, the temperature or the pressure. The composition of the casting solution may be changed by bringing the casting solution in contact with a liquid not well miscible with the material. By changing the solution, the polymer solidifies and the pores are created. Not only by immersion of the casting fluid in a non-solvent, the solidification can be started, also by evaporating a solvent from the casting solution or by reaction of components in the casting solution resulting in a non-solvent agent. Because for almost every polymeric material there is a combination of solvent/non-solvent possible, the process can be used to produce a wide variety of membrane materials.

After solidification the polymers may be cross-linked or cured to improve functional properties, like strength and chemical resistance.

The types of phase separation processes are classified into six categories:

Vapour-Induced Phase Separation (VIPS), which is also known as "dry casting" or "air casting". The evaporation of the solvent in the casting solution will result in a dense or porous product depending on the material and the used solvent/solute mixtures; The vapour may contain water or organic solvent molecules that may be absorbed by the casted film and will influence the porosity of the product.

Liquid-Induced Phase Separation (LIPS), mostly known as "immersion casting" or "wet casting". This process normally yields dense or porous products depending on the material and the used solvents and non-solvent mixtures. This technique may be combined with VIPS for polymer solutions with e.g. two solvents with different boiling points [35,36].

Thermally-Induced Phase Separation (TIPS), frequently called "melt casting"; The material may solidify to a dense or porous product of a material-solvent-additive mixture by varying the temperature of the material-solvent during the casting process.

Colloidal-Induced Phase Separation (CIPS). A phase separated colloidal solution is used to perform structure arrestment on the micro fabricated mould. For example, a polymerizable bicontinuous micro emulsion can be applied onto the mould, or a sol/gel mixture of a ceramic material. The pore morphology can be fine-tuned by the addition of a co-solvent resulting generally into smaller pores.

Pressure-Induced Phase Separation (PIPS), "pressure casting"; The casting solution may contain for example a saturated dissolved gas. Reduction of the pressure (or increase of temperature) will induce growth of gas cells in the casting solution with a closed cell or open cell morphologies and a typical size of 0.01 - 1000 micron.

Reaction-Induced Phase Separation (RIPS), a casting solution containing monomers start to react and initiate phase separation due to for instance increase in molecular weight or production of a non-solvent.

Coating of membranes

Polymeric membranes are often coated to produce composite membranes. On top of an asymmetric porous membrane, a thin layer of a polymeric solution is deposited resulting in a membrane with separation properties determined by the thin layer. A second possibility is coating the complete membrane by a polymer that is not influencing the separation characteristics of the skin layer of the asymmetric membrane (a non-selective polymer). This will plug the voids in the asymmetric membrane and the selectivity of the original membrane is increased.

Interfacial Polymerisation

A coating can also be applied by interfacial polymerisation. In this technique the pores in the membrane are first filled with a liquid A. After immersing the membrane in a bath containing a reactant for liquid A, a polymer is being formed at the interface between the liquid A and the bath. A new selective layer is thus being formed in the pores of the membrane.

3.3   NANOCOMPOSITE MEMBRANES

There are a large number of nanofiltration and gas separation applications with membranes made from a polymeric or an inorganic material. The application of for instance a polymeric material for a separation membrane depends of course upon both the throughput and the purity of the product transported through the membrane. This means that the permeability coefficient and the selectivity should be as large as possible. However, it has been found that simple structural modifications, which lead to an increase in product flux, usually cause a loss in perm selectivity and vice versa [37,38,39,40,41]. This so-called "trade-off" relationship is well described in literature [42,43]. In recent years, the efforts and successes in synthesizing a variety of nanostructured hybrid materials have provided a new degree of freedom for the development of advanced materials with enhanced separation properties [44].

Nanocomposites represent a current trend in developing novel nanostructured materials. They can be defined as a combination of two or more phases containing different compositions or structures, where at least one of the phases is in the nanoscale regime. Kusakabe and his co-workers [45] reported in 1996 that the permeability of CO_2 in a polyimide/SiO_2 hybrid nanocomposite membrane is 10 times larger than in polyimide.

Another application is the use of zeolite nanocrystals (10-100 nm) to combine the unique properties of polymers and zeolites (while overcoming the shortcomings of both). Polymer-zeolite nanocomposite membranes can be used for air separation and is a promising alternative for the conventional energy-intensive cryogenic distillation. Utilization of polymer-zeolite nanocomposite membranes shows a route to achieve a modified polymer matrix offering high molecular sieving selectivities (O_2/N_2>20) while maintaining the ease of polymer processing conditions.

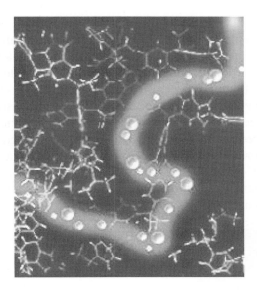

Fig. 13 Gas molecules move through the spaces between the polymer chains. The larger the spaces, the faster and more selective the movement. The finding, which appeared in Science describes a new type of nanoparticle-enhanced filter for separating compounds at the molecular level (Courtesy Science [46]).

For any molecule to move across a membrane, it must go through a solution-diffusion process. The molecule first has to get into the membrane - the solution part of the process - then diffuse through it. Recently, novel gas-separation nanomembranes have been obtained with inorganic silica nanoparticles, with contradictory, but very useful properties (see Fig. 13). It was shown that larger molecules dissolved much faster into the membrane, and once in, move right through to the other side before the smaller molecules had completed the first step.

Silica nanoparticles embedded in a carbon-based membranes may help in the future to produce gases free of impurities. Because of the nanocomposite's ability to trap molecular-sized impurities it could be further used in processes such as biomolecule purification, environmental remediation, sea water desalination and petroleum chemicals and fuel production [46].

3.4   OTHER MEMBRANE STRUCTURES

Track Etched Membranes

One of the most successful attempts to manufacture the ideal membrane (a sieve) is the track-etched membrane. For the production of track etched membranes, dense polymeric films (polycarbonate or polyester) are randomly exposed to a high-energy ion bombardment or particle radiation. This bombardment damages the polymeric chain in the dense film, leaving small 'tracks'. Subsequently the polymer can be etched specifically (in an acid or alkaline solution) along the damaged track. The pores that are being formed are cylindrical channels and very uniform in size. However the membrane has a low porosity, due to the fact that the chance for an overlap between two pores increases with the porosity (the density of the radiation). The membranes are often used in laboratories for analysis.

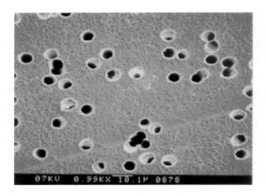

Fig. 14, SEM micrograph of polycarbonate track etched membrane filter (Courtesy Aquamarijn Research).

Anodised Alumina Membranes

The formation of the pores of an anodised alumina membrane is a self-assembling process. By anodisation of aluminium in an acidic solution, a highly ordered structure of pores in the Al_2O_3 matrix can be obtained. Due to lattice expansion by the oxidation of the aluminium, an anisotropic potential distribution and heat development during anodisation, a self-structuring process is induced and creates the shape and interspacing of the pores. The membranes are relatively thick, resulting in long pores with a pore size ranging from 20 nm to 200nm. The pore size is very uniform (see also Fig. 15). However the membranes are unsupported and need, depending on the application, a second support structure.

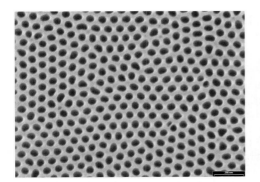 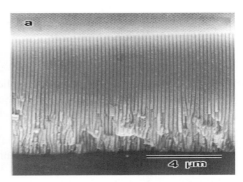

Fig. 15 SEM photographs of an Anopore membrane with a pore size of 50nm. Left: Top view, Right: Side view, note the vertical pore/channel structure (Reproduced by permission of The Electrochemical Society, Inc).

Microsieves

Most of the previously described membranes in this chapter have an sponge-like or a complex (tortuous) and long pore channel structure, which makes it difficult to clean the inside of the membranes. Also the retention of particles, proteins and other coagulates is not only determined by the porous top layer, but also by the pores and cavities in the support structure. Moreover in certain applications, like clarification of beverages, the cleaning agents must be removed completely after cleaning to prevent contamination of the beverage. However, it is nearly impossible to remove cleaning agents that are left behind in dead-end pores and cavities. Until recently the best sieve structures are the track-etched and the anodised alumina membrane. However track-etched membranes have only be made in a polymeric (not chemical resistant) material, whilst the anodised alumina membrane is relatively brittle, only available in a pore range between 20 and 200 nm and has a relatively low flux performance.

In the next chapters the search for the ideal membrane structure and method for manufacturing will be pursued.

REFERENCES

[1] Osmosis is discovered by Abbé J.A. Nollet in 1748 using a pig's bladder membrane.

[2] S. Loeb and S. Sourirajan, Adv. Chem. Ser., 38, 117 (1981).

[3] C.J.M. van Rijn, 'Membrane filter as well as a method of manufacturing the same', PCT Application 95/1386026.

[4] S. Chien, Microvascular Research, 44, 243 (1993).

[5] M.H.V. Mulder, Basic Principles of Membrane Technology, Kluwer Academic Publisher, The Netherlands, (1991).

[6] I. Eriksson, Nanofiltration extends the range of membrane filtration, Env. Progress, 7, 58 (1988).

[7] W. J. Conlon, Pilotfield test data for prototype ultra low pressure reverse osmosis elements, Desalination, 56, 203 (1985).

[8] J. Cadotte, R. Forester, M. Kim, R. Petersen and T. Stocker, Nanofiltration membranes broaden the use of membrane separation technology, Desalination, 70, 77 (1988).

[9] X.-L. Wang, T. Tsuru, M. Togoh, S.I. Nakao and S. Kimura, Evaluation of pore structure and electrical properties of nanofiltration membranes, J. Chem. Eng. Jap., 28, 186 (1995).

[10] X.-L. Wang, T. Tsuru, M. Togoh, S.I. Nakao and S. Kimura, Transport of organic electrolytes with electrostatic and steric-hindrance effects through nanofiltration membranes, J. Chem. Eng. Jap., 28, 372 (1995).

[11] W.-R. Bowen and H. Mukthar, Characterisation and prediction of separation perfomance of nanofiltration membranes, J. Membrane Sci., 112, 263 (1996).

[12] M. Kurihara and Y. Himeshima, The major developments of the evolving reverse osmosis membranes and ultrafiltration membranes, Polym. J., 23, 513 (1991).

[13] L.P. Raman, M. Cheryan and N. Rajagopalan, Consider nanofiltration for membrane separations, Chem. Eng. Progress, 3, 68 (1994).

[14] M. Dekker, R. Boom, TIBTech, 13, 129 (1995).

[15] R.J. Petersen, Composite reverse osmosis nanofiltration membranes, J. Membrane Sci., 83, 81 (1993).

[16] J.K. Mitchell, On the penetration of gases, Am. J. Med., Sci., 25, 100 (1833).

[17] T. Graham, On the absorption and dialytic separation of gases by colloid septa, Phil. Mag., 32, 401 (1866).

[18] J. Haggin, New generation of membranes developed for industrial separations, Chem. Eng., June, 7 (1988).

[19] A. Bos, R.T. Chern and C.N. Provan, Macromolecules, 24, 2203 (1991).

[20] J.H. Petropoulos, J. Membrane Sci., 75, 47 (1992).

[21] M. Wessling, S.Schoeman, Th. v.d. Boomgaard, C.A. Smolders, Gas Sep. & Purif., 5, 222 (1991).

[22] W.J. Koros and I. Pinnau, Polymeric gas separation membranes, CRC Press, 209 (1994).

[23] J.K. Mitchell, On the penetrativeness of fluids, Am. J. Med., 13, 36 (1830).

[24] Y. Mizutani, Structure of ion exchange membranes (Review) J. Membr. Sc., 49, 121 (1990)

[25] T.A. Davis, J.D. Genders and D. Pletcher, A first course in Ion Permeable Membranes, The Electrochemical Consultancy, Romsey, England, ISBN 0-9517307-8-9, 1997

[26] K.N. Mani, J. Membr. Sc., 58, 117 (1991).

[27] G. Pourcelly, Bipolar membrane applications, Chapter 2 in A. Kemperman (ed), "Handbook on Bipolar Membrane Technology." Twente University Press, ISBN 9036515203, (2000).

[28] S. Koter, A. Warszawski, Electromembrane Processes in Environmental Protection, Polish Journal of Environmental Studies, Polish J. Env. Stud., 9, 45 (2000)

[29] B.E Yoldas, Ceramic Bull., 54, 289 (1975)

[30] B. Sea and K.-H. Lee, Molecular Sieve Silica Mambrane Synthesized in Mesoporous □-Alumina Layer, Bull. Kor. Chem. Soc., 22, 1400 (2001)

[31] Y-S. Lin et al, Improvement of thermal stability of porous nanostructured ceramic membranes, Ind. Chem. Eng. Res., 33, 860 (1994)

[32] C. Guizard et al, Nanophase ceramics by the sol-gel process, Mat. Sci. Forum, 154, 152 (1994)

[33] R.S.A. de Lange, J.H.A. Hekkink, K. Keizer and A. Burggraaf, J. J. Membrane Sci. 95, 57 (1995)

[34] N. K. Raman and C. Brinker, J. J. Membrane Sci. 105, 273 (1995)

[35] R.M. Boom, T. van den Boomgaard, T. and C.A. Smolders, J. Membrane Sci., 90, 231 (1994)

[36] R.M. Boom, T. van den Boomgaard, T. and C.A. Smolders, Macromolecules, 27, 2034 (1994)

[37] Z. K. Xu, et al., Polymer, 38, 581 (1997)

[38] Z. K. Xu, et al., Polymer, 38, 1573 (1997)

[39] T. S. Chung, et al., J. Membr. Sci., 75, 181 (1995)

[40] S. A. Stern, et al., J. Polym. Sci., Part-B: Polym. Phys., 31, 939 (1993)

[41] M. Smaihi, et al., J. Membr. Sci., 161, 157 (1999)

[42] L. M. Robeson, J. Membr. Sci., 62, 165 (1991)

[43] G. Maier, Angew. Chem. Int. Ed., 37, 2960 (1998)

[44] Z.-K. Xu, Y.-Y. Xu, M. Wang, J. Appl. Polym. Sci., 69, 1403 (1998)

[45] K. Kusakabe, et al., J. Membr. Sci., 115, 65 (1996)

[46] A. Hill et al., Science, 296, 519 (2002)

Microperforation Methods

1. INTRODUCTION

A review is given of a number of current manufacturing techniques like wet etching, electro-forming, laser drilling and hot embossing to make precision shaped perforations in various materials.

These techniques are bounded by realisable minimum feature sizes and tolerances and in the next chapter on micro engineered membranes it will come forward that micro engineering techniques will considerably bring down those dimensions and tolerances.

2. WET ETCHING OF PERFORATIONS IN METAL FOILS

Wet etching of perforations entails the local removal of metal from one or two sides of a metal foil. Most wet etching processes uses a ferric chloride ($FECl_3$) bath to etch apertures in a stainless steel or a copper foil. The process first entails the production of photo-tools or photographic positives of the required stencil image. These photo-tools, one each for top and bottom, are carefully aligned over e.g. a stainless steel foil covered with a photo-imageable resist, which is then exposed and developed. The developed plate or foil is then placed in an etchant spray, and etching proceeds in the orthogonal direction but also in the planar direction (undercutting). Depending on the spray direction and strength slightly anisotropic profiles can be achieved. The increase in aperture size due to undercutting has to be accounted for during the initial imaging phase. After the etching process part of the ferric chloride solution has been reduced to ferrous chloride ($FeCl_2$) and has to be refreshed.

Wet etching or photo-etching (or chemical milling) technology offers considerable advantages compared to punching, laser cutting and wire-erosion. Those advantages include no burring, no stresses and relatively low tooling costs. Etching is a highly cost-effective method for prototyping, small and the production of (very) large series.

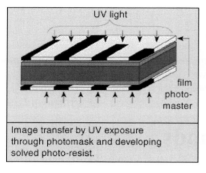

Image transfer by UV exposure through photomask and developing solved photo-resist.

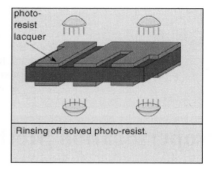

Rinsing off solved photo-resist.

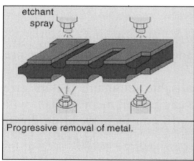

Progressive removal of metal.

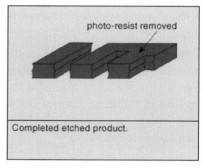

Completed etched product.

Fig. 1 Process flow of double-sided wet etching perforations in stainless steel, courtesy Stork Veco [1].

Precision etching techniques are applicable for material thickness from 10 micron - 2.0 mm. Common metals that can be etched are Cu and Cu-alloys (brass, phosphorbronze, berylliumcopper) Ni and Ni-alloys, stainless steel (301, 302, 304, 316, 430, etc), carbon steel and other alloy steels like Kovar and Invar. The Dutch companies Etchform and Stork Veco specialise in the precision etching of 'exotic' metals such as: beryllium-copper, molybdenum, titanium, tantalum, tungsten, silver, gold and 'special' alloys such as Inconel, Phynox/Elgiloy, niobium and hafnium.

Typical advantages of the precision-etching process are:
High precision
Virtually all metals can be etched

Final product free of burs
Complex shapes are possible
No deformation of the material, no stress
Relatively thick as well as thin material can be processed
Economical tooling and parts

Typical etched products are microsieves, flow elements, springs, lead frames, encoder discs, contact strips and decorative bookmarkers.

Table 1

Specification and tolerances of materials for the wet etching process.

Material thickness (t)	0.01 - 2.0 mm	
Tolerances	depends on material thickness, limit for t: 0.050 - 0.100 mm for t: 0.100 - 0.200 mm for t: > 0.200 mm ± 0.1t	± 0.005 mm ± 0.3t-0.2t ± 0.2t-0.15t
Minimum pore/slot	for t: 0.025 - 0.050 mm for t: 0.050 - 0.100 mm for t: > 0.100 mm	$\cong 1.6t$ $\cong 1.0/1.3t$ $\cong 0.8t$
Minimum bar-width	for t: 0.050 - 0.200 mm for t: > 0.200 mm	$\cong 1.0t$ $\cong 1.0t$
Pitch tolerances	± 0.0025 - 0.010 mm x distance	
Etchprofile (a)	for t: 0.015 - 0,050 mm for t: 0.050 - 0.100 mm for t: 0.100	$a \cong 0.2t$ $a \cong 0.2/0.1t$ $a \cong 0.1t$
Corner radius (r)	$r \cong 0.5 - 0.1t$	
Post treatments	blackened, Au, Ru, Pt, lead-tin plated etc.	

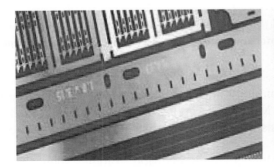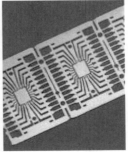

Fig. 2 Left: etchformed microstencil, Right: precision etched lead frame.

3. ELECTROFORMING OF PERFORATIONS

Electroforming entails the local deposition of metal molecules on a lithographic patterned conducting substrate. Main advantages of the electroforming processes are the economical tooling and production costs, the exceptional tolerances (difficult to achieve mechanically), the excellent (part to part) reproducibility and the relatively short processing time (growth rate ±1 μm/min.).

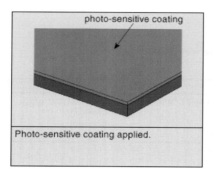

Photo-sensitive coating applied.

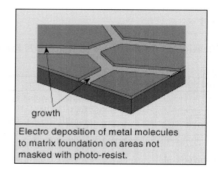

Electro deposition of metal molecules to matrix foundation on areas not masked with photo-resist.

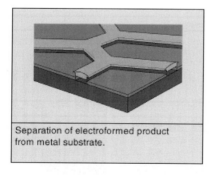

Separation of electroformed product from metal substrate.

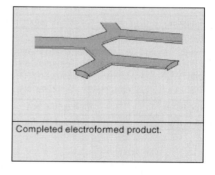

Completed electroformed product.

Fig. 3 Process flow of the electroforming process on a conductive base plate, courtesy Stork Veco [1].

The electroforming technology has applications in several industries: Measurement & Control, Medical Applications, Optical Applications, Automotive, Inkjet Applications, Encoders, Aerospace Industry and Laboratory equipment.

Advantages of this technology are high precision, complex shapes possible, sharp edge definition as defined by the lithographic layer and excellent reproducibility. Typical electroformed products are orifice plates, airflow metering, (fuel) injector applications, diaphragms (field-stops), masks optical, grids for TEMS's, micro precision sieves and screen meshes.

Table 2

Specification and tolerances of materials for the electroforming process.

Materials:	Nickel, (Cu & Au only SEMgrids)
Material thickness (t)	0.015 - 0.2 mm
Tolerances	General ± 0.010 - 0.015 mm limit ± 0.005 mm
Minimum pore/slot	0.005 mm - t = 0.05 mm > 0.05 mm = 0.3 t
Minimum bar-width	0.015 mm + 2t thick resist system 1.5t
Pitch tolerances	general ± 0.01 mm x distance limit ± 0.005 mm x distance
Angular tolerances	general ± 30^2 - 40^2 of arc limit ± 20^2 of arc
Corner radius	inner corner (r) = 90° angle outer corner (r) = t
Post treatments	blackened, gold, Ru, Pt, lead-tin plated etc.

Electroform switch and 3D structures

For the production of perforated membrane structures with a very thin membrane (< 5 micron), small pore sizes (<5 micron) and a reinforced support structure Aquamarijn and Stork Veco have developed a new technique [2] using the effect of electric switching during the electroforming process.

3D mandrels can also be used to produce more elaborate structures. The above depicted mandrel is then first provided with protruding parts.

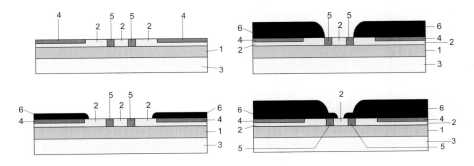

Fig. 4 Electric switching during electroforming to produce a reinforced nozzle structure with nozzles with a small depth/diameter ratio. 1,4,5 are electric conductive layers, 2 and 3 are electric insulating layers and 6 is the electroformed product.

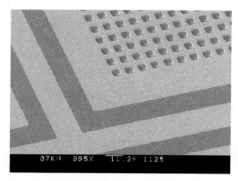

Fig. 5 Top: Overview with an electric switch structure of a 3D mandrel, courtesy Aquamarijn Research.

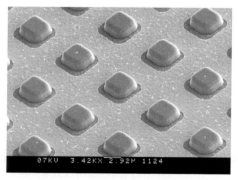

Fig. 6 Detail of the 3D mandrel, courtesy Aquamarijn Research.

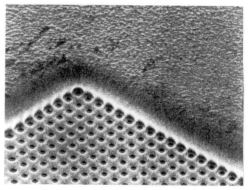

Fig. 7 Reinforced Aquamarijn/Veco microsieve structure obtained with the 3D mandrel in one electroforming step, courtesy Aquamarijn Research.

4. ELECTRO PERFORATION

Electro perforation is a method in which large electrostatic fields are being used to remove (sparking) locally material in a thin foil. For many years' thin foils and papers with area weights of between 20 and 150 g/m^2 have been perforated over large areas or in zones. Among them are to be found packaging foils, cigarette paper, filter materials as well as various non woven and composites. Pore sizes from 5 to 150 microns at pore densities of up to 10^6 /m^2 with a total perforated area of 2% can be reached.

The size and tolerance of the formed pores is difficult to tune, micro perforation is the sole goal of this process, to improve gas and/or liquid (water vapour) permeability. The number of pores is the main determining factor to be matched with e.g. the required respiration rate [3].

Machinery for mass production generally consists of a semiconductor high-level stage, supporting capacitors, high-voltage ferrite transformers, construction unit to build the hybrid drive connection, and their connections in a relatively compact and modular way.

Materials suitable for electrostatic perforation include PTFE, PET, PVC, PE, PP and PSU foils or coatings with thickness' varying between 5-200 micron.

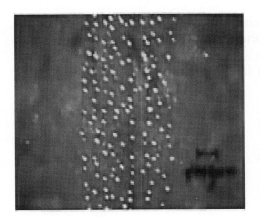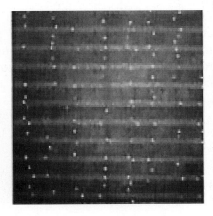

Fig. 8 Some electro perforated foil structures.

5. LASER CUTTING AND DRILLING OF PERFORATIONS

The laser cutting process is a subtractive process like the wet etching process, the only difference being – a laser beam is used to form apertures instead of a chemical bath. The inherent advantages of this process lie in its ability to form very fine aperture sizes with consistent accuracy. Aperture holes are formed one at a time thereby resulting in higher manufacturing costs, which finally depends on the number of apertures in the stencil. Laser cutting is therefore a relatively slow (line writing) and expensive technique to microstructure foils and sheets. Advantageous however is that laser cutting can be applied to many different materials, excluding however e.g. glass and some plastics. Metal sheets up to a thickness of several mm can be processed but it is difficult to obtain smooth edges. Also the material composition due to high temperature gradients induced by the laser changes leading to residual stresses and transformation of crystal structures (e.g. processing stainless steel leads to corroded edges).

The Excimer laser is an ultraviolet laser that can be used to shape a number of materials without heating them, unlike many other lasers, which remove material by burning or vaporising it. The Excimer laser lends itself particularly to the machining of organic materials (polymers, etc).

Precision pore drilling with laser

Precision pores can be made with pulsed laser shots preventing the heating up of the direct surrounding of the intended pore. Laser MicroTools and Lenox laser combine small lasers with precision motion systems, video imaging, custom component fixtures, and proprietary software to carry out precision machining and marking operations on scale sizes in the 2 - 200 micrometer range. These tools allow fabrication of features that are difficult to produce by normal commercially used lasers.

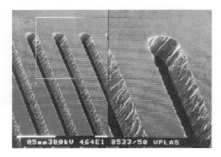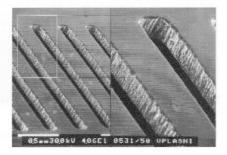
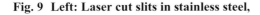

Fig. 9 Left: Laser cut slits in stainless steel,   Right: Laser cut slits in nickel sheet.

Clean, burr-free cutting of patterns in stainless steel, titanium, NiTi, and other metals of thickness up to 250 microns can be achieved with small Nd:YAG and Nd:YLF lasers. Kerf widths and pore diameters as small as a few microns can be produced in thinner foils and films.

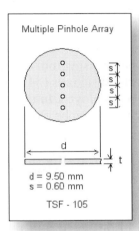

Fig. 10 Precision laser drilled pore in pinhole array.

The array apertures may be used for spatial filters, gas/liquid separators, molecular beam masks, particle counters, pinhole cameras and many other optical or mechanical purposes.

Apert.	Dia. (u)	Tole. +/- (u)	Price
1	2.5	2	$200.-
2	10	3	$200.-
3	17.5	3	$200.-
4	30	5	$200.-
5	50	10	$200.-

Fig. 11 Price indication of precision drilled pores in pinhole arrays.

6. HOT EMBOSSING

In precision technology there is a great demand for low-cost methods for high volume production of small components and systems. On one hand this requires a low-cost substrate material, on the other hand an easy way of microstructuring. Hot embossing of polymers fulfils both criteria, as the raw materials are cheap and the replication character of the method allows rapid processing.

In the first step, a mask with the desired structures is designed, which is used for the fabrication of the embossing tool. This embossing tool can be made with conventional CNC-machining for relatively large structures (>100 µm), for smaller features lithographical techniques have to be employed. In this case, the pattern is transferred to a photo-lacquer layer, which then can be electroformed.

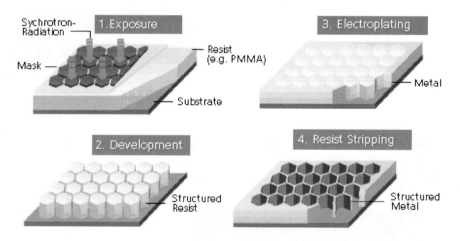

Fig. 12 A conducting substrate is coated with an X-ray resist (usually PMMA) and exposed through a mask with synchrotron radiation (the lithography step). This light is characterized by its short wavelength (in the range < 1nm) and is highly parallel and coherent. After exposure, the resist is developed, the exposed areas dissolve in the developer. The so created voids can now be electroplated and filled with metal (the electroforming step). After stripping of the remaining resist a metal structure with geometric pattern of the mask remains. This metal structure can now be used as a tool in a moulding step, e.g. for hot embossing to replicate the microstructure in polymers.

This process yields a metal tool, usually nickel. Alternatively, the resist structure can be copied into silicon (Aquamarijn/TMP [4]) by means of dry or wet etching. This silicon structure now can be used directly as an embossing tool or it can be electroplated again to yield a nickel tool.

The tool and the substrate are now mounted into a hot embossing machine and are heated separately to a temperature just above the glass transition temperature of the substrate material. The tool is then driven into the substrate with a controlled force, which is kept up for several seconds. Afterwards, the tool-substrate-sandwich is cooled below the glass transition temperature. The polymer material solidifies again and the tool can be taken out of the structure. The big advantage of this method lies in its flexibility, the low internal stresses and high structural replication accuracy due to the small thermal cycle (approx. 40°C), so that structural replications in the micrometer-range are possible [5].

For higher resolution structures or structures with a very high aspect ratio, technologies like LIGA [6,7,8] or laser-LIGA have to be used to fabricate the embossing tool.

The LIGA-technique (abbreviation of the German words Lithographie, Galvanoformung, Abformung, which means lithography, electroplating and moulding) was developed during the 80s at the Forschungszentrum Karlsruhe (Research Center Karlsruhe), originally for the fabrication of nozzles for uranium isotope separation. With this technique it is possible to produce microstructures with very high aspect ratios (up to 100), very small structures (in the submicron range) and with very smooth walls of optical quality (surface roughness < 50 nm).

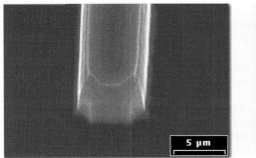 

Fig. 13 Left: apart of an embossing tool fabricated with reactively etched silicon , Right: the structure hot embossed in PMMA [9].

In practice with the method of hot embossing it is very difficult to obtain fully micro perforated product. The basic problem is that the perforated material coming from the pores has to be removed (vanish) from the perforated foil. Research is going on in this field by using a laminated (double) foil consisting of a first part that has to be perforated and a second part that will receive (or absorb) the perforated material. The second part has to have well-tuned elastic properties, it has to be rigid enough to support the first part during the perforation process and on the other hand it has to be elastic enough to make space to receive (or adsorb) the perforated material. It will be clear that this method is not well suitable for perforated products with a high porosity.

REFERENCES

[1] Stork Veco, Eerbeek, The Netherlands.
[2] C.J.Kruithof, H. Knol, W. Nijdam and C.J.M. van Rijn, PCT patent application 20 03 1999, Aquamarijn Micro Filtratrion, Stork Veco, Electroforming method, electroforming mandrel, electroformed product.
[3] Publication summary of the German Magazin PKV-12/98.
[4] C.J.M. van Rijn and J.Elders, NL 1001220, 17-091995, Mould, as well as a method to produce such a mould.
[5] L.Weber, W.Ehrfeld, H.Freimuth, M.Lacher, H.Lehr and B.Pech, SPIE Proceedings Series, Vol 2879, Micromachining and Microfabrication Process Technology II, Austin, 1996.
[6] W.Ehrfeld, D.Münchmeyer, Nucl. Instrum.Methods A 303 (1991) 523-532
[7] M.Heckele, W.Bacher, H.Blum, L.Müller and N.Ünal, Optik und Mikrosystemtechnik, F&M Zeitschrift für Elektronik, Nr.105 (1997)
[8] A.Both, W.Bacher, M.Heckerl, K.-D.Müller, R.Ruprecht and M.Strohrmann: Microsystem Technologies 2 (1996) 104-108.
[9] http://www.jenoptik.de

Chapter 3

Micro Engineered Membranes

> *"The ideal engineer is a composite ...*
> *He is not a scientist, he is not a mathematician,*
> *he is not a sociologist or a writer; but he may*
> *use the knowledge and techniques of any or all*
> *of these disciplines in solving engineering*
> *problems."*
>
> [N. W. Dougherty (1955)]

1. INTRODUCTION MICRO ENGINEERING

Micro engineering refers to the technology and practice of making three dimensional structures and devices with dimensions in the order of micrometers. This technology is also often referred to as Micro System Technology (MST). When considering such small devices, a number of physical effects, such as friction, capillary, electrostatic force etc., have an increased impact on the micrometer scale. The study of these effects on microscopic scales and the adequate design of functional micro systems is the fascinating objective of micro engineers. They further handle with topics such as micromechanics, which focuses on the moving parts of micro engineered devices, and microfluidics, which focuses on the fluidic properties of small devices.

The two constructional technologies of micro engineering are microelectronics and micromachining. Microelectronics, producing electronic circuitry on silicon chips, is a very well developed technology. Micromachining is the name for the techniques used to shape the structures and moving parts of micro engineered devices. In recent years micromachining and micro engineering have become synonyms, mainly because of the advent of nano engineering techniques, i.e. methods capable of shaping structures below the 100 nm scale.

One of the main goals of micro engineering is to be able to integrate microelectronic circuitry into micromachined structures in order to produce completely integrated systems (microsystems). Such systems could have the same advantages, such as low cost, reliability and compact size, as silicon chips produced in the microelectronics industry.

Silicon micro engineering is given most prominence, since this is one of the better-developed micro engineering techniques. Silicon is the primary substrate material used in the production of microelectronic circuitry (silicon chips), and is therefore the favourite, and often the most suitable candidate for the eventual production of micro systems.

Fig. 1 Micromachining: Deep trench etching in silicon (Courtesy Transducer Science and Technology Group, University of Twente).

2. MEMS SILICON MICRO ENGINEERING

The term MEMS is short for Micro Electro Mechanical Systems and originated from the field of miniaturised electromechanical actuators. MEMS is based on integrated circuit (IC) technology combined with thin film deposition, lithographic steps and etching techniques, to fabricate for example micromechanical and microelectronic devices, such as pressure sensors, movable micro mirrors, air bag sensors etc. The most favourite substrate material is mono crystalline silicon because it can be shaped (cf. micromachined) in many ways with anisotropic wet and dry etching techniques. MEMS today is a multidisciplinary field that involves challenges and opportunities for electrical, mechanical, chemical, and biomedical engineering.

Kurt Petersen [1], a MEMS pioneer, who founded Lucas NovaSensors, stated recently that "without exception, every company involved in electronics and small mechanical components should have programs to familiarise themselves with the capabilities and limitations of MEMS. Instrumentation companies that are not fluent in MEMS in the coming years will experience severely threatening competition." Petersen continued that, as MEMS evolves, it is becoming "less an industry unto itself and more of a critical discipline within many other industries." This means that application-specific MEMS processes will undoubtedly evolve as producers discover the best way to use MEMS for their products. Just like production for ICs, processes for MEMS will probably be limited by economic factors, and designers will attempt to satisfy their needs with the simplest, most economical technology.

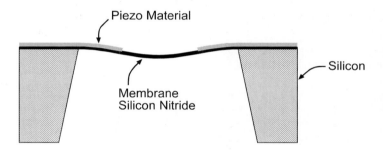

Fig. 2 Silicon micromachined pressure sensor.

3. MICRO ENGINEERED UNPERFORATED MEMBRANES

Silicon nitride membranes with a typical thickness of 1 micron and size of 1x1 mm have traditionally been used to manufacture miniaturised precision pressure sensors. Under application of a pressure difference the silicon nitride membrane will bend and will elongate a little. Piezo material (PZT) present on the membrane will also be elongated and a piezo voltage can be de detected. The relation between pressure difference, elongation and voltage difference can precisely be measured and gauged.

4. MICRO ENGINEERED PERFORATED MEMBRANES

With the former described membrane perforation technology a lower limit for feature sizes is about 2-5 micron in homogeneous foil and sheets with a thickness larger than 5 micron. Although the limits of lithography can be extended to smaller dimensions it is relatively difficult with e.g. wet etching to produce 0.5 micro orifices in a foil or sheet with a thickness larger than 5 micron.

Many applications, especially filtration applications, require pores in a membrane with a low flow resistance, i.e. the length of the pore should be as small as possible. Therefore Aquamarijn [2] introduced the microsieve, a very thin membrane with a specially designed macro perforated support structure to strengthen the thin membrane.

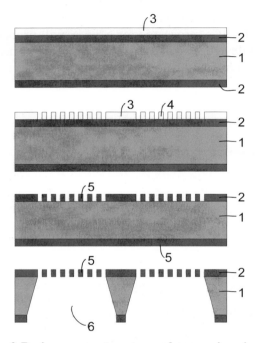

Fig. 3 Basic process steps to produce a microsieve. Layer 2 is low stress silicon nitride that is deposited on a 380 μm thick polished silicon wafer 1 by means of LPCVD (Low-Pressure Chemical Vapour Deposition). On top of this 1 μm thick silicon nitride layer 2 a photosensitive lacquer layer 3 is formed by spincoating. This layer is patterned with small pores 4 by exposing it to UV light through a photo mask, which subsequently can be developed. The pattern in the photosensitive layer 3,4 is transferred into the silicon nitride membrane 5 by means of RIE (Reactive Ion Etching) with a CHF$_3$/O$_2$-plasma.

Using supported microsieve structures and sub-micron lithography techniques form the semiconductor industry it is relatively straightforward to produce microsieves with sub-micron features. A brief review of some standard lithographic techniques will be given. In addition, promising new methods such as laser interference lithography introduced by Van Rijn since 1995 [3] will be discussed.

Fig. 4 SEM micrograph of a microsieve with a membrane and a support structure.

Microsieves produced using silicon micromachining offer new possibilities in microfiltration technology. The pores, which are well defined by photolithographic methods and anisotropic etching, allow e.g. accurate separation of particles by size. The membrane thickness is usually smaller than the pore size in order to keep the flow resistance small (one to three orders of magnitude smaller than other types of filtration membranes).

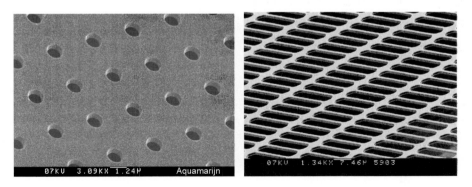

Fig. 5 SEM micrograph of a microsieve with circular and slit-shaped pores (Courtesy Aquamarijn Research).

Various process flows will be discussed for economical production of microsieves with a high effective use of the membrane area.

5. SEMICONDUCTOR LITHOGRAPHY

Contact mask lithography

The contact mask technique makes use of a mask that is in direct contact with the photo lacquer layer during illumination with a suitable UV light source. The patterns in the mask are copied one-to-one to the size of the obtained patterns in the photo lacquer layer. Practically contact masks can be used if the smallest features are larger than 0.5-1 micron, slightly depending on the wavelength of the used UV source.

Fig. 6 shows how the contact mask lithography technique is used to produce a microsieve with a special designed support structure in a silicon <100> wafer. During the KOH etching process silicon is etched from the backside as well as from the topside through the pores of the silicon nitride membrane, both etch techniques have been published by Van Rijn [3].

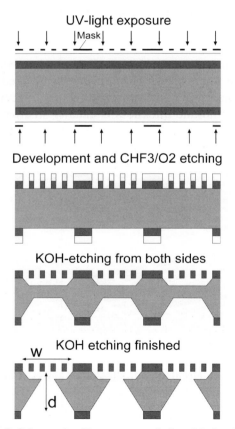

Fig. 6 Schematic diagramme of the fabrication of a microsieve with contact mask lithography.

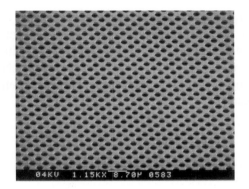

Fig. 7 SEM micrograph of the surface of a microsieve (brochure Aquamarijn 1995) where the silicon is etched and removed via the perforations in the membrane.

Then main disadvantage of using a direct contact mode between the mask and the photo-lacquer layer is that the presence of small particles or contaminants (> 1 micron!) between the mask and the photo-lacquer layer may blur the pattern transfer. The amount of 'blurring' is of coarse directly related by the ratio of the wavelength of the used light and the smallest feature size. Defect free processing of wafer-sized microsieves with small perforations is therefore virtually impossible even in high-class Clean Room environments. A particle with a size of 10-20 micron located between the contact mask (e.g. patterned chromium layer on a glass plate with thickness 2 mm) and the lacquer layer is able to blur or destroy the pattern transfer over an area of 1-100 mm^2. For wafer-sized microsieves or for microsieves with a pore size smaller than 1 micron Aquamarijn therefore made use of more modern lithographic techniques (e.g. ASML wafer stepper's, DIMES Delft, Onstream Eindhoven) to produce microsieves and to be used for research by students and PhD's.

WAFER STEPPER LITHOGRAPHY

In light projection lithography [4,5,6] an optical lens system is used to project the pattern of the mask on the photo-lacquer layer (contact-free technique). The blurring of the pattern is now only restricted to the size of the particles present on the photo lacquer layer. Some particles however can be easily removed if they are loosely attached to the lacquer layer. The projection system often downscales the pattern of the mask between one and ten times with respect to the projected pattern in the lacquer layer. The typical lacquer area that can be projected in one step is about 1-2 cm^2. More steps (of the wafer stepper) are therefore needed to expose the whole wafer (3-12 inch diameter). A typical single step time is 1-2 seconds. This time cannot easily be reduced because of the mechanical repetition step time and the minimum exposure time. Although wafersteppers are very expensive (new generation > 5M euro) they are very economical in mass production. Also, it is relatively easier to get smaller pattern dimension (e.g. 0.12 micron line-widths) than contact mask applications due to the use of high-resolution lenses with a large numerical aperture (NA). Additional advantages are that the mask pattern is relatively large and can be made with relatively cheap laser writers.

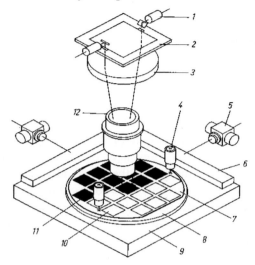

Fig. 8 Wafer stepper. 1 optical adjustment system, 2 reticule 3 tele-centric lens, 4 alignment optic, 5 laser interference meter, 6 mirror, 7 wafer, 8 wafer chuck, 9 x-y table, 10 projected and un-projected pattern fields, 12 objective.

Lithographic down scaling has historically been accomplished by optimising the parameters in the Rayleigh model for image resolution: In this model [7,8], image resolution = k_1/NA, and depth of focus (DOF) = $k_2\lambda/NA^2$, where λ = exposure wavelength and NA = numerical aperture (k_1, k_2 are constants for a specific lithographic process). To pattern devices with decreasing feature sizes, photo-lacquer exposure wavelengths were reduced and numerical apertures were increased [9,10].

Table 1

DUV- and EUV- monochromatic laser light sources.

Wavelength [nm]	Lasertype
248	KrF
193	ArF
157	F_2

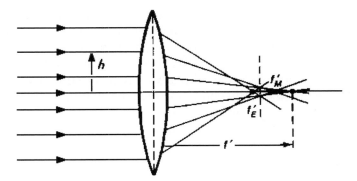

Fig. 9 Chromatic aberration.

Normally lenses are also added for aberration correction [11]. The alignment of the projection optics required a specially developed DUV interferometer to reach diffraction-limited performance. Aberrations are generally described as a variation in the phase of light transmitted by a lens as a function of position within the lens (i.e., as a function of position in the pupil) for a particular point in the field. Since the lens of the pupil is most commonly circular, cylindrical coordinates (r,q) are used. For convenience, this phase error is often fit to a polynomial in radius r and the sinus and cosines of the polar angle q. Rather than using a general polynomial, however, a special polynomial form, called a Zernike polynomial, is used. The advantage of using this Zernike polynomial is that the specific polynomial terms have specific and useful interpretations such as coma, astigmatism, tilt, etc. Since aberrations vary as a function of field position, a separate Zernike polynomial must be determined for various points in the field.

An assumption of the Zernike description of aberrations is that the light being used is mono-chromatic. The phase error of light transmitted through a lens applies to a specific wavelength. The property of a lens element that allows light to bend is the index of refraction of the lens material. Since the index of refraction of all materials varies with wavelength (a property called dispersion), lens elements will focus each wavelengths differently. This fundamental problem, called chromatic aberration, can be alleviated by using two different glass materials with different dispersions such that the chromatic aberrations of one lens element cancels with the chromatic aberrations of the other. As with all aberrations, this cancellation is not perfect meaning that all lens will have some level of residual chromatic aberrations.

The effects of chromatic aberrations depend on two things: the degree to which the Zernike polynomial coefficients vary with wavelength (the magnitude of the chromatic aberrations), and the range of wavelengths used by the imaging tool. For example, a typical i-line stepper might use a range of wavelengths on the order of 10 nm, whereas a KrF excimer laser based deep-UV stepper may illuminate the mask using light with a wavelength range of 2 pm. The obvious difference in the magnitude of the wavelength ranges between these two light sources does not mean, however, that chromatic aberrations are a problem for i-line but not for deep-UV. I-line lenses are designed to use the (relatively) wide range of wavelengths by chromatically correcting the lens with several different types of glass. A typical deep-UV lens, on the other hand, makes no attempt at chromatic correction since only fused silica is used for all the lens elements in the lens. As a result, chromatic aberrations are a concern in deep-UV lithographic lenses even when extremely narrow bandwidth light sources are used.

In principle, every Zernike coefficient is a function of wavelength. In reality, for lenses with no chromatic corrections (i.e., where every lens element is made from the same material), one Zernike term dominates: the term that describes defocus. The plane of best focus shifts with changes in wavelength in a nearly linear fashion. Since the centre wavelength of most excimer lasers is easily adjustable over a reasonably large range, this effect can be readily measured for any given stepper. To account for this behaviour, the Zernike term $Z3$ (the defocus term) can be used.

An excimer laser light source will emit light over a range of wavelengths with a characteristic spectrum (output intensity versus wavelength) that can be roughly approximated as somewhere between a Gaussian shape and a Lorentzian shape. The full width half maximum (FWHM) of the laser output spectrum is called the bandwidth of the laser, with today's KrF lasers having less than 1pm bandwidths. Each wavelength in the laser spectrum will be projected through the imaging lens, forming an aerial image shifted in focus according to the wavelength response characteristic for that lens. The total aerial image will be the sum of all the images from all the wavelengths in the source, resulting in a final aerial image that is somewhat smeared through focus.

For a 180 nm image, the 1pm bandwidth shows a slight degradation of the aerial image, but the 3 pm bandwidth is clearly unacceptable. As one might expect, smaller features will be more sensitive to the effects of chromatic aberrations than larger features. Thus, as deep-UV is pushed to the 130 nm feature size node (pitches of 260 nm, isolated features below 100 nm), even these seemingly minuscule sub-picometer excimer bandwidths will require careful consideration for their chromatic aberration effects.

During the last ten years, image resolution was sufficiently increased to scale minimum dimensions from 1 µm feature sizes for the 1Mb DRAM (computer chip) devices to 0.25 µm features for the 256 Mb DRAM. The depth of focus is proportional to the inverse of the square of the numerical aperture; thus, if increasing NA enhances resolution, the depth of focus becomes very small. If decreasing the wavelength enhances the resolution, the corresponding decrease in depth of focus is less severe. For lithography at the diffraction limit, a shorter wavelength provides more depth of focus at a particular resolution value because the shorter wavelength allows a lower NA photolithography tool to achieve equivalent resolution.

DRAM production of 64 Kb devices utilised scanning exposure equipment operating at a G-line wavelength of 436 nm. These tools were capable of operating at several different exposure wavelengths, including 436 nm, 313 nm, and 245 nm. IBM used these tools for 256 Kb DRAM chips by formulating a resist, TNS, which was functional at the 313 nm exposure region, allowing the critical feature size to be scaled from 2 μm to 1.4 μm. This approach was repeated for 1Mb DRAM chips, and a 245 nm exposure region was used to obtain the 1-μm critical features with the same tool set, which required the development of the first production deep-UV (DUV) chemically amplified, negative-tone resist. At this time, G-line steppers were introduced at a NA of 0.35, and the process for the production of 1Mb DRAM reverted to G-line lithography. For the 4 Mb DRAM generation, stepper technology was extended by scaling the wavelength to the I-line (= 365 nm). High-NA (0.42-0.45) G-line steppers were used to manufacture 4Mb pilot line products at 0.8 μm ground rules, while lower-NA (0.35) I-line steppers were used for final qualification of products with 0.6-0.7 μm critical features. For the 16Mb generation, DUV in the 245 nm exposure region was utilised for development and initial production, while very high NA (0.5-0.6) I-line (365 nm) was used for volume production at 0.4-0.5 μm image sizes. The 64 Mb and 256 Mb DRAM generations use high-NA (0.5-0.6) DUV tools at 248 nm for images at 0.25-0.35 μm resolution.

The most notable transition in recent years however, is one of the most exciting. That is, the transition to the Sub-Wavelength regime. The fact that leading edge device feature sizes are now significantly smaller than the wavelength of light with which they are imaged is transforming the entire semiconductor industry from IC design to wafer fabrication.

The sub-wavelength environment is here to stay until the next generation lithography becomes available. The 1999 SIA International Technology Roadmap for Semiconductors (ITRS) indicates a useful life for leading edge optical lithography for at least the next three technology nodes (130 nm, 100 nm and 70 nm), through the year 2008. Given the previous predictions of leading technologists that the IC industry would not see optical lithography used below 1 mm printed feature sizes, these estimates might be considered conservative.

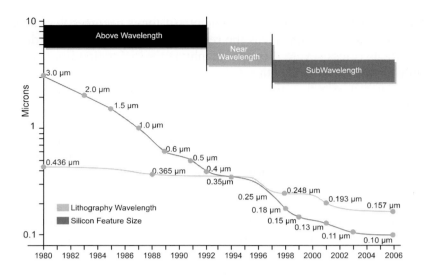

Fig. 10 The Sub-Wavelength Gap. Structures with a feature size below 0.25 micron are made using lightsources with a wavelength larger than the feature size, courtesy EE-design [12].

Reduction of the exposure wavelength from 436 nm (g-line) to 248 nm (DUV) combined with increasing NA from 0.28 to over 0.70 has allowed leading edge fabricators to meet the customer demands and stay on the productivity curve predicted by Moore's Law. Leading edge R&D facilities are now exercising technology based on a wavelength of 193 nm and exploring 157 nm. In parallel, imaging system suppliers are developing higher quality lenses with NA's greater than 0.70. This combined activity ensures that the base technologies necessary to support the extension of optical lithography well into the foreseeable future are firmly in place.

Resolution enhancement techniques

While the traditional technologies continue to advance at a breathtaking speed they are becoming quite costly and are no longer able to provide the resolution and DOF with an acceptable process window on their own. This is a direct result of the Sub-Wavelength environment. Fortunately there is an additional knob the industry can adjust to improve the overall system performance, i.e. the reduction of k_1. The field of low k_1 lithography includes such techniques as optical proximity correction (OPC), phase shift masks (PSM), off-axis or modified illumination, spatial filters and high contrast resists. These techniques, collectively referred to as Resolution Enhancement Techniques (RET's), work in conjunction with the traditional techniques of decreasing wavelength and increasing NA to extract the highest level of performance possible from the advanced lithography systems.

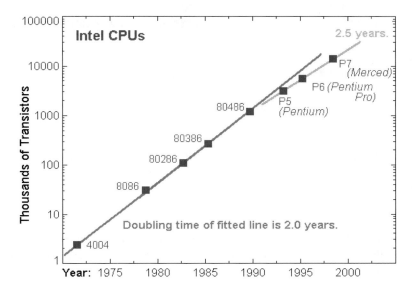

Fig. 11 In 1965 Gordon Moore (Intel) already predicted that the doubling time for chip structures would be about 18 months. The actual doubling time was slightly larger and is of course dependent on the specific devices, the law seems to be more valid for Digital Signal Processor modems of Lucent than for the Computer Process Units of Intel [13].

Central to the resolution enhancement techniques are the Phase Shift Masks. The principle PSMs deployed in the industry include attenuated (usually used for contacts and metal layers) and alternating aperture (used for CD control for gates). The manufacturing environment has seen a steady decline in the value of k_1 from 0.9 to below 0.5. The expectation is that k_1 will come very close, in the next few years, to the theoretical limit of 0.25-0.30. It is interesting to note that if you consider NA as having a "practical limit" of 0.80 ± 0.05 a resolution of 100 nm would drive k_1 values to a range of 0.32 for 248 nm, 0.41 for 193 nm and 0.51 for 157 nm. The technique best suited to function in this environment while providing the best lithographic performance is the alternating aperture PSM.

Alternating Aperture Phase Shift Masks

A critical issue associated with using reticules in a low k_1 environment is the departure from linear behaviour to highly non-linear behaviour. This is manifested by the increase in optical proximity effects as well as the fact that reticule errors and defects will be magnified on the wafer (mask error enhancement factor – MEEF). Fortunately, despite all the pitfalls associated with low k_1 lithography there is a common solution that can increase the overall process window, minimise or even reverse MEEF and provide a solution that has minimal impact to wafer manufacturing, the alternating aperture phase shift mask.

There are several variations of the alternating aperture phase shift mask. These include the multiphase mask, the conjugate twin shifter mask, and the double exposure PSM.

The multiphase mask and the conjugate twin shifter are clear field masks and require one exposure to print the desired pattern. These techniques use weaker phase transitions (such as 0-60°, or 0-90°) in order to avoid printing resist lines, which occur at 0-180° edges. The third technique uses dark field Alternating PSM to define the gates, and clear field Chrome-on-Glass (COG) mask to print the rest of poly layer and trim out unwanted resist lines remaining after the first exposure. The multiphase mask is difficult to design and manufacture, and requires a substantial amount of chip real estate in which to place the phase transitions. The conjugate twin shifter masks demonstrate high sensitivity to defocus conditions and will pose a serious challenge on the design side to compensate for such behaviour. The Dark Field Double Exposure PSM technique offers the best manufacturability solution.

A manufacturable solution

The advantages of the dark field double exposure alternating aperture PSM techniques (including techniques such as Canon's IDEAL) are substantial:

- Stable behaviour through focus
- Available automated design and verification software
- The phase shifting layer can be written using a faster and less expensive laser
- A more manufactureable reticule fabrication process (it is much easier to make "defect free" masks since the area sensitive to the phase defects is greatly reduced compared with alternate PSM techniques)
- An "effective" k_1 of 0.65 for 100 nm features due to frequency doubling
- Less sensitive to lens aberrations under optimum lithographic conditions

The Numerical Technologies, Inc. dark field double exposure technique [14], decomposes the most critical layer into two layers that are less critical than the original layer. This offers a substantial benefit by allowing the stepper settings for the PSM and COG levels to be optimised separately; High NA / low sigma for the phase shift layer and High NA and high sigma for the COG layer. Overlay required for the two levels can be met by any leading edge imaging systems. Trade-offs do need to be made currently with respect to the expected throughput of the imaging systems although this is being actively addressed.

To ensure optimal performance, PSM specific OPC is required to correct new types of optical and process distortions. Two-dimensional OPC effects can be successfully corrected using the NumeriTech's integrated OPC PSM design tool set, iN-Phase.

Technology Adaptation

The increasingly rapid adoption of Dark Field Double Exposure PSM technology is largely because the technology works. The infrastructure that e.g. NumeriTech has helped to develop includes design, mask, lithography, manufacturing equipment, and semiconductor companies working together to provide an integrated solution for the customer.

Results from Lucent Bell Labs attest to the continued adoption of the dark field double exposure PSM and provides real examples of what is capable today in a manufacturing environment.

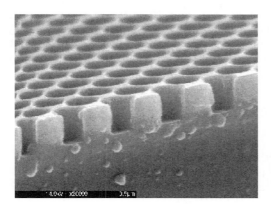

Fig. 12 SEM micrograph of 500 nm pore structures in silicon nitride obtained with use of an ASML wafer stepper (Courtesy Aquamarijn/DIMES-Delft).

Summary

The semiconductor industry is once again changing in attempt to sustain its productivity engine. The focus, as it has been before, is on finding a technically capable, timely, and cost effective lithography solution. This time however, the solution must successfully operate in the Sub-Wavelength environment. Resolution enhancement techniques, particularly Dark Field Alternating Aperture PSMs are here, are real and produce results under those constraints. Those that make the transition will realise the benefits.

As there is an ongoing demand for microsieves with smaller pore sizes the microsieve technology will benefit from the progress and demand for semiconductor products with smaller feature sizes.

E-beam and future semiconductor lithography

Electron beam lithography (EBL) is a specialised technique [15,16] for creating the extremely fine patterns required by the modern electronics industry. Derived from the early scanning electron microscopes, the technique in brief consists of scanning a beam of electrons across a surface covered with a resist film sensitive to those electrons, thus depositing energy in the desired pattern in the resist film. The process of forming the beam of electrons and scanning it across a surface is very similar to what happens inside the everyday television or CRT display, but EBL typically has three orders of magnitude better resolution.

Table 2

New predictions (year node potential solutions) for the ever decreasing line-width's from 1999 ITRS Lithography Roadmap. KrF, ArF und F2 (F2) are lasertypes. RET is Resolution Enhancement Technology, including Phase-Shift-Mask technology for enhanced resolution, EPL is Electron Projection Lithography, EUV is extreme Ultra-Violet, IPL is Ion Projection Lithography, PXL is proximity X-ray lithography and EBDW for electron-beam direct-write techniques.

Year	Line-width	Lithographic tool
1999	180 nm	KrF
2002	130 nm	KrF + RET, ArF
2005	100 nm	ArF + RET, F2, EPL, PXL, IPL
2008	70 nm	F2 + RET, EPL, EUV, IPL, EBDW
2011	50 nm	EUV, EPL, IPL, EBDW
2014	35 nm	EUV, IPL, EPL, EBDW
>2014		Innovative Technology

The main attributes of the technology are:

1) It is capable of very high resolution, almost to the atomic level;
2) It is a flexible technique that can work with a variety of materials and an almost infinite number of patterns;
3) It is slow, being one or more orders of magnitude slower than optical lithography;
4) It is expensive and complicated; E-beam tools cost millions of dollars and require frequent service to stay properly maintained.

The first electron beam lithography machines, based on the scanning electron microscope (SEM), were developed in the late 1960s. Shortly thereafter came the discovery [17] that the common polymer PMMA (polymethyl methacrylate) made an excellent electron beam resist. It is remarkable that even today, despite sweeping technological advances, extensive development of commercial EBL, and a myriad of positive and negative tone resists, much work continues to be done with PMMA resist on converted SEMs.

Currently, electron beam lithography is used principally in support of the integrated circuit industry, where it has three niche markets. The first is in mask making, typically the chrome-on-glass masks used by optical lithography tools. It is the preferred technique for masks because of its flexibility in providing rapid turnaround of a finished part described only by a computer CAD file. The ability to meet stringent line-width control and pattern placement specifications, on the order of 50 nm each, is a remarkable achievement.

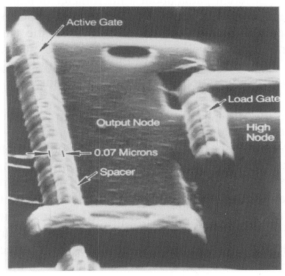

Fig. 13 SEM micrograph of a portion of an integrated circuit fabricated by electron beam lithography. The minimum dimensions are less than 0.1 μm [18].

Because optical steppers usually reduce the mask dimensions by 4 or 5, resolution is not critical, with minimum mask dimensions currently in the one to two μm range. The masks that are produced are used mainly for the fabrication of integrated circuits, although other applications such as disk drive heads and flat panel displays also make use of such masks.

An emerging market in the mask industry is a mask for X-ray lithography. These masks typically have features ranging from 0.25 μm to less than 0.1 μm and will require placement accuracy and line-width control of 20 nm or better.

Should X-ray technology ever become a mainstream manufacturing technique, it will have an explosive effect on EBL tool development since the combination of resolution, throughput, and accuracy required, are far beyond what any single tool today can offer.

The second application is direct write for advanced prototyping of integrated circuits and manufacture of small volume speciality products, such as gallium arsenide integrated circuits and optical waveguides [19,20].

Here both the flexibility and the resolution of electron beam lithography are used to make devices that are perhaps one or two generations ahead of mainstream optical lithography techniques.

Finally, EBL is used for research into the scaling limits of integrated circuits and studies of quantum effects and other novel physics phenomena at very small dimensions. Here the resolution of EBL makes it the tool of choice. A typical application is the study of the Aharanov-Bohm effect [21,22], where electrons travelling along two different paths about a micrometer in length can interfere constructively or destructively, depending on the strength of an applied magnetic field. Other applications include devices to study ballistic electron effects, quantisation of electron energy levels in very small structures, and single electron transistors [23,24]. To see these effects, typically requires minimum feature sizes of 100 nm or less, and operation at cryogenic temperatures.

Alternative Techniques

It is prudent to consider possible alternatives before committing to EBL technology. For chrome-on-glass optical mask fabrication, there are optical mask writers available that are based either on optical reduction of rectangular shapes formed by framing blades or by multiple individually controlled round laser beams. Although at present EBL is technologically ahead of optical mask writers, this may not continue in the future. However, EBL will continue to provide a resolution advantage over the optical mask writers that may be important for advanced masks using phase shift or optical proximity correction. For one-to-one mask fabrication (i.e. X-ray), EBL will continue to be the most attractive option.

Optical lithography using lenses that reduce a mask image onto a target (much like an enlarger in photography) is the technique used almost exclusively for all semiconductor integrated circuit manufacturing. Currently, the minimum feature sizes that are printed in production are a few tenths of a micrometer. For volume production, optical lithography is much cheaper than EBL, primarily because of the high throughput of the optical tools. However, if just a few samples are being made, the mask cost (a few thousand euro's) becomes excessive, and the use of EBL is justified. Today optical tools can print 0.18 μm features in R&D laboratories, and less than 0.12 μm should be possible within a few years.

By using tricks, optical lithography can be extended to 0.1 μm or even smaller. Some possible tricks include overexposing/overdeveloping, phase shift and phase edge masks, and edge shadowing [25]. The problem with these tricks is that they may not be capable of exposing arbitrary patterns, although they may be useful for making isolated transistor gates or other simple sparse patterns. Another specialised optical technique can be used to fabricate gratings with periods as small as 0.2 μm by interfering two laser beams at the surface of the sample. Again, the pattern choice is very restricted, although imaginative use of blockout and trim masks may allow for the fabrication of simple devices [26].

X-ray proximity printing may be a useful lithographic technique for sub-0.25 μm features [27]. Again, it requires a mask made by EBL, and since the mask is one-to-one this can be a formidable challenge. However, if the throughput required exceeds the limited capabilities of EBL, this may be an attractive option. The disadvantage is that X-ray lithography is currently an extremely expensive proposition and the availability of good masks is limited. It also requires either a custom-built X-ray source and stepper or access to a synchrotron storage ring to do the exposures. With care, X-ray lithography can also be extended to the sub-0.1 μm regime [28].

The final technique to be discussed is ion beam lithography. The resolution, throughput, cost, and complexity of ion beam systems is on par with EBL. There are a couple of disadvantages, namely, limits on the thickness of resist that can be exposed and possible damage to the sample from ion bombardment. One advantage of ion beam lithography is the lack of a proximity effect, which causes problems with line-width control in EBL. Another advantage is the possibility of in situ doping if the proper ion species are available and in situ material removal by ion beam assisted etching. The main reason that ion beam lithography is not currently widely practised is simply that the tools have not reached the same advanced stage of development as those of EBL.

Finally, it should also be noted that modern computer simulation tools, together with a detailed understanding of the underlying physics, in many cases allows one to accurately predict exploratory device characteristics without ever having to build the actual hardware.

6. LASER INTERFERENCE LITHOGRAPHY

In 1995 van Rijn [29] proposed for the first time the use of laser interference lithography for the production of micro and nano sieves. Because of the relatively simple periodic structures (orifices and slits) that are needed for the production of microsieves and the fact that laser interference is potentially a very non-expensive patterning technique [30,31,32,33,34], even applicable on non-planar surfaces (very large focus depth), this method will be discussed more in detail. There is little published material on this subject so far [35]. In 1996 in a collaboration of the author [36] with G.J. Veldhuis [37], A. Driessen, P. Lambeck and prof. C. Lodder of the UT, an existing laser apparatus [29] was modified to obtain photo-lithographic patterns for the development of microsieves and isolated magnetic domains. The results were very promising and there were some serious thoughts to patent the laser interference technology for the UT, however an American publication [38] a few months earlier on the production of isolated magnetic dots obtained with a similar method [39,40] jeopardised this plan. Fortunately, the method to produce microsieves with use of a double laser interference technique was already patented in 1995 by Aquamarijn [3]. In a later stage [41], Ph.D's continued with the initial pioneering work. A large Laser Interference Lithography project has recently (2001) been granted to the UT by STW.

Double laser interference exposure technique

When two planar waves of coherent light interfere, a pattern of parallel fringes will appear. These fringes can be used for the exposure of a photosensitive layer.

**Fig. 14 Left: first exposure of the substrate with a photo lacquer layer. Middle: 90°
rotation of the substrate. Right: second exposure of the photo lacquer layer.**

The depth of focus of this method is dependent on the coherence length of the light and can be in the order of metres or more, compared to (sub)microns for conventional optical lithography systems. As a result the demands on substrate flatness and wafer positioning are less critical.

After the first exposure the substrate is rotated over 90° and exposed to laser interference lines again. Now the gratings cross each other and after development a square array of lacquer pores (Fig. 15a) remain. The exposure time of the photo-lacquer layer is a critical factor. In case, a longer exposure time is chosen, a pattern as shown in Fig. 15b will be obtained. Upon further increasing the exposure time, isolated photo-lacquer dots (Fig. 15c) are formed.

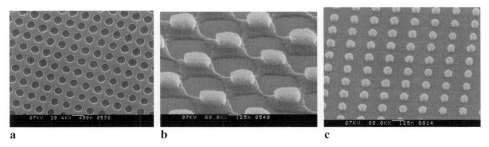

a   b   c

Fig. 15 SEM micrograph of photo-lacquer layer that remains after a double exposure in the laser interference set up. a): short exposure time, b): intermediate exposure time, (left and middle: Courtesy J. Micromech. Microeng.) c): long exposure time (Courtesy G.J. Veldhuis).

6.1   DEVICE FABRICATION WITH SHORT EXPOSURE TIME

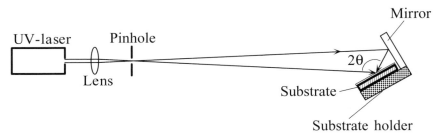

Fig. 16 Set-up for Laser Interference lithography: 'Lloyd's mirror configuration'.

Part of an incoming plane wave is reflected by the mirror and interferes with the undisturbed part of the wave to form an interference pattern (grating) on the substrate surface. To produce the plane wave, TE polarized light of an argon laser with a wavelength $\lambda_{uv} = 351.1$ nm is spatially filtered and expanded by focusing it on a pinhole.

If the light intensity of each beam is I_0, the radiance on the surface is given by:

$$I = 4I_o \sin^2\left(\frac{\pi x}{\Lambda_x}\right)$$

(1)

with Λ_x the fringe period in the x-direction planar to the photo-lacquer layer:

$$\Lambda_x = \frac{\lambda_{UV}}{2\sin\theta}$$

(2)

Here λ_{uv} is the wavelength of the laser light in the medium that surrounds the substrate (usually air) and θ the half-angle between the two beams. The smallest period that theoretically can be obtained is $\theta = 90°$, which is equal to $\lambda_{uv}/2$, and for this configuration $\Lambda=\lambda_{uv}/2=175$ nm. The corresponding smallest pore size (with a porosity >30 %) will be approximately $175/2 = 88$ nm. It is evident that, at low porosity's smaller pore sizes can be made.

Since the beam is only split for a short path length near the substrate, this set-up is very insensitive to mechanical instabilities and no feedback loop [42] is required to stabilise the interference pattern. The thickness of the photosensitive layer needs to be chosen with care to avoid problems with the periodic pattern perpendicular to the substrate surface due to interference between the incoming beam and the one reflected on the substrate surface. Its period is given by $\Lambda_\perp=\lambda_{v\bar{\omega}}/2n\cos\theta_n$ where n is the refractive index of the photo resist and $2\theta_n$ the angle between the beams inside the resist. With $\theta =20°$ and with $n =1.7$ at $\lambda_{v\bar{\omega}}=351.1$ nm one finds $\Lambda=510$ nm and $\Lambda_\perp=105$ nm. Therefore the thickness of the photo resist layer is chosen smaller than 105 nm. The area that can be patterned using a mirror of $2.5 \times 2.5 \text{cm}^2$ equals approximately $9 \times 9\text{mm}^2$ for $\Lambda =510$ nm.

The backside of a single crystalline 3" <100>-silicon wafer with a thickness of 380 µm is pre-etched to a thickness of 15 µm using optical lithography and conventional KOH etching (25%, 70°C), see Fig. 17. On the front side of the pre-etched support (1) a layer (2) of amorphous low stress [43] silicon rich silicon nitride with thickness 0.1 µm is deposited by means of 'Low Pressure Chemical Vapour Deposition' by reaction of dichloresilane (SiH_2Cl_2) and ammonia (NH_3) at a temperature of 850 °C. Except at the area where the microsieve pattern will be formed an etch mask layer (3) of sputtered chromium with a thickness of 30nm is deposited. On top of this chromium layer (3) a layer (4) of positive resist was spun and patterned using interference lithography. A 100 nm thick layer (4) of positive photo resist (Shipley S1800-series) was spun, followed by a 5 min prebake at 90 °C to evaporate the solvent. The resist was exposed to the interference line pattern for 45 seconds. The intensity of the incoming light in the exposed area was measured to be 2 mW/cm^2 for normal incidence ($\theta = 0°$). After rotating the substrate over 90° the exposure was repeated. The resist was developed for 15 seconds in a 1:7 mixture of Shipley-Microposit 351 developer and de-ionised water and dried by spinning.

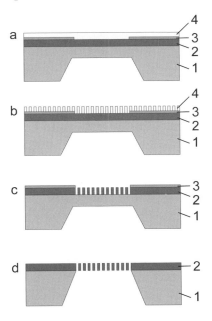

Fig. 17 Schematic representation of the fabrication process of a microsieve. The numbers indicate the silicon support (1), the silicon nitride membrane (2), the chromium etch mask (3) and the photo resist layer (4).

The exposure time was chosen such that only at the crossings of the grid lines (after first and second exposure) the resist receives enough energy to be removed completely after development. Therefore a two dimensional pattern of pores is created in the resist. A SEM image of the exposed (2 × 45s) and developed resist is given in Fig. 15a. The diameter of the pores in the photo resist depends on the duration of exposure, herewith giving a possible tool to vary the pore size at a constant pore-to-pore distance. However, when the exposure time is chosen too long (2 × 75s) the pores in the resist pattern may become too large and will overlap, Fig. 15b.

Next the interference pattern is transferred into the silicon nitride membrane layer (2) by means of CHF$_3$/O$_2$ reactive ion etching at 10 mTorr and 75 Watt for 2 minutes forming the required perforations. Subsequently the silicon underneath the membrane layer (2) is anisotropically etched with an SF_6/O_2 plasma at 100 mTorr and 100 Watt for 10 minutes with an etch rate of $2\mu m/min$ in order to form the macroscopic openings in the support (1), Figure below shows a SEM photograph of the resulting perforated membrane layer (2) showing a very regular pore pattern, the pore size being 260 nm with a pore to pore spacing of 510 nm. The pore size was very uniform over the whole 9×9 mm^2 area.

Fig. 18 SEM micrograph of the microsieve membrane showing pores with a diameter of 260 nm in a 100 nm thick silicon nitride layer (Courtesy Nanotechnology, van Rijn [36]).

6.2 SHORT VERSUS LONG LASER INTERFERENCE EXPOSURE TIMES [41] *

For the fabrication of microsieves an array of pores is needed. Such an array can be obtained by a double exposure with an intermediate rotation over an angle α. For $\alpha = 90°$ the array is square and for $\alpha = 60°$ it is hexagonal [44]. After the second exposure the photo-lacquer layer is developed, where the sum of the two exposure doses determines whether it dissolves in the developer. For a positive resist the areas that receive a dose above a certain threshold dose will completely dissolve. For a certain (short) exposure time only the areas where two intensity maxima overlapped will have received a total dose that exceeds the threshold value. These areas will dissolve during development and an array of

* Partially reprinted with permission from J. Micromech. and Microeng. (2000).

pores will appear in the resist layer. An SEM micrograph of such an array of pores is shown in Fig. 19.

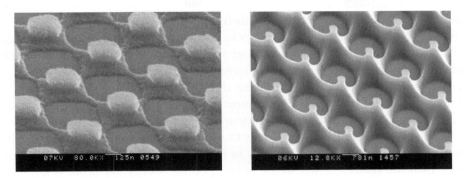

Fig. 19 Photo-lacquer layers after a double intermediate exposure time and subsequent development (courtesy Aquamarijn research).

The picture shows that the resist between the pores exhibits 'saddle points', due to the overlap of a minimum and a maximum. For longer exposure times these saddle points will also dissolve in the development process. The result is then an array of posts on the places where two minima overlap. The transition from pores to posts occurs quite rapidly, as the difference in received dose between the centre of a pore and a saddle point is only a factor of 2. The formation of posts is less critical. In theory the posts will never disappear for increasing exposure times, as the received dose in the centre of a post is always zero. The exposure process for a double-exposed resist layer for $\alpha = 90°$ is explained below.

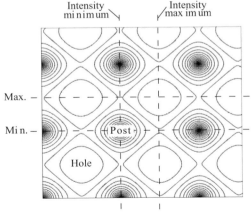

Fig. 20 Contour plot of the received dose of UV-light in photo-lacquer layer exposed twice to a fringe pattern with an intermediate rotation over 90°.

The contour lines in Fig. 20 are equi-dose lines and have been plotted on a logarithmic scale. This implies that, going from a bright region towards a dark region, the difference between two contour lines represents a decrease in received dose by a constant factor (in figure this factor is $\sqrt{2}$). Suppose the first contour in a bright region (where a pore is formed) indicates the threshold dose. The photo-lacquer within this contour line will dissolve and an array of pores will appear. If the exposure time is increased by a factor of $\sqrt{2}$, the next (square shaped) contour line will indicate the threshold dose. An array of large square pores will appear, which are almost interconnected. Another increase in exposure time leads to the next contour line, which represents the contour of a post. A further increase in time leads to smaller posts, but the line density shows that the change in size of the posts is less dependent on a relative increase in dose than the change in size of the pores. In other words, the process latitude for the formation of posts is wider than for the formation of pores. This is important for the uniformity of the array, as in practice the laser light varies in intensity over the surface due to a Gauss distribution profile, the distance to the source, imperfect filtering, drift of the beam and laser noise. In order to overcome the limitations of the pore-formation process, the post-formation process can be used in combination with an image reversal method. Decker et al. [45] used exposure doses for the creation of posts, but applied before development a base-catalysed method to make the exposed areas insoluble. A subsequent flood exposure with a UV lamp made the previously unexposed parts soluble. The result was an array of pores in photo-lacquer produced with the wide process latitude of the post-creation method.

Pattern transfer from a photo-lacquer mask into a silicon nitride layer usually gives rise to tapered walls due to lateral etching of the mask. As tapered pores affect the filtration performance of membranes an alternative process was developed. Posts were created and the pattern was inverted with a chromium lift-off process. 15 nm of chromium is evaporated onto the posts, which are then removed in an ultrasonic acetone bath. The remaining chromium forms a perforated layer. This layer serves as an etch mask for plasma etching. Pattern transfer into the silicon nitride using a chromium mask improves in comparison with a photo lacquer mask, as the plasma hardly attacks the chromium.

During exposure the laser light partly reflects from the substrate and interferes with the incoming light. This causes the creation of an interference pattern in vertical direction. The period Λ_z of this pattern is given by:

$$\Lambda_z = \frac{\lambda}{2n_{res} \cos \theta_{res}} \tag{3}$$

,where n_{res} is the refractive index of the photo-lacquer layer and θ_{res} the angle of incidence in the resist.

As a result of this vertical pattern, the posts will have a rippled sidewall, which makes the lift-off process more effective.

Experimental set-up

The easiest interference lithography system is known as 'Lloyd's mirror configuration' see Fig. 16. Part of an incoming plane wave reflects on the mirror and interferes with the undisturbed part of the wave to form an interference pattern (grating) on the substrate surface. To produce the plane wave, TE polarised light of an argon laser with wavelength λ =351.1 nm is spatially filtered and expanded by focusing it on a pinhole. For large θ (small fringe periods) the system works satisfactory, but for small θ the image of the mirror on the substrate becomes so small that most of the substrate is not exposed to the interference pattern. An increase in mirror size is expensive, as the demands on smoothness and flatness are high. However, the set-up is very useful for research purposes, as it is simple and θ can be changed easily by tilting the substrate holder. For the exposure of large surfaces a second set-up was build. In this set-up the laser beam is split, after which both beams are expanded separately.

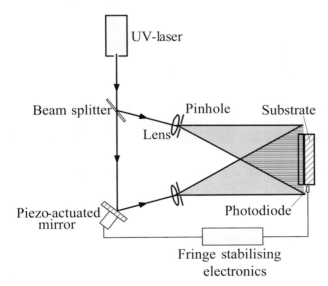

Fig. 21 Interference-lithography set-up for the exposure of large areas.

As both beams travel long separate paths, the effects of vibration and air turbulence can easily disturb them. Therefore a fringe-locking system is necessary. Fringes are detected by a photodiode, after which the signal is used by fringe stabilising electronics to actuate a piezo element. Using a similar set-up, Spallas et al. [46] reported the fabrication of uniform photo-lacquer posts on a 50×50 cm^2 glass substrate. The set-up is less sensitive to dust particles than Lloyd's mirror configuration, as there are no mirrors after filtering of the beams. However, it is time consuming to change θ, because this requires movement of a spatial filter after which the beam has to be aligned on the pinhole again.

Fabrication

A silicon wafer was coated with a 0.5 μm thick silicon-rich nitride layer to obtain a low-stress membrane[43]. On top of this a 250 nm thick positive photo-lacquer layer (1 part Shipley 1805 diluted with 1 part Microposit thinner) was spun at 4000 rotations per minute. Using Lloyd's mirror configuration with $\theta = 20.55°$ a post pattern with period $\Lambda_x = 500$ nm was obtained.

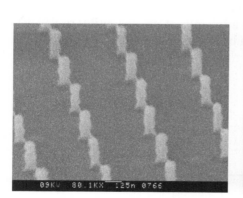 

Fig. 22 Left: SEM micrograph of 80 nm wide posts with rippled sidewalls caused by the vertical interference pattern, Right: SEM micrograph of a membrane after plasma etching through the pores in the chromium layer.

Various exposure times led to posts of different diameters. The smallest posts obtained are shown in Fig. 22. A further increase in exposure time led to posts falling over. The rippled sidewalls show about two periods of the vertical interference pattern in the 250 -nm thick layer. Calculation with Eq. (3) and $n_{resist} = 1.7$ gives a period of 106 nm. After chromium lift-off, i.e. the removal of the posts covered with chromium, the pattern was etched into the silicon nitride membrane using a CHF$_3$/O$_2$ plasma.

The pores in the membrane are significantly larger than the 100 nm pores in the chromium mask. Apparently, an aspect ratio of 5:1 is not possible with this etch recipe. However, microsieves are usually made with a pore size larger than the membrane thickness in order to obtain a small flow resistance and a membrane that is easy to clean. Etching with an aspect ratio of 1:1 would be sufficient for such membranes.

In order to make 100 nm pores, a membrane thickness of about 100 nm is needed. If the grating period is set at 200 nm, θ has to be adjusted to 61.37°. Such a large angle gives a higher reflection from the resist/membrane surface. Moreover, the 100 nm thick silicon nitride membrane absorbs less UV-light than the 500 nm thick membrane used in figures above, which increases the reflection from the membrane/silicon surface. The increased reflection causes a stronger vertical interference pattern, which hinders the formation of posts. The 'waists' of the posts get so thin that the tops fall off. Only the 'foot' formed by the first intensity minimum near the surface remains. The intensity minimum near the resist/membrane surface is caused by the 180° phase shift that arises during reflection on this surface. The incoming and reflected beams are always in anti-phase at the surface.

Fig. 23 Left: SEM micrograph of posts created on a membrane with an improper anti-reflection thickness (100 nm, right: idem with a proper anti-reflection layer.

In order to reduce the total reflection, the reflected light from the silicon/membrane surface should have the opposite phase of the light reflected from the resist/membrane surface. Reflection measurements show this is the case for a membrane thickness around 100 nm. The membrane was covered with a 160 nm thick photo-lacquer layer (1:2 resist/thinner mixture at 5000 rotations per minute) and repeated the exposure conditions. The SEM micrograph in Fig. 23 left shows that the resulting posts have a relatively thin foot, which implies that there has been an intensity *maximum* near the surface. The lift-off process on such small-footed posts is fairly easy. The vertical period for the posts is 121 nm according to Eq.(3). As the resist thickness is only 160 nm, the posts do not show a ripple.

After lift-off and plasma etching a regularly perforated membrane is obtained. The diameter of the pores in the chromium is approximately 65 nm, but the pores in silicon nitride membrane are approximately 100 nm wide due to under-etching. Depending on the application the chromium layer can be left on the surface or be removed by wet etching. An SEM micrograph of the etched membrane with the chromium layer on the surface is shown in Fig. 24.

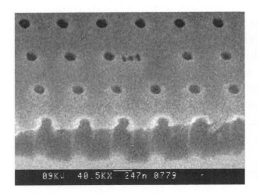 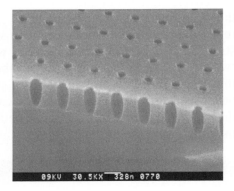

Fig. 24 Pore pattern shapes in an etched membrane. The chromium layer has not been removed in left and right figure.

With the double-beam set-up microsieves were fabricated successfully using 3-inch wafers. A quality check shows that sometimes a pore is missing due to removal of a post during the development process. For filtration purposes this is not a problem. Dust particles form a more serious problem. During exposure a dust particle keeps part of the resist in its shade. This part will not be removed during development and might form a large pore after lift-off. However, a successful lift-off on such a spot is unlikely, as it will probably not have suitable sidewalls (i.e. with a vertical or negative slope). This is confirmed by the fact that no pores larger than the desired pore size are observed, despite the fact the exposure was not carried out in a Clean Room environment.

Microsieves with pore sizes down to 65 nm were fabricated using double-exposure laser interference lithography. The pores are obtained with an inverse process, as the direct process of pore formation in photo-lacquer has narrow process latitude. An array of posts is transferred into an array of pores by evaporating chromium onto the posts, followed by a lift-off in acetone. The resulting patterned chromium layer is used as an etch mask for plasma etching of the silicon nitride membrane. The method is fairly robust, as the lift-off process exploits the rippled sidewalls of the posts to allow the acetone to dissolve the photo-lacquer. Underexposed areas (underneath dust particles) do not have rippled sidewalls and will therefore, in most cases, not result in a large pore. Th removal of a post during development because of a poor adhesion, will only lead to the missing of a pore.

Fig. 25 SEM micrograph of a high-porosity microsieve made with the lift-off method [47].

Conclusions

Microsieves with submicron pore sizes can be made using multiple laser interference lithography. The pore size may practically be chosen in the range between 100 nm and 10.0 μm by adjusting the angle of incidence θ of the incoming laser beams. The exposure method is fast, inexpensive and applicable for large areas.

For use as microfiltration membrane the clean water flux and the maximum allowed transmembrane pressure are important. The clean water flux for the microsieve presented here can be calculated to be 1200 ml/min/bar/cm^2. This value is at least 30 times higher than for most known organic membranes [48,49], such as track etched polycarbonate and stretched PTFE [50], and more than 100 times higher than for inorganic anodic membranes [51], with a typical rating of 10 ml/min/bar/cm^2 at a pore size of 0.2 μm. The maximum allowed transmembrane pressure can be estimated to be 2 bar. This pressure is sufficient for most micro filtration applications.

7. FABRICATION OF MICROSIEVES WITH SILICON WAFERS

Most of the processed silicon wafers today have a <100> orientation. The <111> planes etch approximately 200-1000 times slower than other oriented planes. This means that the <111> planes practically function as an etch stop. A rectangular opening (aligned in e.g. the <001> direction) in a silicon nitride layer covering a silicon wafer will result after wet KOH etching (25%, 70°C) in a pyramid shaped hole, showing the four <111> planes that end in the top of the pyramid.

In order to obtain a silicon microsieve with a high degree of perforation, KOH-etching must be carried out from the backside as well as through the pores on the front side. The etch rate on the front side decreases for smaller pores. For 1 μm pores it is roughly one order of magnitude smaller than the etch rate on the backside. If the wafer thickness is d, the width w of the openings on the backside must be, due to the 54.74° angle between the <111> planes and the wafer surface, approximately $d\sqrt{2}$ (depending on the etch rate underneath the membrane) to etch through the wafer. For a 3-inch wafer with a 375 μm thickness this would result in membrane fields of approximately 530 μm width. Scaling up to 4-inch or 6-inch wafers with a thickness of 525 and 675 μm respectively would give fields of 740 and 950 μm respectively. As large fields are weaker than smaller fields, usually small additional support bars are etched underneath the membrane to reduce w (cf. Fig. 24).

7.1   WET ETCHING THROUGH THE PORES [52][*]

However, for pore diameters below about 1 μm problems occur due to pressure build-up of the hydrogen gas that is created during KOH-etching.

The gas can only escape through the pores in the hydrophilic membrane if its pressure exceeds the bubble-point pressure P_b, given by [53]:

$$P_b = \frac{4\gamma \cos\theta}{d}$$

(4)

where γ is the surface tension of the KOH-solution, θ the liquid-solid contact angle and d the diameter of the pores. Figure below shows that the silicon is etched through the pores.

---

[*] Partially reprinted with permission of Journal Micromech. and Microeng. (2000).

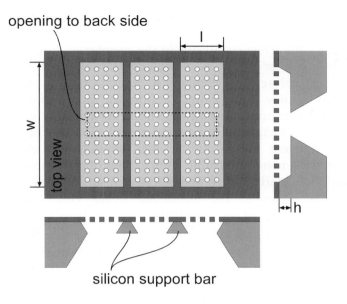

opening to back side

silicon support bar

Fig. 26 Top view of a membrane field that has been split up by small silicon support bars. The dashed line indicates the channel to the backside of the wafer.

After the hydrogen gas has pushed the liquid out of a pore it will form a bubble on top of this pore. When the contact angle of the bubble with the pore wall becomes 0°, the gas pressure reaches a maximum value. For an KOH-solution with an estimated value [54] for γ of 0.075 N/m, this reduces to:

$$p_b = \frac{0.30}{d}$$

(5)

The bubble-point pressure increases for smaller pores. If the pore size is below a certain value, the gas can break the membrane. The maximum pressure p_{max} that an unperforated rectangular membrane [55] can hold can be calculated [56] (See Chapter Membrane Deflection and Maximum Pressure Load, Eq.(16))

$$p_{max} = 0.69 \frac{h\sigma_{yield}^{3/2}}{lE^{1/2}}$$

(6)

Here h is the thickness of the membrane, σ_{yield} the yield stress, l the distance between the silicon support bars and E the Young's modulus. Insertion of some typical values for an unperforated low-stress silicon nitride membrane (h=1.0 µm, l = 200 µm, σ_{yield} = 4.0×10⁹ Pa and E = 2.9×10¹¹ Pa) leads to a calculated maximum pressure of 14 bars. From experience it is observed that during etching the membranes break for pores below approximately 1 µm, which corresponds to a bubble-point pressure of only 3 bar, according to the bubble point equation. The observed difference has several causes. A significant weakening (about 25% depending on the porosity) is caused by the perforations itself [56]. Furthermore, the irregular release of the membrane during etching causes membranes to break far below the maximum allowable pressure for a released membrane. Just before the membrane is entirely released, it will be attached to the silicon by only a few points (besides the side walls). Around these points huge stress concentrations will occur, which can cause the membrane to break. This is in agreement with the observation that the ruptures usually occur more frequent during the release process with alternating perforations and not with non- alternating perforations. Perforations in the microsieves are usually circularly shaped and placed in a square array under an angle of 45° with the edges of the membrane fields. Other shapes or distributions might be stronger, but the correlation between these variables and membrane strength was not known. For membranes with slit-shaped perforations it was found that a 4-5 fold decrease in flow resistance can be obtained in comparison with circular perforations, while the membranes are of comparable strength.

Eq.(6) shows that a decrease in size of the membrane field leads to a stronger membrane. In this way membranes with pores below 1 μm may be released without damage. However, besides that the effective filtration area decreases, another problem arises. For pores around 1 μm it is observed that the etch rate of the silicon underneath the membrane decreases and varies strongly over the wafer surface. Several channels are so shallow that the total sieve resistance increases. The exact mechanism behind this effect is yet not known. Possibly the small pores in combination with the resulting large hydrogen pressure hinders the supply of fresh KOH into the channel. The KOH underneath the membrane continuously creates hydrogen gas, which causes an increasing pressure. This pressurised gas pushes most of the KOH through the pores out of the channel, which reduces the etch rate and thus the rate of pressure build-up. Only after the gas exceeds the bubble-point pressure it can escape and fresh KOH can enter the channel. Obviously this cycle takes much longer for smaller pores and is therefore more sensitive to small variations in pore size. It is also observed that for small pores the channels reach a certain maximum depth, after which the etching almost stops. This effect is possibly caused by the fact that the pressure build-up in a deeper channel takes more hydrogen and thus more time. If, during this time, all the KOH has been pressed out of the channel, the etch process may stop.

The hydrogen gas will not build up a pressure if it is able to escape through the large channels to the backside of the wafer. This is possible if the wafer is etched from the backside first and after that through the pores on the front side. However, with this method it was found difficult to etch the small support bars. Membranes supported by such half support bars turn out to be very vulnerable

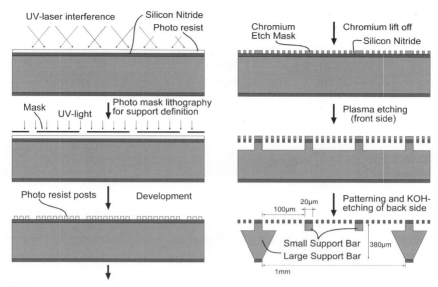

Fig. 28 Process scheme for the release of perforated membranes by plasma etching.

Plasma etching gives such an anisotropy, as the ions can be accelerated into a vertical direction by an electric field. An example is an SF_6/O_2-mixture, as SF_6 etches silicon isotropically while O_2 gives an anisotropic profile by passivating the silicon sidewalls of the trenches LPCVD [57]. Unfortunately the silicon nitride showed a poor etch resistance to the plasma. It was attacked from the inside of the pores (see Fig. 29).

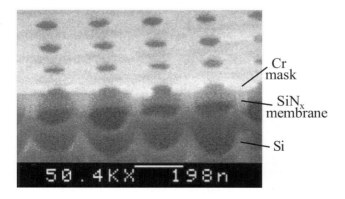

Fig. 29 Silicon etching with an SF_6/O_2 plasma at room temperature through a 100 nm thin silicon nitride membrane covered with a chromium mask. After the removal of only 1 μm of silicon the membrane already shows a significant underetch.

In order to obtain a higher etch selectivity between the silicon support and the silicon nitride membrane, the etch step was repeated in an apparatus with cryogenic substrate cooling (Plasmalab 100, Oxford Plasma Technology [58,59,60]. Using this apparatus an attempt was made to release a membrane with a pore diameter of 1.5 µm and a pitch (pore-to-pore distance) of 4 µm by etching with an SF_6/O_2-plasma at different temperatures. The SEM micrographs of these etch-experiments are depicted in Fig. 30.

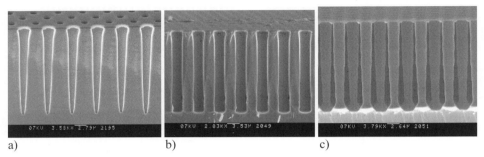

a)   b)   c)

Fig. 30 a) 15 min., -90°C, b) 15 min., -110°C, c) 15 min., -130°C. Results after 15 minutes etching with an SF_6/O_2-plasma at different temperatures.

The micrographs show a vertical profile for a temperature of -110 °C. For -90 °C the profile is positively tapered, for -130 °C it is negatively tapered. Addition of O_2 is not necessary to obtain anisotropy, as the quartz reaction chamber provides for the O_2 while being eroded by the plasma. Bartha et al. [61] explain the changing profile by a temperature-dependent sidewall passivation. Fig. 29 shows a schematic illustration of the mechanism that provides a vertical profile.

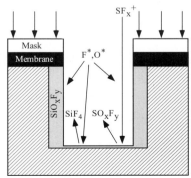

Fig. 31 Schematic illustration of the process of etching with an SF_6/O_2-plasma and cryogenic substrate cooling. Erosion of the quartz reaction chamber also creates oxygen radicals.

The silicon is attacked by radicals (F^* and O^*) creating volatile SiF_4 molecules and a passivating SiO_xF_y layer, see Fig. 31. On the bottom of the trench this layer is removed through sputtering using accelerated plasma ions. A more extensive description of the temperature dependent profile is given by Jansen et al.[62]. The samples do not show enough lateral etch (undercut) of the silicon to connect the trenches. After a second etch of 30 minutes the membranes are still attached to the silicon, although the membrane in sample Fig. 30c has almost been released. However, until now the experiments were merely performed to find the right settings of the process parameters. In practice membranes with large pores like these can still be released by KOH-etching.

For sub micron pores –where KOH-etching gives problems– the release of the membrane by plasma etching is much easier, as less undercut is required to remove all the silicon between the pores. A -130°C recipe was applied on membranes with pore sizes of 400 nm and 70 nm, see Fig. 32a. a)
   b)

Fig. 32 shows that the results are excellent: the membrane is released entirely and the support bars have an acceptably vertical profile. Even the membrane with the 70-nm pores did not show significant etch attack by the plasma.

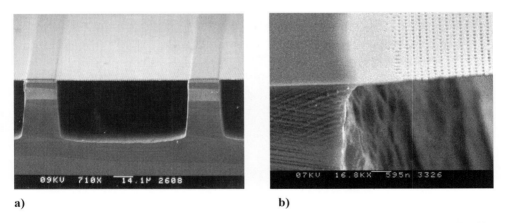

a)   b)

Fig. 32 a) 45 min., -130 °C, 400 nm pores. b) 45 min., -130 °C, 70 nm pores, SF_6/O_2-etching through very small pores.

In contrast with KOH-etching the depth of the channel underneath the membrane does not seem to be influenced by the pore size. The porosity will play a much more important role in the silicon etch rate, as it determines the amount of plasma that can enter the channel.

Fig. 33 SEM micrograph of cryogenic dry etching of silicon beneath a silicon nitride membrane (courtesy Aquamarijn Research).

7.2 <110> SILICON WAFERS [*]

Gas-escape channels in <110> wafers

Pressure build-up of hydrogen gas created during KOH-etching can cause problems when releasing perforated membranes for the production of microsieves. The possibility of creating vertical channels in <110> silicon by KOH-etching allows for the formation of small and therefore strong membrane fields. Build-up of a hydrogen pressure is not observed anywhere, as the gas can escape through the channels. Furthermore, the vertical anisotropy makes it possible to construct very thick and thus strong microsieves. For the new fabrication process chromium etch mask is used instead of photo-lacquer. This chromium mask has the additional advantage of giving better-defined pores, herewith improving the filter performance.

07KV   1.34KX 7.46P 5903

Fig. 34 SEM micrograph of a microsieve with slit-shaped pores (Courtesy of Aquamarijn Research).

KOH-etching on <110> wafers gives vertical walls but unfortunately two oblique <111> planes inhibit the formation of small channels. However, as the shortest width of the fields determines the membrane strength, long but thin slit-shaped channels would still give strong membranes. For <110> wafers the walls of such channels have complementary angles of 70.53° and 109.47°. The walls are vertical, but in the sharpest corners (70.53°) the oblique planes arise with an angle of 35.26° relative to the horizontal <110> plane. A schematic view of such

---

[*] This section has been co-written with S. Kuiper and W. Nijdam.

a channel is given in Fig. 35. In this figure the silicon nitride on the front side has not been patterned yet, in order to etch the silicon only from the backside.

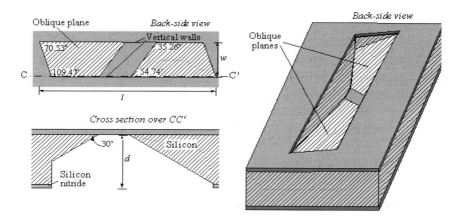

Fig. 35 Schematic view of a slit-shaped pore after KOH-etching of a <110> wafer. The shaded walls inside the channel on the right are <111> planes. Two planes are oblique and four planes are partly vertical.

The channel length l has been chosen such that the two oblique planes intercept at the membrane layer on the top of the wafer before crossing each other. The cross section over CC', see Fig. 35, shows that the angle between the oblique planes and the wafer surface is 30° in the direction parallel to the long vertical walls. This means that in order to etch through the entire wafer the length of the channel has to be larger than $2d\sqrt{3}$. For the channel width w the only restriction that applies is given by flow-resistance requirements. The width should be chosen such that the channel resistance becomes much smaller than the resistance of the attached membrane field. It can be calculated that −even for extremely porous membranes− a width of 100 µm is sufficiently large.

Using the new <110> wet etching method the microsieves can be made much thicker –and thus stronger– than with the conventional <100> wet etching method, as the vertical anisotropy makes the membrane width independent on wafer thickness. Since the aspect ratio of the trenches can be as much as 600 it can be expected that etching of deeper grooves will not cause any problems for the KOH solution (25 %, 70°C) [63]. For concentrations above 30 % the formation of <311> planes has been reported, which inhibits the formation of deep trenches [64]. The maximum etch rate of the <110> planes is found for a KOH concentration of 25 %, and is nearly twice the etch rate of the <100> planes [65].

Fabrication of microsieves on <110> wafers

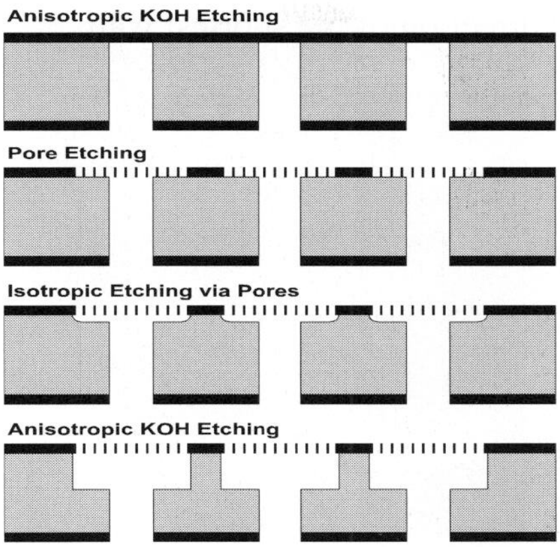

Fig. 36 Schematic view of the microsieve fabrication on a <110> wafer. First, vertical wall KOH etching in <110> silicon, then siliconnitride etching of the pores followed by the release of the membrane with isotropic etching of silicon, and finally KOH etching to make the distance between the perforated membrane and the wafer larger.

As the slit-shaped channels can be made very small, they can be placed closer together, thus creating membrane fields with small widths. Fig. 36 gives a schematic illustration of the fabrication process of a microsieve on <110> silicon.

Instead of a photo-lacquer mask a patterned chromium layer was used as a mask for plasma etching. This layer is obtained either by a lift-off process on photo-lacquer dots or by wet etching of the chromium through pores in a photo-lacquer layer. Before plasma etching of the membrane, vertical escape channels are etched in a KOH-solution. Subsequently the membrane is patterned by plasma etching, using a chromium etch mask. Expected differences in etch rate due to variation in heat conduction between the released and unreleased parts of the membrane were not observed: the pores are uniform over the entire surface. For very small pores or very thin membranes it may be necessary to perform an isotropic HF/HNO_3-etch to release the membrane and make a small space underneath it (third step in Fig. 36). If the HF/HNO_3-solution is chosen in the right composition (less then 1 part 50 % HF on 100 parts 69 % HNO_3), no hydrogen gas will be formed during the release process. The function of the small space underneath the membrane is to give the gas an escape route to the vertical channels. In the discussion a calculation of the required depth for this space will be given.

Using the new method a microsieve was fabricated with slit-shaped perforations and a porosity of 75 %. SEM micrographs of this sieve are shown in Fig. 37 and Fig. 38. The shortest walls underneath the membrane should be partly oblique, but instead they are somewhat vertical with a rough surface. This is either caused by the presence of the perforated membrane, or of non-proper alignment of the mask used to define the direction of the shortest walls.

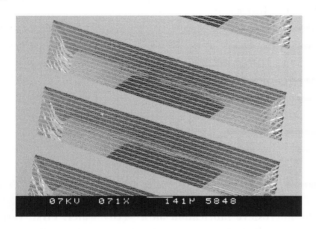

Fig. 37 High-porosity microsieve fabricated on a <110> silicon wafer. Through the slit-shaped pores some typical etch angles can be recognised.

Sub-micron pores in <110> silicon

Sub-micron pores are difficult to make with conventional contact-mask photolithography due to diffraction of the light. Earlier using laser-interference lithography, the fabrication of 0.1 μm pores in a 0.1 μm thick membrane was reported [36]. Using both the method of laser-interference lithography and wet etching, <110> wafers microsieves were fabricated with 0.1 μm thick membranes and 0.1 μm pores. An SEM micrograph of such a sieve is shown in Fig. 38.

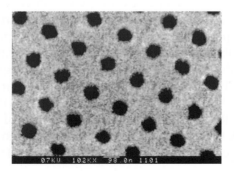

Fig. 38 SEM micrograph of a microsieve with 0.1 μm pores.

Space underneath the membrane

The small space underneath the membrane, formed by HF/HNO$_3$-etching, requires a certain depth to give the gas the opportunity to escape to the large channel. This gas, which is created during the second KOH-step (see Fig. 36), has to push away the KOH-solution underneath the membrane towards the escape channel. The required pressure for this can be calculated by making a force-balance analysis. The force F_γ required to remove the liquid underneath the membrane can be obtained by multiplying the surface tension γ by the wetted perimeter and the cosine of the contact angle θ between liquid and membrane:

$$F_\gamma = 2\gamma(h+w)\cos\theta \tag{7}$$

where h is the depth of the small space underneath the membrane. The gas exerts a force F_p on the liquid, equal to pressure times cross-sectional area:

$$F_p = phw \tag{8}$$

with p the pressure difference between gas and liquid. A balance of the forces in Eqs. (7) and (8) leads to

$$p = \frac{2\gamma(h+w)\cos\theta}{hw} \approx \frac{2\gamma\cos\theta}{h} \tag{9}$$

For the KOH-solution with a contact angle of nearly 0°, it follows that for a depth of 1 μm the pressure underneath the membrane will still be 1.5 bar. For very thin membranes this may be too high, and in that case the depth should be increased by HF/HNO$_3$ etching until a safe pressure level is reached.

Resolution in pattern transfer

The chromium etch mask that is used in the wet etching method with <110> wafers, has an advantage over photo-lacquer regarding the definition of the pores. Chromium can not be etched in the CHF$_3$/O$_2$ plasma and gives therefore a higher resolution in pattern transfer. A photo-lacquer mask gives rise to a positively tapered profile of the pores due to non-vertical photo-lacquer walls and lateral etching of the mask (see Fig. 39). Positevely tapered pores have a negative effect on filtration performance, because particles may more easily be trapped. Moreover, using chromium as an etching mask, a higher porosity can be obtained, which may improve the filtration flux even more.

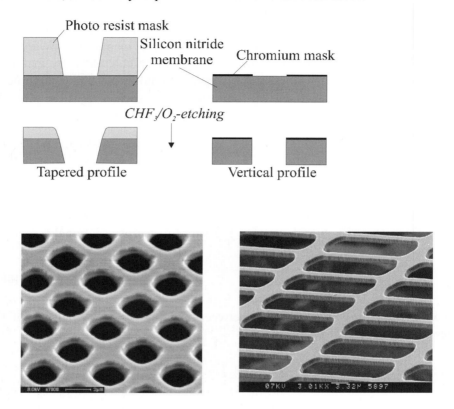

Fig. 39 Comparison of plasma-etch results for a photo-lacquer mask and a chromium mask. The microsieve on the left was made with photo resist, whereas the one on the right was made with chromium.

REFERENCES

[1] K.E. Petersen, Silicon as a mechanical material, Proc. IEEE 70 (1982) 420-457.

[2] C.J.M. van Rijn and M.C. Elwenspoek, Micro filtration membrane sieve with silicon micromachining for industrial and biomedical applications, IEEE proc. MEMS (1995) 83-87.

[3] C.J.M. van Rijn, Membrane filter as well as a method of manufacturing the same, PCT patent application 95/1386026.

[4] M. Rossi, Mikrostrukturierung, Vortrag zu den Übungen in Halbleiterphysik II, Stuttgart, Juni 1999.

[5] Pedrotti, Pedrotti, Bausch and Schmidt, Eine Einführung, Optik, Prentice Hall-Verlag, München (1996).

[6] D. Meschede, Licht und Laser, Optik, Teubner-Verlag Stuttgart, Leipzig (1999).

[7] A. Kroyan, J. Bendik, O. Semprez, N. Farrar, C. Rowan and C.A. Mack, Modeling the Effects of Excimer Laser Bandwidths on Lithographic Performance, Optical Microlithography XIII, Proc., SPIE Vol. 4000 (2000).

[8] C.A. Mack, Inside PROLITH, A Comprehensive Guide to Optical Lithography Simulation, FINLE Technologies (Austin, TX: 1997).

[9] E. Hecht, Optik (2. Auflage), R. Oldenbourg-Verlag, München, Wien (1999).

[10] M. Young, Optik, Laser, Wellenleiter; Springer-Verlag Berlin Heidelberg New York (1997).

[11] D. Widmann, H. Mader, H. Friedrich, Technologie hochintegrierter Schaltungen (2. Auflage), Springer-Verlag Berlin Heidelberg New York (1996).

[12] L. Karklin, Subwavelength challenges, EE-design, 11 september 1999.

[13] M. Seager, The data used to construct this graph have been adapted from the Microprocessor Report 9(6), May (1995).

[14] NumeriTech US pat. 5,858,580.

[15] P.W. Hawkes, E. Kasper, Principles of Electron Optics, Academic Press, London, 1989.

[16] P. Grivet, Electron Optics, Elsevier, Oxford, Pergamon imprint (1965).

[17] M. Hatzakis, Electron resists for microcircuit and mask production, J. Electrochem. Soc. 116 (1969) 1033-1037.

[18] Courtesy of S. Rishton and E. Ganin, IBM.

[19] M.G. Rosenfield, M.G.R. Thomson, P.J. Coane, K.T. Kwietniak, J. Keller, D.P. Klaus, R.P. Volant, C.R. Blair, K.S. Tremaine, T.H. Newman and F.J. Hohn, Electron-beam lithography for advanced device prototyping: Process tool metrology, J. Vac. Sci. Technol. B11 (6) (1993) 2615-2620.

[20] S.A. Rishton, H. Schmid, D.P. Kern, H. E. Luhn, T.H.P. Chang, G.A. Sai-Halasz, M. R. Wordeman, E. Ganin and M. Polcari, Lithography for ultrashort channel silicon field effect transistor circuits, J. Vac. Sci. Technol. B6 (1) (1988) 140-145.

[21] C.P. Umbach, C. Van Haesendonck, R.B. Laibowitz, S. Washburn and R.A. Webb, Direct observation of ensemble averaging of the Aharonov-Bohm effect in normal metal loops, Phys. Rev. Lett. 56 (1986) 386.

[22] V. Chandrasekhar, M.J. Rooks, S. Wind and D.E. Prober, Observation of Aharonov-Bohm Electron Interference Effects with Periods h/e and h/2e in Individual Micron-Size, Normal-Metal Rings, Phys. Rev. Lett. 55 (1985) 1610-1613.

[23] B.J. van Wees, H. van Houten, C.W.J. Beenakker, J.G. Williamson, L.P. Kouwenhoven, D. van der Marel and C.T. Foxon, Quantized conductance of point contacts in a two-dimensional electron gas, Phys. Rev. Lett. 60 (1988) 848.

[24] M.J. Rooks, C.C. Eugster, J.A. del Alamo, G. Snider and E. Hu, Split-gate electron waveguide fabrication using multilayer PMMA, J. Vac. Sci. Technol. B 9 (1991) 2856.

[25] P.H. Woerlee, G.A.M. Hurkx, W.J.M.J. Josquin and J.F.C.M. Verhoeven, Novel method of producing ultrasmall platinum silicide gate electrodes, Appl. Phys. Lett. 47 (7) (1985) 700-702.

[26] E. Anderson, V. Boegli, M. Schattenburg, D. Kern and H. Smith, Metrology of electron-beam lithography systems using holographically produced reference samples, J. Vac. Sci. Technol. B9 (6) (1991) 3606-3611.

[27] R. Viswanathan, D. Seeger, A. Bright, T. Bucelot, A. Pomerene, K. Petrillo, P. Blauner, P. Agnello, J. Warlaumont, J. Conway and D. Patel, Fabrication of high performance 512K static-random access memories in 0.25 um complementary metal-oxide semiconductor technology using x-ray lithography, J. Vac. Sci. Technol. B11 (6) (1993) 2910-2919.

[28] S.Y. Chou, H.I. Smith and D.A. Antoniadis, Sub-100-nm channel-length transistors fabricated using x-ray lithography, J. Vac. Sci. Technol. B4 (1) (1986) 253-255.

[29] C.J.M. van Rijn, Membrane filter as well as a method of manufacturing the same, PCT patent application 95/1386026.

[30] S. Austin and F.T. Stone, Fabrication of thin periodic structures in photoresist, J. appl. Opt. 15 (1976) 1071-1074.

[31] B. de A. Mello, I.F. da Costa, C.R.A. Lima and L. Cescato, Developed profile of holographically exposed photoresist gratings, Appl. Opt. 34 No. 4 (1995) 597-603.

[32] L. Mashev and S. Tonchev, Formation of holographic diffraction gratings in photoresist, Appl. Phys. A 26 (1981) 143-149.

[33] C.J.M. van Rijn, W. Nijdam W, S. Kuiper, G.J. Veldhuis, H. van Wolferen and M. Elwenspoek, Microsieves made with laser interference lithography for micro-filtration applications, Journal of Micromechanics and Microengineering 9 (1999) 170-172

[34] E.H. Anderson, C.M. Horwitz and H.I. Smith, Holographic lithography with thick photoresist, Appl. Phys. Lett. 43 No. 9 (1983) 874-875.

[35] H.Z. Saleem and S.R.J. Brueck,J. Vac. Sci. Technol. B 11 (1993) 658.

[36] C.J.M. van Rijn, G.J. Veldhuis and S. Kuiper, Nanosieves with microsystem technology for microfiltration applications, Nanotechnology 9 (1998) 343-345.

[37] C.J.M. van Rijn, W. Nijdam, S. Kuiper, G.J. Veldhuis, H. van Wolferen and M. Elwenspoek, Microsieves made with laser interference lithography for micro-filtration applications, J. Micromech. Microeng. 9 (1999) 170-172.

[38] A. Fernandez, P.J. Bedrossian, S.L. Baker, S.P. Vernon and D.R. Kania, IEEE Trans. Mag. 32, (1996) 4472.

[39] X. Mai, R. Moshrefzadeh, U.J. Gibson, G.I. Stegeman and C.T. Seaton,Appl. Opt. 24 (1985) 3155.

[40] J.P. Spallas, A.M. Hawryluk and D.R. Kania, J. Vac. Sci. Technol. B 13 (1995) 1973.

[41] S. Kuiper, H. van Wolferen, C.J.M. van Rijn, W. Nijdam and M.C. Elwenspoek, Fabrication of microsieves by laser interference lithography, J. Micromech. and Microeng 2000.

[42] C.O. Boltzer, C.T. Harris, S. Rabe, D.D. Rathman, M.A. Hollis and H.I. Smith, J. Vac. Sci. Technol. B 12 (1994) 629.

[43] J.G.E. Gardeniers, H.A.C. Tilmans and C.G.C. Visser, LPCVD silicon-rich silicon nitride films for applications in micro-mechanics studied with statistical experimental design, J. Vac. Sci. Technol. A, Vol. 14 (1996) 2879-2892.

[44] S.C. Kitson, W.L. Barnes and J.R. Sambles, The fabrication of submicron hexagonal arrays using multiple-exposure optical interferometry, IEEE Phot. Techn. Lett. 8, No 12 (1996) 1662-1664.

[45] J.Y. Decker, A. Fernandez and D.W. Sweeney, Generation of subquarter-micron resist structures using optical interference lithography and image reversal, J. Vac. Sci. Techn. B 15 No. 6 (1997) 1949-1953.

[46] J.P. Spallas, R.D. Boyd, J.A. Britten, A. Fernandez, A.M. Hawryluk, M.D. Perry and D.R. Kania, J. Vac. Sci. Techn. B 14 (1996) 2005-2007.

[47] L.Vogelaar, W. Nijdam, H.A.G.M. van Wolferen, R.M. de Ridder, F.B. Segerink, E. Flück, L. Kuipers and N.F. van Hulst, Large area photonic crystal clabs for visible light with waveguiding defect structures: Fabrication with FIB-assisted Laser Interference Lithography, Advanced Materials 13 (2001) 1551-1554.

[48] Basic Principles of Membrane Technology, Kluwer Academic Publ. 1996, ISBN 0 7923 4248 8, M. Mulder 228.

[49] N.N. Li, J.M. Calo and M. Dekker, Separation and Purification Technology (1992) ISBN 0 8247 8721 8, H. Strathmann 1-17.

[50] K.Scott. Handbook of Industrial Membranes, Elsevier Science Publ. (1995) ISBN 1 85617 233 3, 112&120.

[51] Anopore, alumina membrane filter, Whatmann product guide (1993) cat. No. 9037 9316.

[52] S. Kuiper, M. de Boer, C.J.M. van Rijn, W. Nijdam et al., Wet and dry etching techniques for the release of sub-micron perforated membranes, J. Micromech. and Microeng 10 (2000) 171-174.

[53] M.C. Porter, Handbook of industrial membrane technology, Noyes Publications,1990.

[54] M.C. Porter, Handbook of Chemistry and Physics 1st Stud. Ed., CRC Press Inc.,1988.

[55] For rectangular membranes coefficient is 0.69 instead of 0.58 for square membranes. In J. of Membrane Science 150 (1998) 1-8, these numbers have been intermingled.

[56] C.J.M. van Rijn, M. van der Wekken, W. Nijdam and M.C. Elwenspoek, Deflection and maximum load of microfiltration membrane sieves made with silicon micromachining, J. Microelectromech. Syst. 6 (1997) 48-54.

[57] H.V. Jansen, Plasma etching in micro-technology, Thesis University of Twente, The Netherlands, ISBN 903650810x (1996).

[58] Oxford Instruments, Plasma Technology, North End, Yatton, Bristol BS19 4AP, England.

[59] J.A.Theil, Deep trench fabrication by Si (110) orientation dependent etching, J. Vac. Sci. Techn. B 13(5) (1995) 2145-2147.

[60] H.V. Jansen, M.J. de Boer, H. Wensink, B. Kloeck and M.C. Elwenspoek, The Black Silicon Method VIII: A study of the performance of etching silicon using SF6/O2-based

chemistry with cryogenical wafer cooling and a high density ICP source, Microsystem Technologies, Vol. 6 No. 4 (2000).

[61] J.W. Bartha, J. Greschner, M. Puech and P. Maquin, Low temperature etching of Si in high density plasma using SF6/O2, Microelectronic Engineering 27 (1995) 453-456.

[62] H.V. Jansen, M.J. de Boer, H. Wensink, B. Kloeck and M.C. Elwenspoek, The Black Silicon Method VIII: A study of the performance of etching silicon using SF6/O2-based chemistry with cryogenical wafer cooling and a high density ICP source, Microsystem Technologies, Vol. 6 No. 4 (2000).

[63] A. Hölke and H. Thurman Henderson, Ultra-deep anisotropic etching of (110) silicon, J. Micromech. microeng. 9 (1999) 51-57.

[64] P. Krause and E. Obermeier, Etch rate and surface roughness of deep narrow U-grooves in (110)-oriented silicon, J. Micromech. Microeng. 5 (1995) 112-114.

[65] H. Seidel, L. Csepregi, A. Heuberger and H. Baumgärtel, Anisotropic etching of crystalline silicon in alkaline solutions. I. Orientation dependence and behavior of passivation layers, J. Electrochem. Soc. 137 (11) (1990) 3612-3625.

Chapter 4

Fluid Mechanics

"It is a recognised principle, (...), that, when a problem is determinate, any solution which satisfies all the requisite conditions, no matter how obtained, is the solution of the problem. In case of fluid motion, when the initial circumstances and the conditions with respect to the boundaries of the fluid are given, the problem is determinate."
[G.C. Stokes, Trans. Cam. Phil. Soc VIII (1843) 105]

1. INTRODUCTION

Some relevant elementary Computational Fluid Dynamics (CFD) as well as some analytical fluid flow expressions will be presented for fluid channels and membranes with micro engineered orifices.

2. FLUID FLOW THROUGH A LONG CHANNEL, A SHORT CHANNEL AND AN ORIFICE

Reynolds number

Flow in channels and orifices can be divided in a laminar flow regime, a transition flow regime and a turbulent flow regime. In the laminar flow regime the flow is normally determined mainly by viscous losses. In the turbulent regime the flow is normally determined by mainly inertial (kinetic) but also by viscous effects. In the transition regime unsteady laminar flow is observed with both viscous and kinetic energy dissipation effects.

In order to distinguish between the three flow regimes, the Reynolds number, Re, is an important parameter. The Reynolds number can be considered to be a ratio between kinetic energy (ρv^2) and viscous energy ($\eta v/D$) terms and is a dimension free unit:

$$\text{Re} = \frac{\rho v D}{\eta} \qquad (1)$$

With
  D = hydraulic diameter of the channel and
  v = average velocity of the fluid
  ρ = mass density of the fluid
  η = viscosity of the fluid

At low Reynolds numbers (small fluid velocity and/or small channel diameter), typically less than 1000, the flow in the channel is laminar. In this regime there is a linear (flow resistance) relation between the pressure drop ΔP along the channel and the flow Φ through the channel with a length l.
At high Reynolds numbers, typically larger than 2300, the flow in the channel is turbulent and there is no longer a linear pressure/flow relation.

Table 1

Flow resistance (pressure/flow) and hydraulic diameter of various channel cross sections in the laminar flow regime [1].

Channel crosssection	Flow resistance (Nsm^{-5})	Hydraulic diameter D_h (m)
◯ $\updownarrow$ 2r	$\dfrac{8l}{\pi r^4}\eta$	2r
☐ $\updownarrow$ 2a 2a	$\dfrac{1{,}78l}{a^4}\eta$	2a
b>>h ▭ $\updownarrow h$ b	$\dfrac{12l}{bh^3}\eta$	2h
▽ $\updownarrow h$ b (b=h)	$\dfrac{17{,}4l}{(\frac{1}{2}b)^4}\eta$	0,52b

Fully developed flow in long channels

First, consider a long channel with a fully developed flow (see Fig. 1a). In a laminar flow dominated by viscous losses, energy is needed to compensate for the friction between the (laminar) layers in the fluid flow in order to maintain the flow.
Well known is the Hagen-Poiseuille equation describing the relation between the pressure loss ΔP, the volume flow rate Φ and the viscosity η through a channel with length L ($L>>R$) and radius R:

$$\Phi = \frac{\pi R^4}{8\eta L} \Delta P \qquad (2)$$

If the fluid velocity increases above a certain value, the flow will become unstable and turbulence starts to develop (see Fig. 1b). In turbulent flow great losses due to kinetic effects dominate the flow. The energy necessary to maintain the flow is much larger, due to additional energy losses (inertial losses) associated with the continuous change of the velocity of a fluid element along its fluid path. In both regimes the energy necessary to maintain the flow is obtained from the pressure difference between both ends of the channel.

Because there is no velocity change in a fully developed flow in a long channel, a laminar flow regime dominated by inertial losses does not exist.

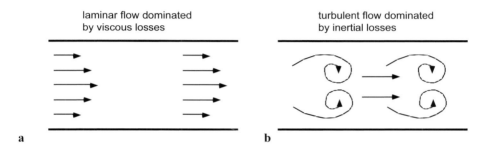

Fig. 1 Fully developed flow in a long channel [2].

At high fluid velocities inertial effects dominate the fluid flow and the relation between pressure loss and flow rate is no longer linear:

$$\Phi = \pi D_h^{\,2} \sqrt{\frac{D_h}{8Lf\rho}} \Delta P \qquad (3)$$

where D_h = hydraulic diameter ($\cong 2R$),

f = friction factor ($\approx 0.14\, Re^{-0.18}$ for fully developed turbulent flow),

ρ = mass density fluid

Fluid flow in short channels and orifices

Now consider a short channel (see left Fig. 2). Let the velocity field be fully developed before the short channel has been entered. The fluid velocity in the channel differs not much from the fluid velocity before the channel. And hence the flow is mainly determined by viscous losses.

and hence the flow is in the laminar viscous regime.

The openings in a microsieve are orifices if $L/Dh < 0.5$. When the openings are circular, it follows that $L/2R < 0.5$, thus if the thickness of the microsieve is smaller than the diameter, the opening may be considered as an orifice.

The volume flow rate Φ of a fluid with viscosity η through a circular orifice ($L<<R$) with radius R is proportional with the pressure difference ΔP (Stokes flow, Happel and Brenner [6])

$$\Phi = \frac{R^3}{3\eta} \Delta P \tag{6}$$

Dagan et al. [7] considered the case of a short channel with a length L comparable with the radius R ($L \approx R$). A good approximate value for the flow rate Φ is given by them

$$\Phi = \frac{R^3 \Delta P}{3\eta} \left[1 + \frac{8L}{3\pi R} \right]^{-1} \tag{7}$$

For L>>R this relation will scale to the Hagen-Poiseuille relation Eq.(2). At higher fluid velocities the flow rate Φ will also be determined by inertial (kinetic) effects.

In this regime an important equation is the Bernoulli equation stating the energy conservation of the sum of external work (pV), potential energy (mgz), and kinetic energy ($1/2mv^2$) of a fluid element with mass m and volume V(=m/ρ) along a streamline:

$$\frac{p}{\rho} + gz + \frac{\bar{v}^2}{2} = \textbf{constant along a streamline.} \tag{8}$$

where $z =$ the height,
                     $p =$ the hydrostatic pressure,
                     $\rho =$ the fluid density,

g = gravity and
$\bar{v}$ = fluid velocity field.

Restrictions:
a.   Steady flow
b.   Incompressible flow
c.   Frictionless flow (η=0)
d.   Flow along a streamline

Eq.   (8) gives a non-linear relation between pressure and velocity. For laminar flow at higher flow rates a minimum pressure drop therefore will arise at the entrance of an orifice (velocity before the entrance is negligible) to create a fluid element with a kinetic energy contribution.

$$\Delta P = \frac{\rho \overline{V_{orifice}^2}}{2} - \frac{\rho \bar{V}_{orifice}^2}{2} \cong \frac{\rho \bar{V}_{orifice}^2}{4} = \frac{\rho \Phi^2}{4\pi^2 R^4} \tag{9}$$

The total pressure drop is in approximation then the addition of the viscous and the minimum kinetic contribution. The relation between pressure drop and flow rate for viscous flow with a correction for the kinetic contribution then becomes:

$$\sum \Delta P = \frac{3\eta}{R^3}\left[1 + \frac{8L}{3\pi R}\right]\Phi + \frac{\rho}{4\pi^2 R^4}\Phi^2 \tag{10}$$

One can calculate that for orifices with a radius of about 1 micron the viscous contribution equals the minimum kinetic contribution for a water flow at a $\Sigma\Delta P$ pressure of about 6 bar and scales with $\eta^2/\rho R^2$. It should be emphasized that at very high flow velocities the actual kinetic contribution is often larger than the required minimum kinetic contribution. The ratio between those contributions is often empirically determined.

Table 2

Overview of flow rate and pressure drop relations in different flow regimes and different channel length/diameter ratio's.

Type of restriction	Definition	Viscous losses dominate laminar flow ($Re \ll Re_t$)	$Re_{transition}$	Inertial losses dominate (turbulent) ($Re \gg Re_t$)
Orifice	$L/D_h < 0.5$	$\Phi = \dfrac{D_h^{\,3}}{24\eta}\Delta P$	15	$\Phi = \pi D_h^{\,2}\sqrt{\dfrac{\Delta P}{8\rho\xi}}$
Short Channel	$2 < L/D_h < 50$	$\Phi = \dfrac{D_h^{\,3}\Delta P}{24\eta}\left[1+\dfrac{16L}{3\pi D_h}\right]^{-1}$	$30\,L/Dh$	orifice: $\xi \cong 2.6$ channel: $1 < \xi < 1.5$
Long Channel	$L/D_h > 100$	$\Phi = \dfrac{\pi D_h^{\,4}}{108\eta L}\Delta P$	2300	$\Phi = \pi D_h^{\,2}\sqrt{\dfrac{D_h}{8Lf\rho}}\Delta P$

where D_h =hydraulic diameter ($\cong 2R$),
ξ = empirical kinetic contribution constant
f = friction factor ($\approx 0.14\,Re^{-0.18}$ for fully developed turbulent flow)

Computational Fluid Dynamics

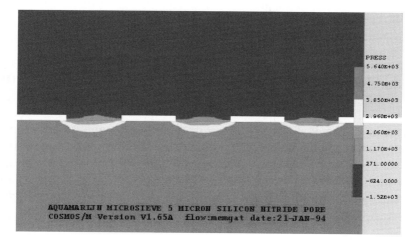

Fig. 5 Cosmos computer simulations [8] of the pressure profile of the flow through a microsieve with pore size 5 micron.

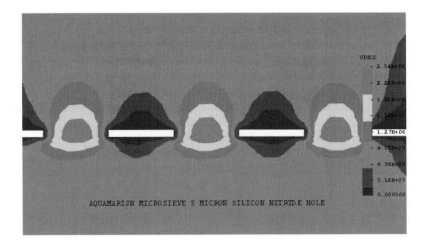

Fig. 6 Cosmos computer simulations of corresponding flow profile microsieve with pore size 5 micron.

Generally the motion of a fluid in as well the viscous and the kinetic regime with constant ρ (incompressible) and constant η can be described by the Navier-Stokes differential momentum (force balance) equation:

$$\rho \frac{d\bar{v}}{dt} = \rho g - \nabla p + \eta \nabla^2 \bar{v} \tag{11}$$

density * acceleration/unit volume = gravity force + pressure force +viscous force

The Navier-Stokes equation can also be extended to compressible flow and is an excellent control element for fluid computer simulations and will be more discussed in detail below.

The development of efficient and accurate computational schemes for optimal control of fluid flows is an extremely complex undertaking [9]. Most fluidic problems are highly non-linear and require much computing time. First one has to define properly all the variables, e.g., velocity, pressure, viscosity, temperature, density, internal energy, etc. Next one has to deal with the boundary constraints or design variables, e.g. shape of obstacles and walls, amount of fluid injected, heat flux along a portion of the boundary, etc. Next a full system of equations of fluid flow e.g., the Navier-Stokes system, the Euler equations, the potential flow equations, etc. has to be translated in a computational algorithm.

Lattice Boltzmann method*[10]

In the last years the Lattice Boltzmann method has proven to be very successful in the modelling of physical transport phenomena [11]. Contrary to traditional modelling methods that have a continuum approach with discretization of differential equations, the Lattice Boltzmann method is based on the kinetic theory. The Lattice Boltzmann method describes the movement of discrete particles over a regular lattice, assuming a truncated Maxwell-Boltzmann distribution of the velocities in a system. Such a group of particles with a velocity distribution can be regarded as a fluid package. The size of a Lattice Boltzmann fluid package lies typical in the meso-scopic length scale (1µm to 1cm). Sometimes, the method is used on the macro scale (from 1cm).

The driving force for transport is a density gradient in the conserved quantity. The ideal gas law is taken as a constitutive equation to relate the density difference to a pressure difference. Near the compressibility limit (in case of low Mach numbers), the rules for ideal gases can also be applied to incompressible fluids. Many transport phenomena, which can be described by the kinetic theory can be described by Lattice Boltzmann. Lattice Boltzmann particularly shows its strength when two different length scales are involved and the system can no longer assumed as a continuum:

flow through porous media [12]
multiphase flow of immiscible fluids [13]
behaviour of emulsions and suspensions [11]

Convection-diffusion phenomena are described by:

$$\left(\frac{\partial}{\partial t}+\bar{u}\cdot\nabla\right)\rho=-D\nabla^2\rho \tag{12}$$

in which D is the Fickian diffusion coefficient, $\bar{u}$ the convective velocity, and ρ the concentration.

Fluid flow is described by:

$$\left(\frac{\partial}{\partial t}+\bar{u}\cdot\nabla\right)\rho\bar{u}=-\nabla\cdot\vec{\Pi} \tag{13}$$

This complete Navier Stokes equation is a momentum balance with $\bar{u}$ the velocity, ρ the "density" of the fluid and Π the stress tensor defining flow properties in terms of bulk viscosity and shear viscosity.

By performing the Chapman-Enskog expansion of the Lattice Boltzmann equation it can be proven that this will lead to a discretitized version of the Navier Stokes equation [14].

---

* Contribution of R. van der Sman, Food Process Technology, University of Wageningen.

The simplest and currently widely used Lattice Boltzmann model is the so-called BGK (Bhatnagar-Gross-Krook) model, where the collision operator is based on a single-time relaxation toward the local equilibrium distribution. Well described schemes are the D2Q9 scheme with 2 dimensions and 8 velocities in 8 directions and a rest particle and D3Q19 scheme in 3 dimensions with 18 velocities and also including a rest particle. For 3D also the D3Q15 scheme is frequently used. Systems with more than one velocity (e.g. D2Q17) can be used to describe supersonic flows.

The particle velocities and directions are discretisations, such that the particles move to the neighbouring cells within the integration time interval, except for the rest particle that does not move.

The particle movement consists of subsequent collision and propagation steps. Each particle distribution in time can be regarded as a distribution consisting of an equilibrium and non-equilibrium part. During the collision step, the new equilibrium distribution is calculated from the existing particle distribution. Omega, the collision operator or inverse relaxation parameter controls the update of the particle distribution from the calculated equilibrium distribution. In the propagation step, the particles are propagated into their direction.

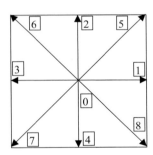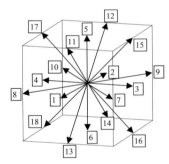

Fig. 7 Particle transport directions of D2Q9 (left) and D3Q19 schemes (right).

The D2Q9 system is further defined as follows:

Mass balance: $\quad \sum_i f_i(x,t) = \rho(x,t)$

Impulse balance: $\quad \sum_i f_i(x,t)\bar{c}_i = \rho(x,t)\bar{u}(x,t)$

Equilibrium distribution function for fluid flow:

$$f_{i,eq} = w_i\rho[1 + \frac{\bar{c}_i \bullet \bar{u}}{c_s^2} + \frac{(\bar{c}_i \bullet \bar{u})^2}{2c_s^4} - \frac{\bar{u}^2}{2c_s^2}] \tag{14}$$

in which $w_1 = w_2 = w_3 = w_4 = \dfrac{c_s^2}{3c_i^2}$, $w_5 = w_6 = w_7 = w_8 = \dfrac{c_s^2}{12c_i^2}$ and $w_0 = 1 - 4w_1 - 4w_5$

Collision and propagation steps:

$$f_i(x + \bar{c}_{ix}\Delta t, y + \bar{c}_{iy}\Delta t, t + \Delta t) = f_i(x, y, t) - \omega(f_i(x, y, t) - f_{i,eq}(x, y, t))$$

The viscosity (for a given value of omega) is defined as:

$$v = c_s^2 \Delta t \left(\dfrac{1}{\omega} - \dfrac{1}{2} \right) \tag{15}$$

Table 3

Displacements, velocities and weight factors for the D2Q9 system.

i	Δx	Δy	$C_{i,x}$	$C_{i,y}$	C^2	w_i
0	0	0	0	0	0	4/9
1	+1	0	+C	0	1	1/9
2	0	+1	0	+C	1	1/9
3	-1	0	-C	0	1	1/9
4	0	-1	0	-C	1	1/9
5	+1	+1	+C	+C	2	1/36
6	-1	+1	-C	+C	2	1/36
7	-1	-1	-C	-C	2	1/36
8	+1	-1	+C	-C	2	1/36

The D3Q19 scheme can be derived in a similar way with the displacements and velocities in x, y and z direction and weight factors $w_i=1/3$ for i=0, $w_i=1/18$ for i=1,..,6 and $w_i=1/36$ for i=7,..,18.

Stability criteria (omega equal to 1)

For diffusion, the grid Fourier (Fo^*) criterion must be met and in case of diffusion-convection also the Courant number (Cr) must be taken into account. In literature often the grid Peclet number (Pe^*) is found, which is a combination of Fourier and Courant.

$$Fo^* = \dfrac{D\Delta t}{\Delta x^2} \le 0.5 \qquad Cr = \dfrac{\bar{u}\Delta t}{\Delta x} < 1 \qquad Pe^* = \dfrac{Cr}{Fo^*} = \dfrac{\bar{u}\Delta x}{D} < 2 \tag{16}$$

With diffusion coefficient D, gridsize Δx, time interval Δt and $\bar{u}$ being the convective velocity.

When the Peclet number increases the value of 2, a so-called Lax-Wendroff scheme is formed. Characteristic for this method are the numerical oscillations (also known as numerical dispersion or wiggles).

To describe fluid flow the system should comply with Courant and Euler number (Eu) constraints and additionally the Peclet number in case of diffusion-convection when a dissolved component is present in the fluid flow.

$$Cr = \frac{\bar{u}_{max}\Delta t}{\Delta x} < 1 \qquad \text{with} \quad \bar{u}_{max} = \frac{\Delta Ph^2}{2\eta L} \quad \text{and} \quad Eu = \frac{\Delta P}{\rho_0 C_s^2} < 0.01 \qquad (17)$$

$\bar{u}_{max}$ is the maximum fluid velocity for Poiseuille flow in a slit form channel with height h and length L. ΔP is the pressure drop over the channel length and η is the viscosity of the fluid.

Fourier, Courant and Peclet criteria are also applicable to central difference finite volume methods. The Euler number is typically for Lattice Boltzmann, because it describes the incompressibility limit of the fluid. Additional requirements for fluid flow are $1/3c^2 < c_s^2 < 2/3c^2$ and a square form lattice. For diffusion phenomena a square form lattice is already sufficient.

Symmetry is required for the Lattice Boltzmann method in general, in order to get isotropy of the tensor of the system [14]. In practice often a square or hexagonal lattice is defined. When omega goes from 1 to 1.99 the stability criteria become less strict (especially for Pe*).

The influence of the value of omega

When the collision operator omega equals 1, the Lattice Boltzmann scheme approximates the finite volume scheme [15]. When omega is 1, this can sometimes cause oscillatory behaviour in the form pressure waves. In practice Lattice Boltzmann is somewhat more accurate than second order finite volume, because when omega is higher than 1, it can work with steeper gradients, smaller physical dimensions and higher Reynolds numbers. Normally Lattice Boltzmann goes to Re$\cong$1000; with turbulence model up to Re$\cong$5000. Some drawbacks of the LB method are the slow convergence when omega is close to 1.99 and possibly stability problems after very long integration times [16].

Choosing the right system parameters

The Lattice Boltzmann method offers several possibilities to convert a physical problem to a model. In general omega, Δx and Δt offer enough degrees of freedom. For fluid flow c_s^2 is normally chosen $1/3c^2$, whilst for single diffusion processes the ratio c_s^2/c^2 is even less fixed.

The physical dimensions of the system are converted to the amount of gridpoints that is required for the accuracy. Then the physical parameters are introduced and finally a integration time interval is chosen that meets the stability criteria. When a diffusive component is present in the fluid flow (convection-diffusion) the ratio c_s^2/ c^2 for diffusion is adapted to get a good value of omega for the diffusion process.

Boundary conditions

Depending on the system several different boundary conditions can be applied:

Static walls: Particles that leave the system are reflected ("bounce-back") and introduced again into the opposite direction. At the wall the velocity is zero, therefore this boundary is also known as the no-slip condition. Usually an imaginary wall is located between the gridpoints ("half-way-wall bounce-back"). Bounce-back can be used to define obstacles in the flow and gives quite accurate results.

Moving walls: A moving wall is treated according to the bounce-back scheme, but with a compensation for the wall velocity (also a no-slip condition). Moving walls are used to create a shear field or to decrease the grid by assuming symmetry in the system.

Open inlet or outlet with a fixed pressure: e.g. according to Zou and He [17]. By applying fixed pressures, flow velocities and mass transport can be better controlled.

Periodic boundary conditions: particles that leave the system are introduced again at the opposite side of the system. The advantage of periodic boundary conditions is that a small system can be simulated by assuming it is repeating space. Periodic boundary conditions are easy to implement and therefore they are used very often.

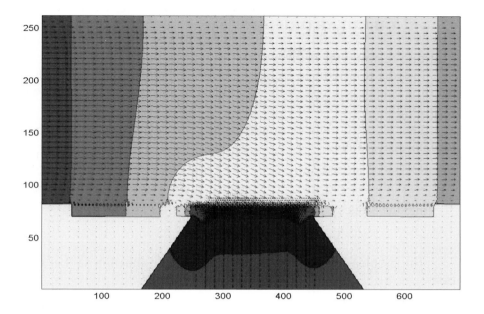

Fig. 8 Lattice Boltzmann computer simulation of the (cross flow) pressure profile above the microsieve Courtesy of [18]. Beneath the microsieve a pressure loss can be observed in the left and right channel, in case of low channel depth/ channel length ratios.

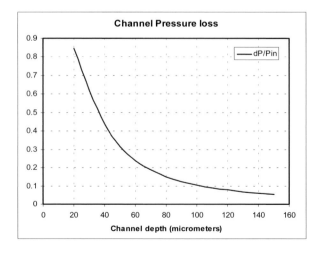

Fig. 9 Pressure loss as a function of the channel depth beneath the microsieve membrane, courtesy of [19].

3. FLOW THROUGH MICROSIEVES WITH RECTANGULAR AND CIRCULAR ORIFICES

Multiple rectangular orifices [20]

In order to determine the pressure-flow relation for a microsieve with slits, the flow is assumed to be fully determined by the viscous effects (the Reynolds number being very small).

The velocity field $\vec{v}$ should satisfy the conservation of mass for an incompressible flow:

$$\nabla \bullet \vec{v} = 0 \tag{18}$$

and the Stokes equation:

$$\nabla p = \eta \nabla^2 \vec{v} \tag{19}$$

The problem will be solved with a harmonic function Ω. Let the velocity field

$$\vec{v} = \zeta \, \nabla\Omega - \Omega \, \hat{e}_\zeta \tag{20}$$

and the pressure

$$p = p_c + 2\eta \frac{\partial \Omega}{\partial \zeta} \tag{21}$$

where ζ is a coordinate axis in the y-direction and p_c a constant.

The function Ω will be solved by solving Laplace's equation:

$$\nabla^2 \Omega = 0 . \tag{22}$$

The slits (width $2R$) are periodically distributed on the microsieve as shown in Fig. 10. The flow is driven by a transmicrosieve pressure Δp. Let the velocity field be

$$\vec{v} = (u,v) = \left(\zeta \frac{\partial \Omega}{\partial x}, \, \zeta \frac{\partial \Omega}{\partial y} - \Omega \right) \tag{23}$$

The boundary conditions are:

$y = 0$, on microsieve: $u = 0$, $v = 0$

$y = 0$, in slits: $\qquad u = 0$, $\quad \dfrac{\partial^2 u}{\partial y^2} = 0$

$y \rightarrow \pm\infty$: $\qquad\qquad u = 0$, $v = \text{const}$, $p = \text{const}$

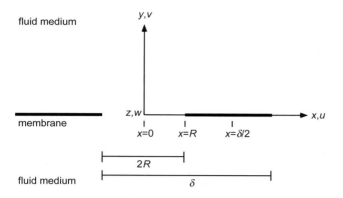

Fig. 10 A microsieve with parallel slits (width 2R, thickness infinitesimal small) in an infinite fluid medium.

With the formulas for the pressure and the velocity field, the Laplace's equation needs to be solved:

$$\frac{\partial^2 \Omega}{\partial x^2} + \frac{\partial^2 \Omega}{\partial y^2} = 0, \quad -\infty < x < \infty, \quad -\infty < y < \infty$$

(24)

With the boundary equations:

$y = 0$, on the microsieve: $\quad\quad\quad\quad \Omega = 0$

$\quad\quad\quad\quad\quad\quad\quad\quad\quad\quad\quad \frac{\partial \Omega}{\partial y} = 0$

$y = 0$, in the slits:

$y \to \pm \infty :$ $\quad\quad\quad\quad\quad\quad\quad \frac{\partial \Omega}{\partial x} \to 0$ faster than $\frac{1}{y}$

$\quad\quad\quad\quad\quad\quad\quad\quad\quad\quad\quad y\frac{\partial \Omega}{\partial y} - \Omega = \text{const}$

$\quad\quad\quad\quad\quad\quad\quad\quad\quad\quad\quad \frac{\partial \Omega}{\partial y} = \text{const}$

Because of the symmetry in $y=0$ and the periodicity in x (see Fig. 10), a possible solution is [20]:

$$\Omega = -Ky + \frac{1}{2}a_0 + \sum_{n=1}^{\infty} a_n e^{-2n\pi y/\delta} \cos(2n\pi x/\delta)$$

(25)

where $K = \text{const}$. The far field pressure is equal to

$$P_{y=\infty} = P_{slit} - 2\eta K$$

(26)

The pressure drop from $y = -\infty, y = \infty$ is equal to Δp and because of the symmetry in y it can be shown that

$$K = \frac{\Delta p}{4\eta}$$

(27)

The fluid velocity for $y = \infty$ is equal to $-a_0 / 2$ (see Eq.(25)). With

$$a_0 = \frac{2K\delta}{\pi} \ln \cos (\pi R / \delta)$$

(28)

it follows that

$$\dot{v}_{y=\infty} = - \frac{R\Delta p}{2\eta\pi\kappa} \ln \cos (\pi\kappa / 2)$$

(29)

where $\kappa = 2R / \delta$ is the fraction of perforated area.

For small κ (Eq.(29)) can be expanded to

$$\dot{v}_{y=\infty} = \frac{\pi R\Delta p}{16\eta} \kappa \left(1 + \frac{\pi^2}{24}\kappa^2 + \frac{\pi^4}{360}\kappa^4 + ...\right)$$

(30)

The volumetric flow rate Q through the wall is

$$Q = \dot{v}_{wall} \cdot A = \frac{\dot{v}_{y=\infty}}{\kappa} \cdot 2lR = \frac{\pi l R^2 \Delta p}{8\eta} \left(1 + \frac{\pi^2}{24}\kappa^2 + ...\right)$$

(31)

with l the length of the slit.
For $\kappa = 0$ (a single slit in a plate), the result reduces to

$$Q = \Delta p \frac{\pi l R^2}{8\eta} .$$

(32)

The velocity in the slit is $v_{slit} = -\Omega$ ($x < R$, $y = 0$) since $u = 0$ in the slit and $\zeta = y = 0$.

$$v_{slit} = \frac{R\Delta p}{4\eta} \left\{\left[1 - \left(\frac{x}{R}\right)^2\right]^{1/2} + \frac{\kappa^2\pi^2}{24}\left[1 - \left(\frac{x}{R}\right)^2\right]^{1/2} + O(\kappa^4)\right\}$$

(33)

In Fig. 11 the velocity profile in the slit, v_{slit}, has been calculated for different values of κ.

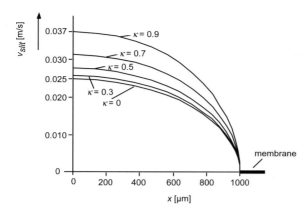

Fig. 11 Velocity profile in a slit of fluid flows through a microsieve with slits ($\Delta p = 100$Pa).

Multiple circular orifices
Tio and Sadhal [20] have also calculated the flow through a microsieve with a regular pattern of circular orifices. They show that the flow rate for a microsieve with a square array of orifices (where fraction κ of perforated area) is

$$Q = \frac{R^3}{3\eta}\Delta p\,\frac{1}{1-f(\kappa)} \tag{34}$$

with $f(\kappa) = \sum_{i=1}^{\infty} a_i \kappa^{(2i+1)/2}$ and $a_1 = 0.344$, $a_2 = 0.111$ and $a_3 = 0.066$.

The correction terms in Eqs. (7) and (34) are considered to be independent [21]. Thus the relation between the pressure and the flow rate becomes:

$$Q = \frac{R^3}{3\eta}\Delta p\,\frac{1}{1+8L/3\pi R}\frac{1}{1-f(\kappa)} \tag{35}$$

Flow rate through decreased openings in a thin plate
The volumetric flow rate without correction terms Q_C through a circular opening with radius R is given by:

$$Q_c = \Delta p\frac{R^3}{3\eta} \tag{36}$$

with Δp the pressure difference across the plate and η the viscosity of the fluid.
The volumetric flow rate Q_S through a slit with length l and width $2R$ is given by:

$$Q_s = \Delta p \frac{\pi l R^2}{8\eta} \tag{37}$$

In order to compare the volumetric flow rate of both type of openings the open area of a number of circular openings must be equal to the open area of a single slit, thus

$$x \cdot \pi R^2 = 2lR \tag{38}$$

$$x = \frac{2lR}{\pi R^2} = \frac{2l}{\pi R} \tag{39}$$

with x the number of openings.

Comparing these two openings, the ratio of Q_S and Q_C is

$$\frac{Q_s}{Q_c} = \frac{3\pi l}{x8R} = \frac{3\pi^2}{16} \tag{40}$$

When the open area of a number of circular openings with radius R is equal to the open area of a single slit with a width $2R$ it follows that the flow ratio is:

$$Q_s/Q_c = 1.85 \tag{41}$$

This means that the flow rate through a microsieve with slits is 1.85 times larger than the flow rate through a microsieve with pores at equal porosity's of the microsieves.

Note that this value is only valid for orifices (openings without a channel length), for openings with a certain channel length a flow ratio can be calculated of 32/12 using the presented formulas.

Also the advantage of a slit over a circular opening can be seen because the volumetric flow rate of a microsieve with slits scales with R^2 and the volumetric flow rate of the sieve with circular openings scales with R^3. So the ratio

$$\frac{Q_s}{Q_c} \sim \frac{1}{R} \tag{42}$$

will increase when the openings are made smaller (e.g. deposition of an additional anti-fouling or chemical inert layer after the processing of the actual microsieve).

In Fig. 12 the ratio is shown as a function of the thickness of the deposited layer with 3 slits (2μm * 10μm) in an area versus 16 circular pores ($R=1$μm) in an area of equal size.

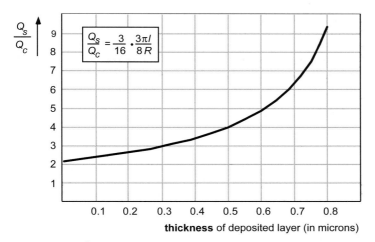

Fig. 12 The ratio $\dfrac{Q_s}{Q_c}$ **when the openings are made smaller with an additional layer.**

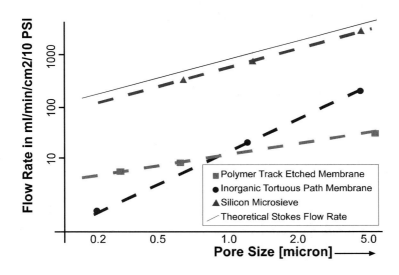

Fig. 13 Typical flowrate in ml/min/cm2/10 psi of microsieve with circular pores and other current filter types. Calculated water flux at room temperature for a microsieve with a square array of circular perforations, a porosity of 20% and a thickness equal to the pore diameter.

The graph shows that very high fluxes (up to two orders of magnitude higher than for other types of microfiltration membranes such as track-etched membranes, asymmetric membranes or tortuous path membranes) can be obtained by using a high-porosity membrane with circular perforations. Moreover changing the circular perforations with diameter D to slits with a width D, at an equal perforated fraction, an at least 1.85 increased flow rate will be obtained (Eq.(41)).

REFERENCES

[1] R.K. Shah, A.K. London, Laminar flow forced convection in ducts, Academic Press New York 1978.

[2] R.W. Fox and A.T. McDonald, Introduction to fluid mechanics, John Wiley & Sons, New York, 1985.

[3] P. Gravesen. et al., Microfluidics - A Review, Micro Mechanics Europe, Workshop Digest, (1993) 143-164.

[4] R.K.Shah and A.L. London, Laminar flow forced convection in ducts, Academic Press, New York, 1978.

[5] Z. Dagan. et. al., An infinite-series solution for the creeping motion through an orifice of finite length, J. Fluid Mech., Vol. 115 (1982) 505-523.

[6] J. Happel and H. Brenner, Low Reynolds number hydrodynamics, Martinus Nijhoff, Dordrecht, 1986.

[7] Z. Dagan, S. Weinbaum and R. Pfeffer, Chem. Eng. Sci., 38 (1983) 583-596

[8] J. van Kuijk, forAquamarijn 1995.

[9] Succi S. (2001) The Lattice Boltzmann equation for fluid dynamics and beyond. Oxford University Press, England.

[10]R.G.M. van der Sman, Lattice Boltzmann schemes for convection-diffusion phenomena; application to packages of agricultural products. PhD thesis (1999) Wageningen University.

[11] A.J.C.Ladd and R. Verberg, Lattice Boltzmann simulations of particle fluid suspensions, Journal of Statistical Physics 104 (2001) 1191.

[12] J. Bernsdorf, G. Brenner and F. Durst, Numerical analysis of the pressure drop in porous media flow with lattice Boltzmann (BGK) automata, Computer Physics Communications 129 (2000) 247.

[13] F.M. van Kats, P.J.P. Egberts and C.P.J.W. Kruisdijk, Three-phase effective contact angle in a model pore, Transport in Porous Media 43 (2001) 225.

[14] D.A. Wolf-Gladrow, Lattice-Gas cellular automata and lattice Boltzmann models, Springer Verlag, Berlin, Germany (2000)

[15] M. Junk, A finite difference interpretation of the Lattice Boltzmann method, Numerical Methods for Partial Differential Equations 17 (2001) 383.

[16] D.Kandhai, A. Koponen, A. Hoekstra, M. Kataja., J. Timonen and P.M.A. Sloot, Implementation of 3D lattice-BGK: boundaries, accuracy and a new fast relaxation method, Journal of Computational Physics 150 (1999) 482.

[17] Q. Zou and X. He, On pressure and velocity boundary conditions for the lattice Boltzmann BGK model, Phys. Fluids 9 (1997) 1591.

[18] G. Brans and R.G.M. van der Sman, Lattice Boltzmann Computer Simulation 2002.

[19] U. Dobberstein, FCDF research centre 2002.

[20] K.Tio and S. Sadhal, Boundary Conditions for Stokes flow Near a Porous Membrane, Applied Sc. Research, Vol. 52 (1994) 1-20.

[21] C.J.M. van Rijn and M.C. Elwenspoek, Micro Filtration Membrane Sieve with Silicon Micro Machining for Industrial and Biomedical Applications, IEEE Micro Mechanical Systems, (1995) 83-87.

Chapter 5

Membrane Deflection and Maximum Pressure Load

> *"Paper is strongest at the perforations."*
> [Correy's Law]

1. INTRODUCTION

The perforated membrane fields of a microsieve are much thinner (< 2 micron) than most commonly used homogenous perforated membranes (e.g. thickness of track etched membranes are typical 10 micron). The microsieve therefore has always a proper designed support structure to subdivide the actual sieve field into smaller membrane areas that are supported at regularly spaced intervals to increase the pressure load of the whole microsieve structure. An analytical expression will be given together with an experimental verification between the relation of membrane size, membrane shape and maximum pressure load for unperforated membranes based on earlier work. This work cannot be found in textbooks, but experiments supports the validity of the presented theory.

Perforations in the membrane will of course generally lower the maximum pressure load of the membrane. For filtration membranes the porosity of a membrane should be as high as possible. However, a large increase in porosity will lead to a further decrease in membrane strength. For the design of microsieves it is important to know the correlation between strength and porosity. Unfortunately analytical expressions[1] to describe this correlation do not exist, but computer simulations [12] and experiments support the proposed theory.

---

* Timoshenko [1] only considers the effect of one perforation in a membrane.

Unperforated membranes

A thin unperforated membrane will bend on applying a pressure difference. The stress in the membrane at each point is the sum of the local tensile stress and the local bending stress. The bending stress reaches a maximum near the edge of the membrane where it is attached to the support.

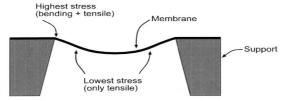

Fig. 1 Bending of a membrane under an applied pressure difference.

In case the pressure load is further increased the membrane will break initiated from small ruptures from the edge. It will be clear that if the width of the support distance (width of membrane field) is reduced the membrane can withstand a larger pressure load. If the length of the membrane field is much larger than the width of the membrane field, the maximum pressure load will be dependent mainly on the width as will be shown.

2. ANALYTICAL PRESSURE LOAD EXPRESSIONS FOR AN UNPERFORATED RECTANGULAR MEMBRANE[*][11]

For silicon micromachined membranes it is advantageous to subdivide the actual sieve field into smaller rectangular membranes. Sieve fields with a very small width (<200 micron) can easily be made. Based on the work of Timoshenko [1] new analytical formulas will now be derived taking into account the bending stress near the edge of the membrane. In the here presented model the assumption is made that the rectangular membrane is only clamped to a rigid support along the long edges and not on the short edges. This model of course is then more valid for rectangular membranes with a large length to width ratio.

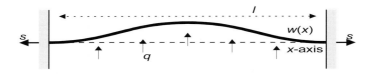

Fig. 2 Deflection w(x) of a two edge clamped membrane stretched with axial distributed force S and uniformly loaded with pressure q.

[*] Partially reprinted with permission of J. Microelectromech. Syst. (van Rijn 1997).

The deflection curve $w(x)$ of a rectangular plate with dimension $l \cdot b \cdot h$ clamped at two edges $x=0,l$ stretched by an axial distributed force S and uniformly loaded under a pressure q is given by the well known differential equation

$$D\frac{d^4w(x)}{dx^4} - S\frac{d^2w(x)}{dx^2} = q \qquad (1)$$

with flexural rigidity

$$D = \frac{Eh^3}{12(1-v^2)} \qquad (2)$$

($v=$ Poisson's ratio, $E=$ Young's modulus).

The general solution symmetrical in $x=l/2$ following Timoshenko [1] is given by

$$w = \frac{ql^4}{16u^3 D\tanh u}\left\{\frac{\cosh\left[u\left(1-\frac{2x}{l}\right)\right]}{\cosh u} - 1\right\} + \frac{ql^2(l-x)x}{8u^2 D} \qquad (3)$$

with definition $u^2 = Sl^2/4D$.

The deflection near the centre of the membrane is mainly determined by the parabolic term, whereas the deflection near the edge of the membrane is mainly determined by the first term. The points of inflection ($d^2w/dx^2=0$) of the deflection curve are determined by the dimensionless parameter u. For small values $u \ll 1$ the inflection points are almost independent of u and given by

$$x = l/2 \pm \frac{l}{2\sqrt{3}}.$$

At a large axial force S the points of inflection will move toward the edges. For $u \gg 1$ these points are located at

$$x = l/2 \pm \left(l/2 - \frac{l\ln u}{2u}\right) \qquad (4)$$

For very thin plates or membranes u will be already large at moderately values of the axial force S. The general solution is fully determined by the constants D, S and q. For large deflections the axial force S will increase due to elastic extension Δl of the plate clamped between the edges. It will be shown that S may be expressed as a simple function of D, q, l and v in the limit for large values of u. In this limit S is still a constant (dependent only on other constants) and is independent on x, so the deflection curve $w(x)$ is still a solution of the differential equation. The increment ΔS is related to the increment Δl:

$$\Delta l = \varepsilon l = \frac{\Delta S\left(1-v^2\right)l}{hE} = \int_0^1 dl - l = \int_0^1 \sqrt{dx^2 + dw(x)^2} - l \tag{5}$$

It can be shown that the latter term is almost independent on $w(x)$ for all type of deflections under the condition $dw(x)/dx \ll 1$, or $w_{max} \ll l$. Δl Scales then with w_{max}^2/l. For a parabolic deflection curve, $\Delta l = (8/3)\, w_{max}^2/l$. Using

$$u^2 = (S + \Delta S)l^2 / 4D = u_0^2 + \Delta S l^2 / 4D \tag{6}$$

one obtains

$$\Delta l = \frac{4D\left(u^2 - u_0^2\right)\left(1-v^2\right)l}{l^2 hE} = \frac{8}{3}\frac{w_{max}^2}{l} \tag{7}$$

Alternatively w_{max} is determined by the deflection curve, for $u \gg 1$ at $x = l/2$, $w_{max} = ql^4/32u^2 D$. For large deflections the inflection parameter u_0 related to the initial axial force density S may be neglected, i.e. $u \gg u_0$, hence the following relation using Eq.(2) and Eq.(7)is found for u:

$$u^6 \underset{u \gg 1}{=} \frac{9\left(1-v^2\right)^2 q^2 l^8}{8E^2 h^8} \tag{8}$$

The maximum deflection w_{max} at $x = l/2$ is given by

$$w_{max} \underset{u \gg 1}{=} \frac{ql^4}{32 Du^2} = 0{,}361 l\sqrt[3]{\frac{ql\left(1-v^2\right)}{Eh}} \underset{v=0.25}{=} 0{,}35 \cdot l\sqrt[3]{\frac{ql}{Eh}} \tag{9}$$

The constant tensile stress in the plate is then estimated for large values of $u \gg 1$

$$\sigma_{tensile} = \frac{S}{h} = \frac{4u^2 D}{hl^2} = \frac{Eu^2}{3\left(1-v^2\right)}\left(\frac{h}{l}\right)^2 \underset{\substack{u \gg 1 \\ v=0.25}}{=} 0{,}37\sqrt[3]{\frac{q^2 l^2 E}{h^2}} \tag{10}$$

The maximum bending stress at the edge of the membrane is given by

$$\sigma_{bend} = \frac{E}{1-v^2}\frac{d^2 w}{dx^2} z \tag{11}$$

The above expressions are valid for a rectangular plate clamped at two edges, and may be valid for a thin membrane plate under a substantial load at large deflections $w_{max}/h \gg 1$. Equation (10) gives a good scaling relation for $\sigma_{tensile}$ for all type of deflections under the condition $w_{max} \ll l$, whereas σ_{bend}

$$\sigma_{bend} = \frac{E}{1-v^2}\frac{d^2w}{dx^2}z \bigg|_{\substack{u\gg l \\ x=0 \\ z=h/2}} = \frac{3q}{2u\tanh u}\left(\frac{l}{h}\right)^2 \bigg|_{v=0.25} = 1{,}47\sqrt[3]{\frac{q^2l^2E}{h^2}} \tag{12}$$

gives after calculation [2] a similar scaling relation at the edge of the membrane.

The actual case to be considered here is a rectangular membrane clamped at all four edges. With the principle of virtual work Timoshenko [1] has calculated the deflection w_0 of the centre of a square membrane clamped at four edges

$$w_0 = 0.802a\sqrt[3]{\frac{qa}{Eh}}\bigg|_{a=l/2} = 0.318 l\sqrt[3]{\frac{ql}{Eh}} \tag{13}$$

and corresponding tensile stress σ_0 in the middle (centre) of the membrane

$$\sigma_0 = \frac{E}{1-v}0.462\frac{w_0^2}{a}\bigg|_{v=0.25} = 0.396\sqrt[3]{\frac{q^2a^2E}{h^2}}\bigg|_{a=l/2} = 0.29\sqrt[3]{\frac{q^2l^2E}{h^2}} \tag{14}$$

The calculated values for w_0 and σ_0 correspond reasonably well with the values found for w_{max} and $\sigma_{tensile}$ for the two clamped case. The value found for the maximum deflection w_{max} in the two edge clamped case is slightly larger than in the four clamped edge case due to the extra constraint and thus limiting the value of the deflection in the middle of the membrane.

A fortiori it is assumed that the ratio between σ_{bend} and $\sigma_{tensile}$ remains unchanged for the two-clamped and four-clamped case, because both stresses scale identical at the edge of the membrane [3]. Moreover in the four-clamped case the maximum stress [4] is found near the middle of the edges. The deflection curve will resemble then the two-clamped case. The total stress at the edge is the addition of the constant tensile stress due to stretching and the bending stress near the middle of the edge of a square membrane:

$$\sigma_{total} = \sigma_{tensile} + \sigma_{bend} = 0.29(1+1.47/0.37)\sqrt[3]{\frac{q^2l^2E}{h^2}}\bigg|_{\substack{u\gg l \\ x=0}} \tag{15}$$

A similar relation is also valid for a rectangular membrane, in which the long side is much larger than the short side:

$$\sigma_{total} = \sigma_{tensile} + \sigma_{bend} = 0.37(1+1.47/0.37)\sqrt[3]{\frac{q^2l^2E}{h^2}}\bigg|_{\substack{u\gg l \\ x=0}} \tag{16}$$

In the above mentioned equations (15) and (16) internal stresses of the materials are not taken into account. It is well known that stochiometric silicon nitride membranes fracture at relatively low pressure due to high intrinsic tensile stress of the order of 1 GPa [5]. The intrinsic tensile stress σ_0 in a silicon rich silicon nitride membrane with thickness 1μm is much smaller, from recent data [6], it ranges from $0.8 \cdot 10^8$ to $1.6 \cdot 10^8$ Pa. The maximum tensile stress σ_{yield} (before rupture occurs) for silicon rich silicon nitride [7] is approximately $4.0 \cdot 10^9$ Pa. The intrinsic tensile stress may, therefore be neglected in calculating the maximum pressure q_{break} before fracture occurs ($u \gg u_0$). Using Young's modulus E for silicon nitride $2.9 \cdot 10^{11}$ Pa and $v = 0.25$ one finds $q_{break} = 2.7$ bar for a dense square silicon nitride membrane with width $l = 1000$ μm. The inflection parameter then is $u = 40$. The inflection points of the membrane are then located at 25 μm from the edge.

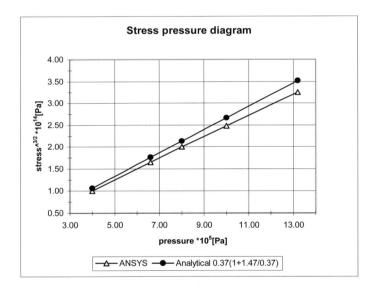

Fig. 3 Comparison of the stress near the edge of the long side of an unperforated rectangular silicon nitride membrane (200 x 600μm) with a thickness of 1 μm as predicted by an ANSYS computer simulation and analytical equation (16) derived by van Rijn et al. [11].

In Table 1 some theoretical estimates for the maximum load q_{break} have been given for some inorganic materials, metals and polymers for square membranes. For non-ductile inorganic materials the ultimate stress $\sigma_{ultimate}$ at which the membrane breaks at pressure q_{break} coincides with the pressure q_{yield} when the stress in the membrane reaches in the middle of the edge σ_{yield}

($\sigma_{ultimate} = \sigma_{yield}$). The maximum load q_{break} for these materials is calculated here with Eq.(15) and Eq.(16) using both $\sigma_{tensile}$ and σ_{bend} in the middle of the edge of the membrane because for non-ductile (brittle) materials there is no stress regime above σ_{yield} for plastic deformation.

Table 1

The maximum load of square membranes (with l=1mm, h=1μm). Many given values for E, σ_{yield}, $q_{ultimate}$ are bulk values from Timoshenko Mechanics of Materials.

Material	E [G Pa]	σ_{yield} [M Pa]	$\sigma_{ultimate}$ [M Pa]	q_{tens} tens. [bar]	q_{yield} tens.+bend. [bar]	q_{break} calculated [bar]	q_{break} measured [bar]	u	Bending Point from edge [μm]
Al	70	50	70	0.1	0.01	> 0.21	1.3	>28.7	< 58
Polyimide	2.5	215				> 9.6			
Polycarbonate	2.5	63				> 1.85			
Ni	210	400	500	1.10	0.10	> 1.1		>74	< 29
Cu	110	330	380	1.14	0.10	> 1.14	5.3	>93	< 24
Stainless Steel	200	450	800	1.35	0.12	> 1.35		>81	< 27
Ti	110	400	500	1.54	0.14	> 1.53	1.5	>103	< 22
W	360	2500	2500	13.3	1.2	1.2		64	32
$Si_{rich}N_{poor}$	290	4000	4000	30.1	2.7	2.7	2.5	89	25
Si	190	7000	7000	86	7.8	7.7		146	17
Al_2O_3	530	15400	15400	168	15.2	14.9		130	19
Si_3N_4	385	14000	14000	171	15.5	15.4		145	17
SiO_2	73	8400	8400	182	16.5	16.4		258	11
SiC	700	21000	21000	233	21.0	20.9		132	18
Diamond	1035	53000	53000	768	69.4	68.9		72	15

For ductile metals like Al, Ni, Cu, Stainless Steel and Ti there is a linear relation between the strain and the applied stress up to σ_{yield}. The membrane will not break when the pressure q_{yield} is reached, in fact the stress in the middle of the edge may increase unto $\sigma_{ultimate}$. Between σ_{yield} and $\sigma_{ultimate}$ the strain of the membrane strongly increases. In this region E can not be considered as a constant, because E will decrease as a result of plastic deformation. This means that q_{break} will strongly increase in this region. Therefore, for a more elongated membrane a higher load may be applied, resulting in $q_{break} \gg q_{yield}$. For ductile materials therefore only an under-estimate of the maximum load q_{break} can be given. The under-estimate for the maximum load q_{break} for these materials is calculated here with Eq.(15) and Eq.(16) using only $\sigma_{tensile}$ in the middle of the edge of the membrane. For ductile materials the local stress originating from bending will be released due to local plastic deformation (in the middle of the edge) when this local stress reach a value above σ_{yield}.

Applying a higher load such that the lateral strain increases above σ_{yield} will cause the membrane to break when $\sigma_{ultimate}$ is reached. Eq.(15) and Eq.(16) are still valid although the reduced value of E is not known. The under-estimate of the maximum load q_{break} in this case is therefore higher than the pressure q_{tens}

defined as the pressure necessary for creating plastic deformation due to tensile stresses only, i.e. q_{tens} is calculated from Eq.(15) and Eq.(16) leaving out σ_{bend} .

For a perforated membrane eforementioned equations may be used choosing a different value for E and σ_{yield}. In a first order approximation both E and σ_{yield} are smaller and proportional with the unperforated fraction of the membrane [8]. This will result in a smaller maximum load, also proportional to the unperforated fraction as can be obtained from above scaling equation Eq.(15) and Eq.(16) for σ_{total}.

3. EXPERIMENTS

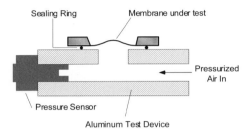

Fig. 4 Test set-up.

To determine the maximum load of membranes, a small test-device has been made, in which membranes can be clamped. With pressurised air, the membranes can be deflected to a certain load. This maximum pressure can be measured with a pressure sensor (Honeywell, 24PC, 7Bar) connected to a digital multimeter with a peak-hold function to memorise the maximum applied pressure before fracture occurs.

Silicon nitride membranes

The dependence of the maximum load q_{break} on the membrane width, membrane thickness, membrane perforation, membrane shape and some membrane materials has been determined.

In Fig. 5 to Fig. 7 the dependence of the maximum load q_{break} on the membrane width l is shown for various membrane thickness' h. q_{break} seems to be reasonably well inversely proportional to l in accordance with the earlier presented theory, see Eq.(15).

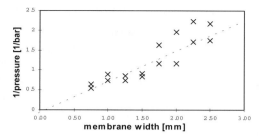

Fig. 5 Results obtained from maximum load measurements (silicon nitride membrane: 0.5 μm thick) for square membranes with width *l*. Drawn line corresponds with σ$_{yield}$=4000.

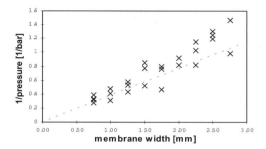

Fig. 6 Results obtained from maximum load measurements (silicon nitride membrane: 1.0 μm thick) for unperforated square membranes with width *l*. Drawn line corresponds with σ$_{yield}$=4000.

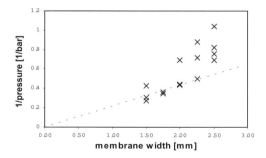

Fig. 7 Results obtained from maximum load measurements (silicon nitride membrane: 2.0 μm thick) for unperforated square membranes with width *l*. Drawn line corresponds with σ$_{yield}$= 4000.

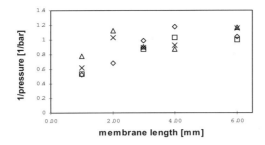

Fig. 8 Dependence on membrane shape; maximum load data of a rectangular membrane (silicon nitride, 1.0 μm thick), with a fixed width of 1 mm.

In the three figures straight lines have been drawn corresponding with σ_{yield}= 4000 MPa. These lines correspond quite well in all three figures with the data, herewith verifying that also the thickness h scales according to Eq.(15).

In Fig. 8 it is seen that the maximum load values are nearly constant for unperforated rectangular membranes with size 1 mm x 2 mm, 1 mm x 3 mm, up to 1 mm x 6 mm. The square membrane with size 1 mm x 1 mm has a q_{break} roughly 30 % higher than the more elongated versions. Qualitatively this difference may easily be explained as the difference in maximum load for a membrane clamped at four edges (good model for a square membrane) and a membrane clamped at two edges (good model for a rectangular membrane with an infinite length and a fixed width; an approximating model for a square membrane). Comparing Eq.(15) and Eq.(16) it can be calculated that the theoretical ratio between the two maximum load pressures for a rectangular and a square membrane with a width l is $(0.29/0.37)^{3/2} = 0.69$.

Ductile membranes

Maximum load values for breaking of the membranes of the ductile materials titanium, aluminium and copper are depicted in Fig. 9 to Fig. 11. The data indicate also a linear relation between the maximum breaking load and the width l.

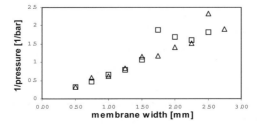

Fig. 9 Results obtained from maximum load measurements (titanium membrane: 1.0 μm thick) of an unperforated square membrane with width l.

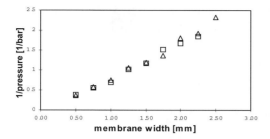

Fig. 10 Results obtained from maximum load measurements (aluminium membrane: 1.0 μm thick) of an unperforated square membrane with width *l*.

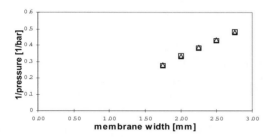

Fig. 11 Results obtained from maximum load measurements (copper membrane: 1.0 μm thick) of an unperforated square membrane with width *l*. The pressure of our system (4 bar) was not high enough to break membranes smaller than 1.75 mm.

As discussed in the former section for ductile materials the maximum breaking load may be much higher than the maximum load q_{yield} for reaching a stress σ_{yield} at the middle of the edge in the membrane.

In comparing the experimental found maximum breaking load for these ductile materials with the calculated q_{break} as presented in Table 1 (membrane width = 1 mm) it is found that the maximum breaking load for titanium membranes is slightly higher, and that for aluminium and copper membranes it is much higher than $q_{break.}$. Apparently the materials aluminium and copper have a broader plastic deformation regime than titanium. Also the grain size as defined by the sputter process may have an effect, especially for less ductile materials, such as titanium. It is also known that the reactive ion plasma may attack titanium, whereas copper and aluminium are not sensitive for this process.

If the yield stress of a ceramic membrane is exceeded, it will rupture. For metals and polymers the calculated maximum pressure may be used as a lower limit for the really applicable pressure, as the material will plastically deform at the places of maximum stress, thus redistributing the stress over a larger area.

Suppose a 1 μm thick polyimide membrane is supported by bars with an intermediate distance of 1000 μm. If values σ_{yield} = 215 MPa and E = 2.5 GPa [9] for the cured polyimide are used in Eq.(15), a lower limit for p_{max} of 9.6 bar is obtained. Polycarbonate is significantly weaker, as it does not contain cross-links like the cured polyimide.

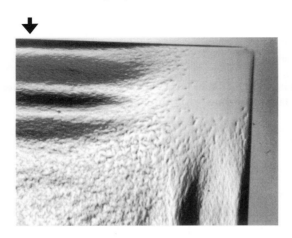

Fig. 12 An aluminium membrane 1 mm × 1 mm with thickness 1μm after loading above q_{yield} (0.03 bar) showing plastic deformation. The arrow in the figure indicates the middle of the edge of the membrane.

With σ_{yield} = 63 MPa and E = 2.35 GPa [10] a lower limit for p_{max} of 1.85 bar is obtained. Experiments show that for metals the pressure at which the membranes break is approximately one order of magnitude higher than the lower limit estimated with Eq.(15). For polymers a similar difference is expected. If one incorporates a 50 % strength reduction for the perforations in the membrane [11], and if one assumes that the support bars are much stronger than the membrane with a separation width of 200 μm and that the sieve rests on a rigid support, one obtains a very rough estimation for the maximum pressure of 15 bar for the polyimide and 3 bar for the polycarbonate membrane.

Summary
Maximum load values q_{break} for silicon nitride membranes have been experimentally determined on varying the width, the thickness and the shape of the membranes. The results are in accordance with the theoretical predictions for q_{yield} based on the derived Eqs.(15) and (16). The maximum load values q_{break} for perforated membranes are found to be smaller for the unperforated membranes.

The maximum load values q_{break} for unperforated membranes composed of a ductile material are higher than the estimated values for q_{yield}. For ductile materials q_{yield} may be considered as a safe under-estimate for q_{break}.

4. PERFORATED MEMBRANES AND MAXIMUM PRESSURE LOAD

Suppose a membrane is fully perforated except near the edges. Such a membrane will be more flexible than an unperforated one. Therefore it will show a larger deflection when a pressure is applied.

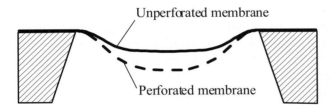

Fig. 13 Deflection of a perforated and unperforated membrane for equal pressure loads.

A larger deflection leads to a stronger bending at the edge with a resulting increase in the bend stress. Therefore an unperforated membrane will be able to withstand larger pressures than a perforated one. The distribution of the pores over the surface will influence the stiffness and herewith the strength of the membrane, which has been studied by Nijdam [13]. This stiffness is dependent on the number, size and shape of the 'bars' in-between the pores. This is illustrated for a rectangular membrane with slits.

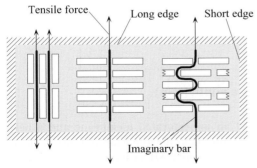

Fig. 14 Different slit configurations in a rectangular membrane from Nijdam [13].

As the membrane is rectangular, the tension perpendicular to the long edges will be larger than the tension perpendicular to the short edges. The tension pulls on the imaginary bars between the slits. The structure on the left will probably be stronger than the one in the middle, as the number of bars is higher. The structure on the right has the same number of bars as the one in the middle, but since these bars are curved (spring-shaped), they will be less stiff. Therefore a membrane with the structure on the right will show a larger deflection for equal pressures. A larger deflection leads to a larger bending stress on the edges and it can be expected that the membrane with the alternating structure will be the weakest. In order to verify this hypothesis, microsieves with several different patterns were tested.

Pressure-load tests on microsieves
As can be seen in Fig. 15 the maximum load of a perforated membrane is nearly two times smaller than the maximum load of an unperforated membrane. The perforations are circular pores with a diameter of 5 μm. The total perforated area is approximately 25 %. The mean yield strength and the mean Young's Modulus of the material silicon nitride will decrease within the perforated area with a fraction (1-κ) with κ the as the perforated area. The rupture starts at the edge of the membrane and at this place the membrane is not perforated and both the Young's Modulus and the yield strength remain constant. However due to a lowered mean Young's Modulus the membrane will have a larger deflection and a fortiori a larger bending near the edge of the membrane.

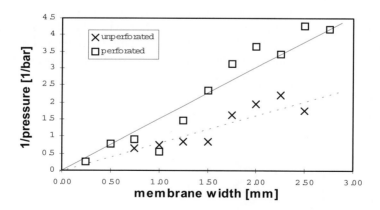

Fig. 15 Difference in maximum load of an unperforated membrane and a perforated membrane (silicon nitride membrane 0.5 μm thick) for unperforated square membranes with width l. Drawn lines correspond to σ_{yield} = 4000 MPa (dotted) and σ_{yield} = 2500 MPa (solid).

An estimate of this effect at the edge of the membrane may be derived from:

$$\sigma_{total} \approx \frac{E_{edge}}{1-v} \cdot \frac{w_0^2}{l} \approx E_{edge} \sqrt[3]{\frac{q^2 l^2}{h^2 E_{mean}^{2}}} \approx \sqrt[3]{\frac{q^2 l^2 E_{edge}}{(1-\kappa)^2 h^2}} \qquad (17)$$

This relation implies a linear relation between the maximum pressure load q and the unperforated fraction $(1-\kappa)$.

Extensive pressure load tests on microsieves with slit-shaped perforations and with circular-shaped perforations have been performed by Brink et al. [12].

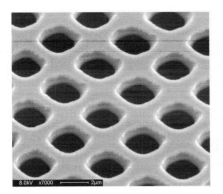 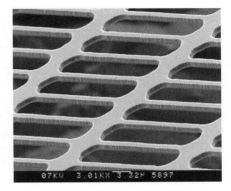

Fig. 16 SEM micrographs of microsieves with circular pores and with slits.

Microsieves were constructed by deposition of a 1 µm thin low-stress silicon nitride layer on a 380 µm thick silicon wafer using LPCVD (low-pressure chemical vapour deposition). On top of that a 1.8-µm thin photosensitive layer was spun. This layer was patterned with UV-light through a photo mask that contains slits (size 3 x 15 µm) and circles (diameter 3µm) in several different patterns. These patterns were transferred into the silicon nitride by RIE (reactive-ion etching). Finally, the silicon nitride membrane was released by removing selectively a part of the silicon underneath by anisotropic KOH-etching.

The 9×9 mm^2 sieves have a perforated free hanging membrane of 1×3 mm^2. As the membrane is not square, a strength dependency on the orientation of the perforations is to be expected. Sixteen different patterns for slits and sixteen for circular pores were designed and made. Fig. 17 shows how the perforations of the thirty two different sieves are oriented.

As the highest stress in the membrane occurs at the edge, the first 20 µm from the edge of the membrane was not perforated. The sieves are placed in a holder and subjected to a pressurised yeast-cell suspension. The yeast cells block the pores and thus prevent the water from flowing through the membrane. This assures an even distribution of the pressure force over the membrane area and makes determination of the pressure difference easier. The pressure on the top-side of the membrane is slowly increased until rupture occurs.

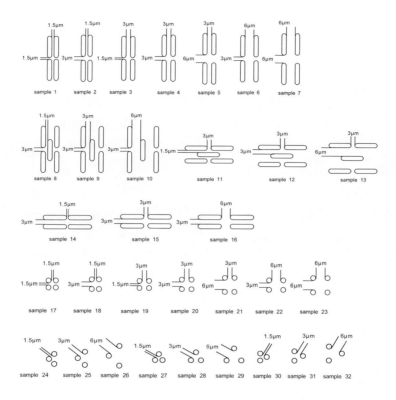

Fig. 17 Various pore configurations used for the strength tests. The long side of the membrane is horizontal with respect of the given pore configurations.

Unperforated membranes were also tested and a mean pressure load of 9.6 bar was measured. The yield stress of the silicon nitride of this membrane was estimated to be 11 GPa from Eq.(16), a value that was also confirmed ($\pm$ 1 GPa) with ANSYS computer simulations. This value is remarkably higher than the value of the yield stress in former experiments (4 GPa), however the batches of those membranes originated from a different supplier. The ANSYS computer simulations for the maximum pressure load of the perforated membranes are higher than the actual measured values. The analytical expression Eq.(17) predicting a linear relation between the maximum pressure load and the unperforated fraction of a perforated membrane, seems to underestimate slightly the actual measured values. The difference between the square and the brick pattern is difficult to observe, both from experiments and computer simulations.

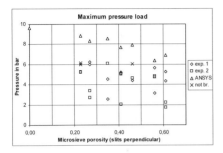 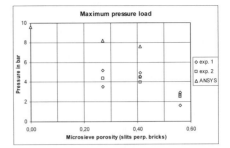

Fig. 18 Maximum pressure load of slits perpendicular to the length of the membrane in a square pattern (samples 1-7) and in a brick pattern (samples 8-10). Data (Δ) from ANSYS computer simulations are also plotted.

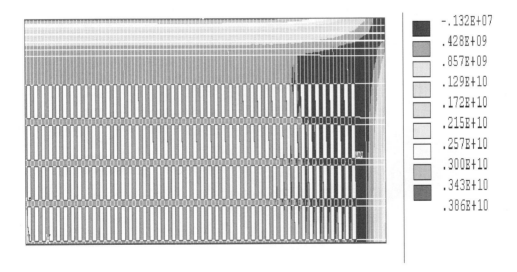

Fig. 19 Indication of the stress distribution on the surface of the membrane. Only a quarter of the membrane is shown [12].

An example of a stress calculation with ANSYS is shown in Fig. 19. It gives an indication of the stress distribution over the membrane surface. The plot gives only a rough indication of the actual stress values, as the mesh size is not small enough to show stress peaks that should occur at the corners of the slits.

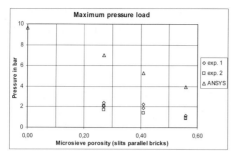 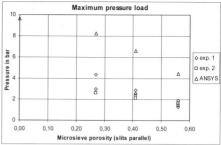

Fig. 20 Maximum pressure load of slits parallel to the length of the membrane in a brick pattern (samples 11-13) and in a square pattern (samples 14-16). Data (Δ) from ANSYS computer simulations are also plotted.

The square pattern of parallel slits seems to be stronger (ANSYS and experimental) than the brick pattern of parallel slits as in confirmation with the earlier work of Nijdam (see also Fig. 14). Perpendicularly placed slits are much more stronger than parallel placed slits along the long side of the membrane.

Fig. 21 Indication of the stress distribution on the surface of the membrane, Nijdam [13]. Only a quarter of the membrane is shown.

The ANSYS computer simulations for the maximum pressure load of the perforated membranes and the analytical expression Eq. (17) both over estimate here the actual measured values. The maximum deflection at the middle of the membrane according to the ANSYS simulation is 135 μm for the brick pattern (porosity 0.56), 70 μm for the square pattern (porosity 0.56) of parallel slits, 45 micron for an unperforated membrane, and 45 -55 μm for most other pattern configurations with different porosity's at a pressure of 9.6 bar.

Nijdam [13] already in 1995 performed 3D computer simulations with a fine mesh size, here the stress peaks at the corner of the parallel placed slits can be observed.

ANSYS simulations and experiments indicate that the parallel square and the diagonal square pattern of pores are similar in strength, see Fig. 22. Also according to ANSYS, the calculated maximum deflections in the middle of the membrane are similar for both structures independent of the amount of porosity.

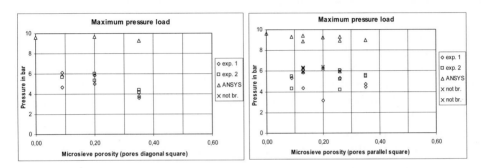

Fig. 22 Maximum pressure load of membranes with cylindrical pores in a parallel square pattern (samples 17-23) and in a diagonal square pattern (samples 24-26). Data (Δ) from ANSYS computer simulations are also plotted.

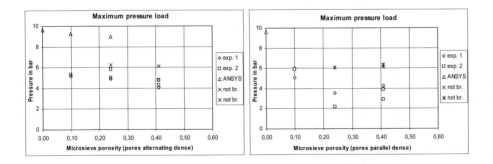

Fig. 23 Maximum pressure load of membranes with cylindrical pores in a parallel dense pattern (samples 27-29) and in a alternating dense pattern (samples 30-32). Data (Δ) from ANSYS computer simulations are also plotted.

With a dense 'pore packing' higher porosity's are achievable than with the square pore pattern, see Fig. 23.

The alternating dense pattern has the advantage that along the long side of the membrane not all the pores close to the edge of the membrane are at the same distance from the membrane. In the past experiments showed that an alternating pattern facilitates a more defect free processing during the free etching of the membrane from the silicon substrate.

A defect free membrane manufacturing process is a prerequisite for all further operations. Experiments indicate, see Fig. 23, that the alternating dense pattern of cylindrical pores seems to be slightly stronger than the parallel dense version.

Conclusions

Both experiments and ANSYS computer simulations support the analytical expressions for the maximum pressure load derived by Van Rijn based on formulas by Timoshenko for unperforated membranes. For perforated membranes the ANSYS computer simulations appear to overestimate the measured values. It is shown that a simple linear analytical expression eq. (17) can also be used for the prediction of the maximum pressure load and gives a prediction more in line with the experiments than the ANSYS simulations.

A very strong maximum pressure load dependency for slit shaped configurations was found. Membranes with slits placed perpendicular to the long edges are significantly stronger than those with parallel placed slits. Experiments and ANSYS simulations with cylindrical pore configurations for a given porosity, give similar results. Alternating dense cylindrical pore structures have a number of advantages, such as a higher maximum pressure load for the same porosity and a defect free fabrication process.

Membrane filtration applications using membranes with slits offer an advantage over membranes with circular pores. The flow resistance of a membrane is nearly a factor 2 (chapter 4.3) smaller for a configuration with slits (with a width d) than for circular pores (with a diameter d) at the same porosity. Furthermore, the porosity of the slit shaped configurations of the membrane can potentially be chosen higher (measured up to 0.61) while the experimental maximum pressure load values stay relatively high (> 5 bar).

REFERENCES

[1] S.P. Timoshenko and S. Woinowsky-Krieger, Theory of Plates and Shells, McGraw-Hill Book Company, New York, 1959.

[2] This calculation is to author's knowledge not known in literature.

[3] For round membranes: M.P. Di Giovanni, Flat and Corrugated Diaphragm Design Handbook, Marcel Dekker, ISBN 0-8247-1281-1, ch 14.

[4] C.Y. Chia, Nonlinear Analysis of Plates, McGraw-Hill, New York, 1980.

[5] M. Heschel and S. Bouwstra, Robust compliant silicon nitride membranes, MME 1995, 3-5 september 1995, Copenhagen, Denmark (1995) 84-87

[6] V.L. Spiering, S. Bouwstra, M. Elwenspoek and J.F. Burger, J. Micromech. Microeng., Vol. 3 (1993) 243-246.

[7] S. Bouwstra, R. Legtenberg and T. Popma, Sensors and Actuators, March 1990, Poster Eurosensors, Nov. 88, Enschede.

[8] D. Bynum and M.M. Lemcoe, Stresses and Deflections in Laterally Loaded Perforated Plates, Welding Research Council Bulletin, Vol. 80 (1992).

[9] Arch Chemicals N.V., Keetberglaan 1A, Havennummer 1061, B-2070 Zwijndrecht, Belgium.

[10] General Electric Plastics, Distributor: B.V. Snij-Unie HiFi, Zoutketen 23, 1061 EX, Enkhuizen, Nederland.

[11] C.J.M. van Rijn, M. van der Wekken, W. Nijdam and M.C. Elwenspoek, Deflection and maximum load of microfiltration membrane sieves made with silicon micromachining, J. Microelectromech. Syst. 6 (1997) 48-54.

[12] R. Brink, Aquamarijn, Thesis, Hoge School van Utrecht, 1998.

[13] W. Nijdam, Aquamarijn, Masters Thesis, University of Twente, 1995.

Chapter 6

Polymeric Membranes

"Life in plastic, it's fantastic"
[Barbie, song by Aqua]

1. INTRODUCTION

Silicon micromachined microsieves with a silicon nitride membrane layer are durable filters with an outstanding chemical inertness. These properties make them especially suitable for applications where a long lifetime and an easy cleanable filter are required, for example crossflow filtration of milk and beer. For applications where disposable filters are preferred, for example the medical or microbiological analysis market, the use of these high-quality microsieves may not always be economically feasible. For such applications a low-cost sieve would be highly desirable. As most disposable filters are made of polymers (e.g. track etched membrane filters), it is worthwhile to investigate the production feasibility of polymeric microsieves.

Lithographic techniques

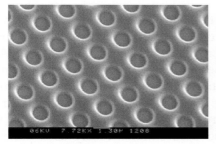 

Fig. 1 Left: Polyimide polymeric microsieve structure on a silicon substrate, Right: microsieve released from the substrate.

A very simple technique is to spin coat a photo lacquer layer (e.g. polyimide) on a dense substrate, expose this layer to a micro sieve pattern, develop the photo lacquer layer, cure the photo lacquer layer and remove the created polymeric microsieve from the substrate [1].

Nijdam [2] originally proposed to use a double exposure technique, in order to create a polymeric microsieve with a thin membrane layer together with a polymeric support structure. Kuiper [3] has studied this method in more detail and published results on this in his PhD work. The method will be described later. The medical company Baxter also filed patent applications [4,5] related to afore mentioned methods for the production of polymeric microsieves for leukocyte removal from blood cell concentrates.

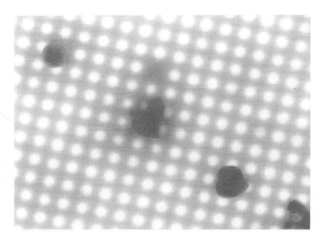

Fig. 2 Leukocytes retained on an Aquamarijn microsieve [6].

Deposition techniques on patterned substrate structures
Another very simple technique is to deposit a polymeric (or inorganic) material on a substrate structure with vertically or negatively tapering sidewalls. The microsieve structure is then formed on the substrate. After a curing process the created microsieve structure is then removed from the substrate. After cleaning, the substrate structures may be reused. In order to avoid sticking, a Teflon layer can first be put on the substrate before the deposition of the micro sieve material. Instead of an anti sticking layer a dissolvable sacrificial layer can also be used. In Fig. 3 a photo lacquer layer has been used for this purpose with the additional advantage, that it can be used simultaneously as a patterned substrate structure [1]. Also the use of other sacrificial layers have been proposed to facilitate release and to avoid distortion of the high aspect ratio

micro-structures [7]. In the latter case sacrificial layers are applied between the mould and the product in subsequent process steps. The removal and re-deposition of such sacrificial layers is however complicated and time consuming, which renders the process less suitable for application on an industrial scale.

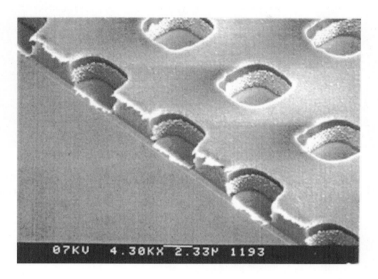

Fig. 3 SEM micrograph of a silicon substrate with a microsieve on top of a patterned photolacquer layer. The sacrificial photolacquer layer can be dissolved easily in acetone hence releasing the microsieve. This method is applicable for a wide variety of organic and inorganic materials.

Hot embossing and other moulding processes

Moulding techniques for the production of microsieves also have a long history. Originally LIGA techniques (see chapter 2) were proposed for the production of macro sieve structures with pore size at least larger than 10 micron. Nickel stamps were fabricated and used as the mould with this technique for the hot embossing or injection moulding of polymeric foils. In 1995 van Rijn [1] proposed to use silicon moulds (see figure below) obtained with deep aspect ratio etching techniques as originally developed by the silicon wizard Jansen [8] for the production of polymeric micro sieves.

Elders and Van Rijn originally described [9] the transformation of the silicon mould (see Fig. 4) to a nickel mould for the production of perforated polymeric and metal foils (DEEMO process).

Some recent PhD work [3] on this matter will be discussed later including it pitfalls and drawbacks.

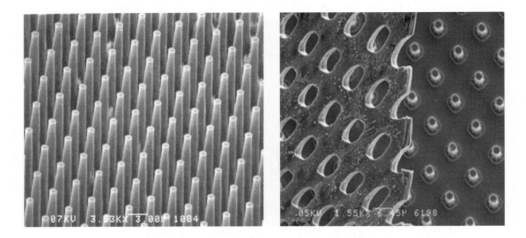

Fig. 4 SEM micrographs, Left: Silicon mould for the production of polymeric membranes, van Rijn 1995 [1]. Right: silicon mould with released polymeric membrane.

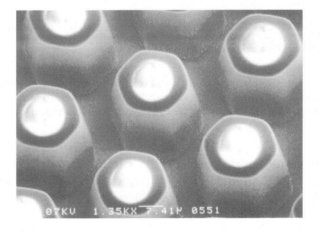

Fig. 5 Silicon mould for the production of polymeric and metallic perforated membranes.

Phase Separation Micro Moulding

This relatively new 'soft embossing' technique has been developed by the Membrane Techology Group of the University of Twente and Aquamarijn in 2001 and appears to be a very economical production method for the production of polymeric micro sieves (see Fig. 16).

2. PHOTOLITHOGRAPHIC POLYIMIDE MEMBRANES

Polymeric microsieves can be fabricated using a negative-tone photo-sensitive polyimide. Polyimide is an extremely strong polymer. After curing, polyimide (Durimide™ 7510) has a yield strength [10] of 215 Mpa, which is approximately half of the tensile strength of steel. A method for the fabrication of microsieves with this polymer is described below (see Fig. 6).

A flat and smooth substrate is covered with a 2 μm thin layer of photosensitive polyimide by means of spin coating. As a substrate a polished silicon wafer was used, but other substrates like metals, ceramics or glass are also suitable. The polyimide layer is exposed to a micro pattern of UV-light. This pattern was obtained using a contact mask, but alternative exposure methods like wafer stepping or laser-interference lithography could also be employed. The exposed areas polymerise, so that only the unexposed areas will be etched and dissolved in cyclopentanone. The result is a 2 μm thick perforated polyimide membrane attached to a flat substrate. This vulnerable membrane must be reinforced in order to be able to withstand sufficiently high transmembrane pressures. Spinning a second polyimide layer over the perforated layer at a lower angular speed can make such reinforcement. A 50 μm thick second layer was spun, and exposed it through a photo mask containing a macro pattern for the creation of the support bars. After dissolving the unexposed areas, a polyimide microsieve is obtained still attached to the substrate.

The polyimide is annealed for one hour at 350°C in an oxygen-depleted environment to evaporate the remaining solvent and to increase its strength by cross linking. During this step the thickness decreases by a factor of 2, leaving a 1 μm thick membrane. The final step is the release of the sieve. Depending on the substrate, the polyimide may be peeled off or has to be released in a stripping solution. Other possibilities are the use of an etchable sacrificial layer underneath the membrane (e.g. aluminium), or the use of an anti-sticking layer (e.g. fluorocarbon). The etch or stripping solutions should not damage the substrate, as multiple use of the substrate will decrease the total costs significantly. The release step was not investigated further and the microsieve was simply released the by etching the silicon substrate in a HF/HNO_3 solution.

The resolution of negative-tone photosensitive polymers is usually lower than that of positive-tone polymers. According to the supplier, the resolution limit of the polyimide used is approximately 0.5 μm. For applications that require smaller pores, other (high-resolution) photosensitive polymers may be employed (e.g. SU8).

Instead of a second polyimide support layer, the macroperforated support may also be made in the substrate by etching the substrate through the pores in the membrane. Using a metal substrate the result would be a micro-perforated polyimide membrane attached to a macro-perforated metal support. For this method a good adhesion of the polymer to the metal is required.

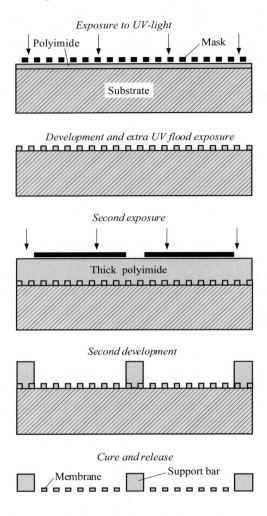

Fig. 6 Schematic illustration of the production process for a polyimide microsieve.

Using this method of double-layer photolithography a microsieve with a 1 μm thick membrane and a pore size of 4 μm was obtained. SEM micrographs of this microsieve are shown in Fig. 7.

Fig. 7 SEM micrographs of the back side of a polyimide microsieve with a pore size of 4 μm and a pitch of 10 μm. The membrane thickness is about 1 μm.

3. HOT MOULDING TECHNIQUES [3]

Analogous to the production of compact discs, it might be possible to make microsieves with a moulding process as described earlier [1]. The pattern in a compact disc contains features on the (sub)micrometer scale, and is obtained with injection moulding. Similar techniques for the replication of a master into a polymer copy are hot embossing or imprinting. The effects of light diffraction, scattering, interference and reflections on a substrate do not limit the resolution of such methods. Chou et al. [11] investigated the imprinting of nanopatterns for lithographic purposes. They produced an array of 60 nm deep pores with a diameter of 10 nm with a pitch of 40 nm. The smallest pore obtained had a diameter of 6 nm. It is potentially a high-resolution and high-throughput method for low-cost mass production. During imprinting or hot embossing a thermoplastic polymer is heated above its glass-transition temperature such that it can be shaped mechanically without damaging the mould. After conforming to the master, the polymer is cooled down and subsequently released from the mould. The technical feasibility of the hot embossed imprint process for the production of microsieves is described below.

Mould fabrication

The mould should contain at least two levels. First a shallow level with approximately 1 μm tall posts to obtain the perforated membrane, and second a deeper level for the formation of large support bars. If necessary an intermediate level for small support bars can be made. Using silicon micromachining technology it is fairly easy to make such a mould. Anisotropic etching with an SF_6/O_2 plasma allows for the creation of structures with vertical or tapered sidewalls [12]. The fabrication process is shown in Fig. 8. The chromium dots on top of the substrate were obtained by shadow evaporation [13,14], although standard lithographic methods like lift-off or chromium etching may be used as well.

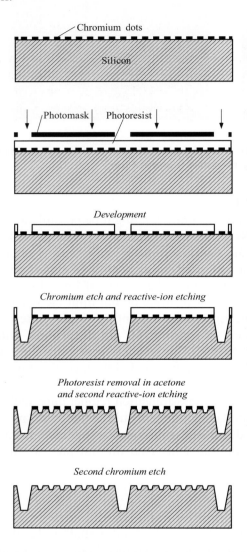

Fig. 8 Schematic illustration of the fabrication of a mould for the production of polymeric microsieves.

In order to prevent adhesion of the polymer the mould was covered with a 25 nm thick fluorocarbon layer by deposition from a CHF_3-plasma [15]. Jaszewski et al. [16] have shown that using such a layer hundreds of embossing steps can be performed without depositing a new layer.

The hot imprint (melt) process

Many thermoplastic polymers can be structured e.g. polymethylmethacrylate, polyethylene, and good results have been obtained with polycarbonate [17]. Well known filtration membranes like track-etched membranes [18] are made of polycarbonate or polyester. An advantage of polycarbonate and polyester is the possibility to etch it in a controlled way, which is a crucial step in the production of track-etched membranes with well-controlled pore sizes.

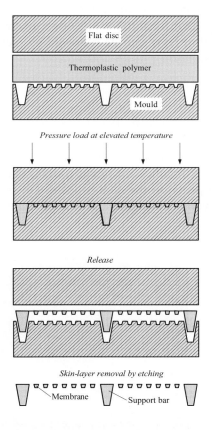

Fig. 9 Schematic illustration of the melt-imprint moulding process for the fabrication of microsieves.

The fabrication process for hot (melt) imprinting is schematically described in Fig. 9. A 125 μm thick sheet of polycarbonate (Lexan® 8B35 film [19]) is placed between the mould and a flat disc. The whole is placed in a vacuum oven and heated to a temperature of 250°C, which is 100° C above the glass transition temperature of polycarbonate. A continuous pressure is applied by placing a weight on the disc. The 1 kg weight on the 0.5×0.5 cm^2 mould area gives a pressure of approximately 4 bar. Part of the molten polymer is pushed into the pattern and the excess escapes to the sides. After 1 hour the oven is cooled down and the polycarbonate is released from the disc and the mould.

Most papers on imprint lithography report the remaining of a very thin layer of polymer (PMMA) on the places where the mould touched the disc. Jaszewski et al.[16] found a thickness between 20 and 50 nm. Chou et al.[11] deliberately keep the mould from touching the disc in order to prolong its lifetime. In imprint lithography the thin layer is usually removed by oxygen-plasma etching, while the imprinted layer is still fixed to the substrate. In case of imprinting for microsieve production the thin layer can be removed by etching the entirely released structure in a caustic solution.

Results and discussion

A mould was fabricated for a microsieve with 0.7×3 µm^2 slit-shaped perforations and a honeycomb-shaped support structure [20]. Sometimes not all of the posts are perfectly slit-shaped, caused by defects in the used evaporation mask. Furthermore the sidewalls of the deep trenches are rather rough, which is a result of the influence of the chromium posts on the etch process. On places where no chromium was present, the honeycomb structure had smooth walls.

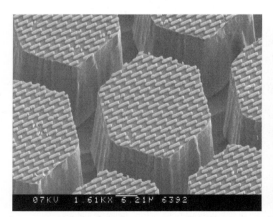 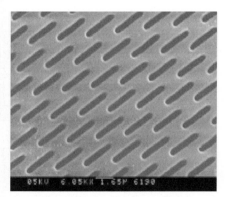

Fig. 10 SEM micrographs. Left: Mould used for making microsieves with slit-shaped perforations. Right: Obtained polycarbonate microsieve.

Despite the rough walls and the rather high aspect ratio of the trenches the imprinted polycarbonate can be peeled off the mould. The imprinted microsieve is a relatively good replica of the mould, although the membrane appears to have been stretched and damaged at several places.

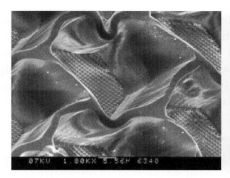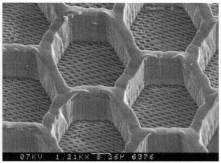

Fig. 11 SEM micrographs. Left: Distorted polycarbonate microsieve after direct release from the mould. Right: Released polycarbonate microsieve with slit-shaped (distorted) perforations.

But regarding the fact that no optimisation with respect to an anti-sticking layer and release temperature was carried out the results are encouraging for the future. It is generally known from the LIGA imprint process that replica's on a larger scale than the presented 2 x 2 mm trials here [21] will show up more distortion on larger scale moulds, e.g. 200 x 200 mm. A remarkable observation was the sometimes locally absence of the skin layer that was expected to have been formed between the top of the mould and the flat disc at a few places. This absence can well be seen in Fig. 12, where an SEM micrograph shows the situation during the release of a membrane with 4 μm wide square pores.

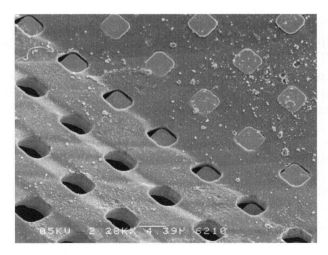

Fig. 12 Release of the polycarbonate membrane from a mould with 4 μm wide square pedestals.

Possibly the skin layer is so thin that it ruptures during release and remains adhered on top of the posts. If this is true, one might exploit this effect by removing the fluorocarbon layer from the top of the posts prior to imprinting, thus enhancing the adhesion of the skin layer. After the release the ruptured layer will simply mingle with new polymer in a next imprint process. Another explanation may be thermal shrinking of the polycarbonate.

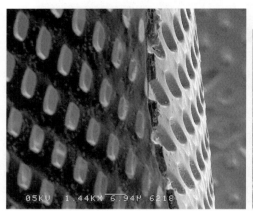 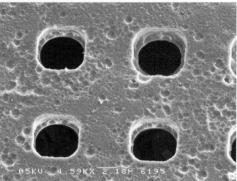

Fig. 13 SEM micrographs. Left: Cross-section of a released membrane with a thin skin layer covering some pores. Right: The skin layer has been removed with KOH etching, but etch pits have been created.

With its coefficient of thermal expansion of $6.8 \cdot 10^{-5}$ °C^{-1} [22] the poly-carbonate will shrink approximately 1.5 % during cooling to room temperature. The resulting stress will concentrate at the edges of the skin layer and may cause rupture. However it is possible that the layer may not have been formed at all. The cohesive forces between the polycarbonate molecules may have caused the molten polycarbonate to withdraw from the narrow space between mould and disc.

In order to investigate smaller pore dimensions a silicon mould with circular posts with a diameter of 0.4 μm was made. Fig. 14 shows SEM micrographs of the mould and the resulting perforated membrane.

Scaling up

Concerning scaling-up of the two discussed methods, the method of exposing photosensitive polyimide to UV-light is preferred. However the method of imprinting would be quite new for the sub-100 nm feature size.

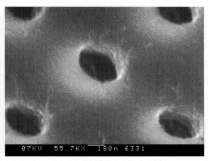

Fig. 14 SEM micrographs of a mould with 0.4 μm diameter posts and the resulting microsieve.

The first article on nano-imprint lithography was published in 1995 [23]. Imprinting with submicron resolution is already routinely used for reproducing refractive optical microstructures [17], such as holograms and gratings [24]. It is even thinkable that the imprint process may be applied in a continuous way with a roll-to-roll method. Commercial roller systems achieve imprint speeds of up to 1 m/s in polycarbonate foil up to a width of 2 m. Using a roller technique Tan et al. [25] obtained imprints with sub-100 nm resolution. However, for high aspect ratio imprint depths in excess of about 1 μm problems occur when using a roller method or a stamping hot press because of the perforation and sticking problems of thin membrane films with a thickness in the order of 0.5 micron or larger.

4. PHASE SEPARATION MICRO MOULDING[*] [26,27]

The moulding of perforated membranes with micro-structures featuring high aspect ratio's in the micron and sub-micron range is relatively difficult because of the mechanical fragility, distortion and anchoring of those structures during release of the products from the mould. Injection moulding, pressure moulding, hot embossing, slip casting and hot (melt) imprinting of the product on the mould are known techniques, partially described earlier, but all face problems with respect to the release of products with high aspect ratio structures. Accordingly, there remains a need for economical moulding of micro-structured products that can be made out of a wide variety of polymeric and inorganic materials with a high aspect ratio.

Phase separation processes have been extensively used to produce micro porous membranes for water production and for separation processes in chemical, pharmaceutical, medical and food industry, from which it has been proven in practice that they can be made out of a wide variety of materials and can be applied in an economical and reproducible manner.

 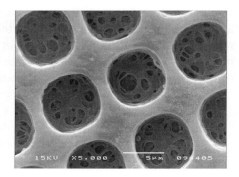

Fig. 15 SEM micrographs of a micro porous phase separated polyimide and polyethersulfone microfiltration membranes (source, MTG University of Twente and Aquamarijn).

Phase separation process

Phase separation (some times also denoted as phase inversion) entails the process of changing a one-phase casting fluid (solution) into at least two separate phases. One phase is defined to be uniform in chemical composition and in physical structure throughout the material. In all phase separation processes, a casting solution is precipitated into at least two phases: a solid

---

[*] Collaboration between Matthias Wessling, Membrane Technology Group of Twente University and Aquamarijn Research.

material-rich phase that forms the matrix of the product and a material-poor phase that may form the inner product pores. The phase separation is generally a thermodynamically driven process and may be induced on the mould (or shortly prior or after casting) by changing e.g. the composition of the solution, the temperature or the pressure of the casting solution. The composition of the casting solution may be changed by bringing the casting solution in contact with a non-solvent, which is normally a gas (vapour) or a liquid not well miscible with the material, or by evaporating a solvent from a casting solution containing a non-solvent, or by reaction of components in the casting solution resulting in a non-solvent agent. The material of the product may be polymeric as well as inorganic or a combination of both.

Prof. M. Wessling, head of the Membrane Technology Group (MTG) of the University Twente having much expertise in phase separation processes proposed in 2000 to study phase separation processes to manufacture micro structured flat sheet micro filtration membranes with moulds and moulding technology from Aquamarijn (since 1995). The method will now be presented.

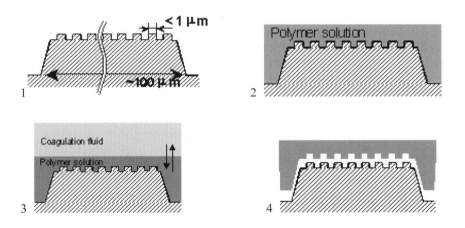

Fig. 16 Preparation route of micro-architectures by phase separation micro moulding. The illustration shows the preparation Liquid-Induced Phase Separation (LIPS).

The phase separation micro moulding process comprises four steps:

1. Preparation of a micro master mould based on technologies derived from microelectronics and photolithography
2. Casting an appropriate polymer solution onto the master
3. Solvent and non-solvent liquid exchange initiating phase separation until the polymer solution contains sufficient non-solvent to precipitate
4. Solidification and final release of the micro to nanosized three-dimensional structure

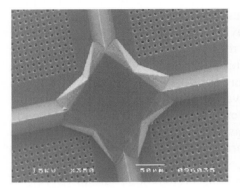 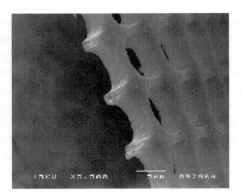

Fig. 17 SEM micrographs of polymeric microsieves obtained with phase separation micro moulding. Left: Microsieve fields with support structure. Right: Cross section of a microsieve.

The process was relatively easy and fast with good results, showing a very good replica of the mould. There was some hope to manufacture microsieves (fully perforated micro filtration membranes) without a closing flat disc (cf. Fig. 9) and with a special mould produced by Aquamarijn (W.Nijdam) having very high aspect ratio pins. J. Barsema succeeded in a first trial (late summer 2000 [28]) to make the first phase separated polymeric microsieve in polyimide (see Fig. 17).

Three types of phase separation processes are well known:
1. Vapour-induced phase separation (VIPS), also called "dry casting" or "air casting". Evaporation of the solvent will result in a dense or porous product depending on the polymer and the used solvent(s) mixtures; The vapour may contain water or organic solvent molecules that may be absorbed by the casted film and will influence the porosity of the product.
2. Liquid-induced phase separation (LIPS), mostly referred to as "immersion casting" or "wet casting". This process normally yields dense or porous products depending on the polymer and the used solvents and non-solvent mixtures; This technique may be combined with 1) for polymer solutions with e.g. two solvents with different boiling points.
3. Thermally induced phase separation (TIPS), frequently called "melt casting"; The polymer may solidify to a dense or porous product of a polymer-solvent-additive mixture by varying the temperature of the polymer-solvent during the casting process.

Moulding of a microsieve with Vapour and Liquid Induced Phase Separation

A polymeric micro sieve **6** is made by casting a thin film **4** with a thickness of 40 micron of a polyethersulfone (PES) solution with the aid of a casting knife over a mould **3** having a large number of small orthogonal protruding cylindrical rods **1** with a diameter of 2 micron and a height of 8 micron on MESA's **2** with a height of 30 micron, Fig. 18. The polyethersulfone solution contains 3.75 g PES (BASF, Ultrason E 620p), 25 g N-methylpyrilidone (NMP) and 20 g acetone per 48.75 g. The acetone is a solvent with a low boiling point and the NMP is a solvent with a high boiling point. After the evaporation of the acetone and shrinkage of the casting solution for 10 seconds in a water vapour environment the mould together with the PES precipitate is immersed in a water bath at a temperature of 20°C for 30 seconds, which induces a further shrinkage of the casting solution **5**.

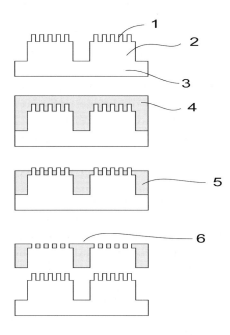

Fig. 18 Process steps for the production of a polymeric microsieve obtained with a phase separation process.

The NMP in the PES solution diffuses to the water bath while the water diffuses into the NMP-rich PES solution. The water is a non-solvent for the PES and phase separation is induced resulting in a microporous PES film still containing water and NMP in the pores. The resulting product **6** is easily released from the product and is dried at ambient temperature. A fast solvent exchange may first be applied by first washing with ethanol and subsequently hexane before further drying steps. This may be important when the polymers used are rubbery and tend to collapse upon drying.

Due to the shrinkage of the PES film partially by the evaporation of the acetone and mainly during the immersion in the water bath, a very thin (thickness 3 micron) crack-free microporous microsieve is obtained after the drying process. Due to the shrinkage of the PES solution the perforations (perforation diameter 6 micron) are somewhat larger than the diameter of the rodlike protrusions on the mould (Fig. 17 shows a SEM cross-section and a topview of the obtained supported microsieve). Using a higher initial polymer concentration the amount of shrinkage is reduced. The addition of surfactants (SPAN 80, TWEEN) might enhance the perforation through the rods 1 of the casting solution. Moulds with strong rods 1 have also been made with nickel electroplating techniques from Stork Veco. A great advantage of the described technique is that the casting knife does not have to touch the mould during casting so that abrasion of the mould is negligible in time.

Phase separated polymeric microsieves can be made in a broad range of pore sizes and from many different polymers like polyimide, polypropylene, polyamide, Teflon®, polyurethane, bio and non-bio degradables, etc. Liquid induced phase separation is very well accomplished by immersion in an immersion bath, in which the concentration of the non-solvent in the bath is larger than 5 % and preferably larger than 95%, in order to prevent the formation of thin and porous skin layers. Vapour induced phase separation is very well accomplished, in which the concentration of the non-solvent in the vapour is less than 25% and sometimes depending on the aspect ratio of the micro perforated structures, less than 1%. Vapour induced phase separation may be quickened, in that the non-solvent is already partially dissolved in the fluid with a concentration less than 2%, in above example with a water concentration of e.g. 0.6%. Good results have been obtained with e.g. perforated and non perforated structures and a dense or open skin layer covering all sides of the product in that the casting solution contains at least two solvents in which a lowest boiling solvent and a highest boiling solvent have a difference in their respective boiling points of approximately 50°C or greater; removing a predominant amount of said lowest boiling solvent by evaporation; contacting the solution with a non-solvent for the material, but which is miscible with said at least two solvents to induce phase separation and solidification; releasing the solidified product from the mould. Good results can be obtained with a non-solvent selected from the group consisting of water, methanol, ethanol, isopropanol, toluene, hexane, heptane, xylene, cyclohexane, butanol, cyclopentane, octane, higher alcohols, glycols and miscible mixtures of these non-solvents.

Fig. 19 SEM micrographs of the mould (left), mould with polymeric microsieve (middle) and detail (right) showing the intrinsic shrinkage of the polymer allowing a defect free release.

It has surprisingly been found that due to the phase-separation a relatively dense and highly flat skin layer is formed at the interface with the mould surface that is conformal to the surface. Moreover the resulting product tends to shrink isotropically to a certain extent due to the phase separation process. These two circumstances greatly improve the release of the eventual product from the mould without notable distortion of the microstructure.

The polymer phase separation processes performed on mould shows unique features:

1. It allows rapid processing: solidification of the polymer occurs within seconds.

2. Shrinkage of the polymeric architecture is an intrinsic part of the solidification process and subsequent release. It allows zero-defect fabrication in contrast to for instance simple spin coating or hot embossing requiring subsequent pealing-off. The architectures produced show a high regularity over large areas: hence the defect concentration is very low.

A mould with micro fluidic structures has been prototyped in silicon with standard lithographic and anisotropic reactive ion etching techniques.

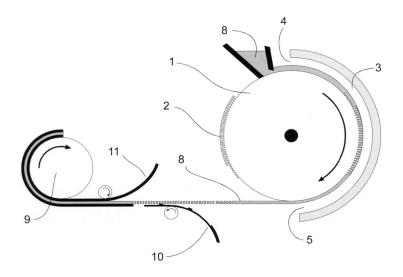

Fig. 20 Schematic diagram of a roll-to-roll apparatus for the production of polymeric microsieves and fluidic boards using a phase separation process.

Production of large membrane areas

Moulds made of a microstructured nickel foil with a thickness of 50-200 micron can easily be wrapped around a drum for continuous production of polymeric microsieves and or partially perforated fluidic boards (see Fig. 20).

With such a drum **1** (e.g. diameter 9.6 cm) having a 100 micron thick microstructured nickel foil **2** continuously polymeric moulded micro fluidic products can be made roll-to-roll with the VIPS technique. In this example a polymer solution containing a high concentration of polyimide, in a 40/60 NMP/acetone solution has been casted with a mean thickness of 300 micron on the nickel foil **2** with an insert at the highest point of the drum **1**. In this example the casting solution **8** is already pre-saturated with a small amount of water (<1%) to speed up the vapour induced phase separation process. The rotation velocity (clock-wise) of the drum is set at 1 cm/sec. Around the drum a ring shaped air gap **3** is provided with a mean channel height of 1 cm in which air saturated with water vapour is fed (anti clock-wise), at an elevated temperature depending on the casting settings between 30 and 90 ºC, from an inlet **5** to an outlet **4** and next to a condenser to collect the solvent deposit. The mean interaction time of the casting solution with the air is 60 seconds. The produced phase-separated foil **8** with micro fluidic structures is subsequently collected by another roll **9**.

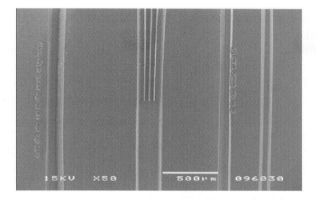

Fig. 21 SEM photograph of a top view of a phase separated fluidic board, also suitable for the implementation of polymeric wave guide channels (middle, optical splitter structure), (Courtesy Aquamarijn Research/Membrane Technology Group, University of Twente).

Alternatively the microstructured polymeric foil **8** is first adhered to a support foil **10** bearing the phase separated foil and in a second stage to a cover foil **11** covering the fluidic structures before the pick up by the collecting roll **9**. Support foil **10** may contain fluidic structures like in- and output ports and liquid channels to form functional components together with the phase separated microstructured foil **8**. Instead of a drum also extended belts with the micro-structured mould may be used for the roll-to-roll manufacturing of these microfluidic structures or other methods according to the phase separation micro moulding process.

Microstructured products made according to the phase separation method may also be used as a (micro porous) mould to produce other microstructured products. Non-solvent vapour or liquid may be applied through pores of a microporous mould to induce phase separation of the moulded article. Fluids may also be extracted through the pores e.g. in case of colloids suspended in the fluid to induce solidification.

Monolithic polymeric microsieves may be used for e.g. beer and wine clarification, cold sterilisation of beer and milk, blood plasma filtration, leukocyte or plasma filtration, microbiological analytical applications, Polymerase Chain Reaction (PCR) probe filters, shadow masks for e.g. electronical applications, fotonic crystals, nano stencilling methods, micro contact printing, prefilters, mixers, as a tool to produce light emitting displays etc. The microsieves may be used without a support or with one or more supports. The obtained product may be elastomeric or non-elastomeric depending on the chosen casting solution and the intended application. Also reinforced nozzle plates for a variety of applications can be made such as ink jet printing, (cross flow) emulsification nozzles for the production of single, double and multiple emulsions, atomisation nozzles for fuel injection, (deep pulmonary) drug delivery, foaming purposes, spraydrying, spraycooling, spraycoating, and as an electrically insulated but perforated spacer (having e.g. an electrolyte or another medium within the perforations) for e.g. battery applications . Of coarse many other perforated articles such as encoder disks, micro and nano stencils for e.g. the evaporation of isolated magnetic domains, SAM's, and many other organic and inorganic materials can be made according to the phase separation method.

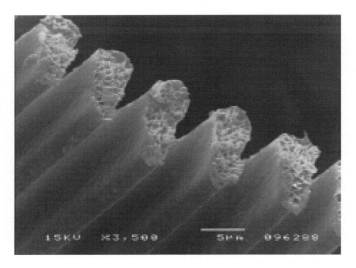

Fig. 22 SEM photograph of a cross-section of a prototyped phase separated polyethersulfone encoder disk, (Courtesy Aquamarijn Research/Membrane Technology Group, University of Twente).

Despite the low costs of polymeric microsieves, they will not replace the silicon nitride microsieves in all filtration areas. For applications such as clarification of beer [29] the silicon nitride sieves have important advantages concerning durability, hydrophilicity, particle adsorption, chemical inertness and thermal resistance. However, for applications where disposable filters are desired, like in blood filtration, the polymeric sieves may become a good alternative for the present microsieves. Moreover, in any area where polymeric membranes with pore sizes in the order of a micron are used, polymeric microsieves may be a welcome improvement concerning flow resistance and pore uniformity.

4.1 MOULDING OF MICRO ARRAYS, MICROWELLS AND MICRO TITRATION PLATES [27]

On a microarray, a plurality of regions (called features) is defined on which different probes e.g. nucleic acids, amino acids, oligonucleotides, antibodies, antigens, etc. are immobilised. The microarray is placed into a reaction container together with a sample containing for instance DNA material to induce hybridisation with the probes immobilised on the respective features of the microarray. Thereafter, the microarray is irradiated with light to record fluorescence or fluorescence activity of each feature. Based on the measured fluorescent intensities, the type and place of bonding between the respective probes and the sample DNA is obtained and converted into desired information.

Probes can conveniently be applied with ink-jet technology on solid supports such as glass slides or porous supports like nitrocellulose, polyvinylenedifluoride, polypropylene, polysulfone and nylon. Ink-jet technology also allows for the accurate deposition of defined volumes of liquid.

On porous products often splattering (diffusion) of the probes is observed, a problem a problem to becomes more serious when more densely packed array probes are made. Porous products with orthogonal pore channels may be used (like anodised alumina membranes) to inhibit splattering. Another method is to introduce a grid of hydrophobic material on the porous product, or laying down stripes of an uncured or otherwise flowable resin or elastomer solution and allowing the material, al least partially, to infiltrate the porous array product to prevent splattering.

Phase separation micro moulding method may also be used for a new type of (polymeric) micro array structures to prevent splattering, especially in view of the need for further densification of probe spots. First of all polymeric phase separation techniques may be employed to provide more orthogonal directed pore channels like macro voids in the porous product. Orthogonal oriented macropores can be made with PMMA/butylacetate as material/solvent and n-hexane as a non-solvent. Added surfactants in general (e.g. TWEEN 80) may also influence the forming of orthogonal channels. Secondly a mould may be used for heat-seal stamping of the porous support sealing locally pores and in addition forms a water-impervious barrier between different probe regions. This mould may also be used for cold seal stamping during or short after a phase separation process to obtain both porous and dense areas. A microporous mould saturated with an alcohol or glycol (delayed mixing) may also be used to pattern locally regions with a very thick skin layer. Thirdly, phase separation can be used to obtain porous MESA's (probe regions) with a grid of thin and deep gaps between the different probe regions. Fourthly, moulding may be employed to obtain micro-well structures, in which the walls of the wells are microporous, the skin layer is dense and the bottom of the wells are microporous (see Fig. 25, right picture). Each probe could be deposited in a different well. The probe material will be absorbed mainly by the bottom of the well. A surplus of probe material will be more absorbed by the microporous walls of the well through the bottom of the well. Using a mould in which the different parts have different phase separation conditions such products can easily be made.

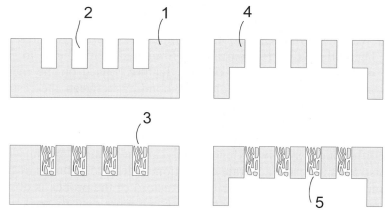

Fig. 23 Different layouts for a microarray/multiwell structure.

Another elegant way is to construct the microarray in two steps (see Fig. 23). Step one entails the formation with e.g. phase separation of a dense or hydrophobic grid/multiwell structure 1,4 with chambers 2 and step two entails the filling of these chambers with a phase separation process with a microporous probe binding material 3,5 (see

Fig. 24). The filling of these chambers 2 with the microporous material may be done with inkjet printing together or sequential with the probe material 3,5. The filling may also be done by normal phase separation casting techniques especially when using a fully perforated grid. The material of the grid may be dense, hydrophobic and/or microporous with a closed cell structure (non-connecting pores). After applying a casting film for phase separation e.g. with a thickness of 20 micron on a grid 1,4 with a thickness of say10 micron, due to the large shrinkage of this film during precipitation in the non-solvent bath, the microporous polymer (with an open cell structure/connecting pores) will mainly reside within the perforations of the grid. The material of the grid 1 may be e.g. polymeric or ceramic. The grid 1,4 may be microporous and a treatment may be given to fill or to make it hydrophobic before the filling of the chambers 2. Preferentially the walls of the chambers 2 are tapered and/or have a sufficient micro roughness to lock the probe material 3,5. A number of different hybrid microarrays have been manufactured with polyvinyldifluoride (PVDF) and polypropylene as grid material and nitrocellulose, PVDF and nylon as probe binding material 3,5.

Fig. 24, shows a SEM picture of a topview of a micro array with 200 micron sized probe regions according to the phase separation micro moulding method. Of coarse many other combinations of grid and probe binding materials as well as other dimensions are possible.

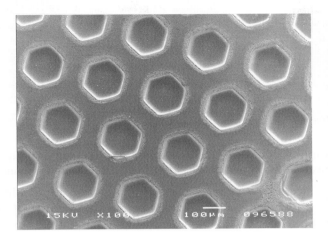

Fig. 24 SEM micrograph. Topview of a prototype microarray structure with non-splattering probespots (Courtesy Aquamarijn Research/Membrane Technology Group, University of Twente).

The thickness of the hybrid or non-hybrid microarray film is preferably between about 10 and 100 micron. In order to improve alignment of the microarray/well plate with a suitable, probe filling device alignment marks can be provided in the product. Also the distance (e.g. 5 micron) between the different probe regions can be made as small as possible with respect to the size of the probe regions (e.g. 100 micron), which can be obtained easily with high aspect ratio structures. Without alignment techniques the probe material can also be deposited in more than one multiwell, provided that the different multiwells have a size (e.g. 2 micron) are much smaller than the probe spot (e.g. 20 micron). In a preferred embodiment, each microarray contains about 10^3-10^8 chambers for 10^3-10^6 distinct polynucleotide or polypeptide biopolymers on a surface area less than 1 cm^2. The biopolymers in each microarray region normally are present in a defined amount between about 0.1 femtomoles and 100 nanomoles. Each probe region may be provided with an imprinted determination-code originating from the used mould. The ability to form high-density arrays of biopolymers, where each region has a well-defined amount of deposited material, can be achieved in accordance with the microarray production method described above.

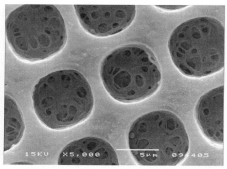

Fig. 25 SEM photographs of a microporous multiwell and a flow-through microarray, (Courtesy Aquamarijn Research/Membrane Technology Group, University of Twente).

Products fabricated according to the phase separation micro moulding method may also be used to manufacture dense/microporous micro well and titration plates, preferably provided with an unique identification or bar code for each well or titration section.

Fig. 25 shows SEM pictures of microporous micro multiwell plate structures of a hydrophilic polyethersulfone/polyvinylpyrrolidone, which can be used for instance for cell encapsulation and to investigate cell growth. The feed solutions can be applied simply through the microporous material. Each surface section may have a different surface morphology (with or without a locally smooth skin layer) depending on the exact liquid-liquid demixing and liquid-solid demixing conditions during phase separation.

A preparation method to manufacture a multiwell plate will be described in more detail. The micro moulded multiwell plate of polystyrene is here made by thermally induced phase separation. Polystyrene (Dow) was dissolved in diisodecylphthalate (Merck) to a 23 wt % casting solution and heated to a temperature of 120 °C. With a casting knife the solution was casted 300 micrometers thick on a mould provided with cylindrical pillars with a diameter of 5 micron and a height of 30 micron. Both the casting knife and the mould had a temperature of 120 °C. The solution was gradually cooled down to a temperature of 20 °C and a porous layer formed at the imprint side of the structure. Alternatively the solution was cooled down very rapidly in a water bath of 20 °C, leading to a dense skin layer on the moulded and topside of the structure. Afterwards, the polystyrene multiwell plate was rinsed in ethanol.

4.2   FLAT SHEET AND TUBULAR POLYMERIC STRUCTURES[*] [27]

Flat sheet and tubular polymeric structures are well known in membrane technology, where they serve as membrane or barrier, mostly for separation purposes. Flat sheet polymeric structures can also be found in packaging, laminates, etc., but are then produced by a different production process like melt extrusion, blow moulding, vacuum suction, etc. These films are produced from a polymer melt, while most filtration membranes are normally produced from a polymer solution using a phase separation process.

Flat sheet polymeric membranes are prepared as self-supporting sheets or as composite structures where the polymeric film is supported by a woven or non-woven cloth. The woven or non-woven cloth provides additional mechanical support to the membrane and may serve as an interspacing to transport the feed and permeate solution. Flat sheet membranes may be stacked in a plate and frame module. Long flat sheet membrane structures together with the supporting cloth may be winded in a so-called spiral wound membrane filtration module. Narrow strips can also be winded around a cylindrical template while the edges of the strip are being contacted and welded on each other, to construct a tubular polymeric structure.

In order to increase the effective surface area of a filtration membrane and hence the filtration performance (flux) three dimensional micro-structures can be introduced on the surface of a the filtration membrane (see Fig. 26).

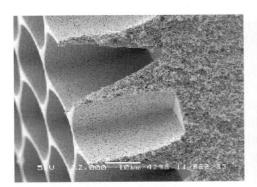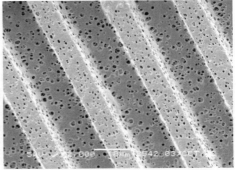

Fig. 26 SEM photographs of two phase separated polyethersulfone microfiltration membranes with an enhanced effective filtration area, (Courtesy Aquamarijn Research/Membrane Technology Group, University of Twente).

Micro-structures on the membrane surface (e.g. ridges along the permeate side with a height between 50 en 500 micron) may also serve as a spacer

---

[*] This section is co-written with G.H. Koops, European Membrane Institute, University of Twente.

between two membrane sheets or as a turbulent promoter in e.g. flat sheet plate and frame modules and spiral wound modules. A pattern of ridges and placed along, oblique or perpendicular relative to the mean cross flow direction may enhance the mass transfer coefficient and may counteract the build-up of a cake layer. Ridges contributing to turbulent promotion may differ in e.g. height with respect to the height of the spacer. Such ridges may be provided with an oblique upper plane in order to promote turbulence or to guide fast laminar flow streams towards the membrane layer to inhibit the build up of a cake layer. Ridges may also be provided on the membrane layer orthogonal to the cross flow direction to induce turbulent flow. This can be accomplished by structuring the casting knife or roll, which is in contact with the top side of the film or by structuring the support layer (this could be a plate or roller) on top of which the polymeric film is casted, rolled or sprayed.

4.3   CAPILLARY POLYMERIC STRUCTURES *[27]

Hollow fibre and capillary polymeric structures are typically found in the field of membrane technology, where they serve as membranes, mostly for separation purposes. Hollow fibre membranes and capillary membrane differ in dimensions, but are both self-supporting cylindrically shaped membranes with typical inside diameters of 10 - 3000 micrometer. These membranes can either have asymmetric or symmetric structures, which consist of at least one skin layer and a support layer. The skin layer is normally very thin (typically between 0.05 – 10 micrometer) compared to the support layer (typically between 30 and 500 micrometer) and are present at the shell side, the bore side or both sides of the hollow fibre/capillary membrane. The skin layer is responsible for the separation, while the support layer only serves as mechanical support with a flow resistance as low as possible for the transported species. Hollow fibre/ capillary membranes are applied in all kind of separation processes, such as gas separation, reverse osmosis, nanofiltration, ultrafiltration, microfiltration, membrane extraction, membrane contactor applications, supported liquid membrane applications, but is certainly not limited to these applications. The membranes used today are characterised by either a porous skin or a non-porous dense skin. The surface of these skin layers are relatively smooth, but might contain some surface roughness, which is caused by the membrane formation process itself (e.g. phase separation) and the polymer used as membrane matrix material in combination with the process conditions, like e.g. the spinning/extrusion rate, the dope composition, the bore liquid composition and/or the flow rate and of the bore liquid. The spinnerets or extrusion nozzles used to produce these hollow fibre/capillary membranes consist of one or more small cylindrical tubes in one or more cylindrical orifices, with smooth tube and

---

* This section is co-written with G.H. Koops, European Membrane Institute, University of Twente.

orifice walls without specific microstructures. Normally, a rough surface is undesired since in most applications this will enhance membrane fouling. One example in membrane filtration that forms an exception is the so-called low-pressure thin film composite reverse osmosis membrane, which has an extremely rough skin surface, applied on purpose to increase the total surface area and thus enhancing the membrane flux. Since the feed for reverse osmosis is relatively clean membrane fouling due to an increased membrane roughness is negligible. Polymeric hollow fibre/capillary membranes are most of the time produced by a spinning technique in combination with a phase separation process. The initial starting dope solution contains at least one membrane matrix forming polymer and a solvent for that polymer. Often other components like non-solvents, a second or even a third polymer, salts, etc. are added to manipulate the membrane structure. This dope solution is shaped into a hollow fibre or capillary by the use of a spinneret or nozzle and the shape is formed and fixated by a phase separation process. After leaving the spinneret/nozzle the nascent fibre/capillary can either pass through a so-called air gap before the fibre/capillary is immersed in a non-solvent or solvent/non-solvent bath or can enter or being contacted by a non-solvent or non-solvent/solvent bath immediately. The first process is called dry-wet spinning and is normally performed using a tube-in-orifice spinneret/nozzle Fig. 27a), the latter process is called wet-wet spinning and is normally performed by a triple layer spinneret/nozzle (Fig. 27b).

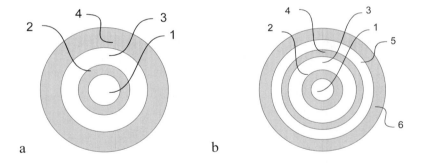

Fig. 27 Examples of conventional spinneret's for production of capillary membranes.

An example of the production of an ultrafiltration membrane by a dry-wet spinning process is as follows:

16 wt % polyethersulfone (Ultrason E 620p, BASF) is dissolved at room temperature in 38.5 wt % N-Methyl pyrrolidone, 38.5 wt. % polyethylene glycol (PEG 200) and 7 wt. water.   The polymer solution has a viscosity of 9350 Cp at 25 °C. This solution is extruded through a tube-in-orifice spinneret/nozzle

(needle OD = 0.6 mm, orifice OD = 1.2 mm) at a rate of 5 m/min at 22.5 °C. After passing through an air gap of 10 mm the nascent fibre enters a coagulation bath filled with water of 20 °C. As bore liquid a mixture of 40 wt % water, 30 wt% PEG 200 and 30 wt % NMP was applied. After thorough rinsing with water the fibres are immersed in a 10 wt % glycerol solution for 24 hours after which the fibres are dried in air. These hollow fibres have an outside diameter of 0.97 mm, an inside diameter of 0.67 mm, a pore size of 15 nm, a pure water flux of 1400 L/m^2/hr/bar and a BSA retention of 91%.

Phase separation can be established by changing the polymer solution composition (VIPS, LIPS) and/or by changing the polymer solution temperature (TIPS). This can be accomplished by penetration of a non-solvent vapour into the polymer solution, penetration of a non-solvent liquid into the polymer solution, evaporation of solvent from the polymer solution, diffusion of solvent out of the polymer solution, increasing or decreasing the temperature of the polymer solution.

Using moulded extrusion nozzles or spinnerets in combination with a phase separation process according to the phase separation micro moulding method it has been proposed by van Rijn [27] to provide any shell or bore surface layer with 3D microstructures (Fig. 28a,b) such as micro grooves and ridges **10-16** along the length of the fibre/capillary.

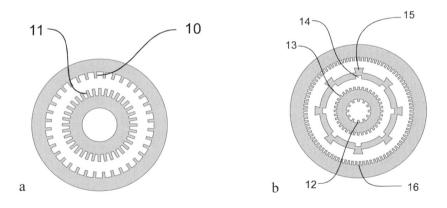

Fig. 28 Cross-section of spinneret's provided with microgrooves and ridges.

Typical dimensions of these microstructures are between 1 and 200 micron with aspect ratio's up to 10. Other dimensions are also possible demanding on the specific application. According to the phase separation micro moulding method it will be shown for the first time that such micro-structures with a high aspect ratio can be made, and more important, that the actual membrane surface can be completely covered by micro-structures with a substantial uniform thickness of the skin layer. A clear advantage of such microstructures is of course a surface enlargement of the membrane layer and hence an improved filtration flow rate, analogue to the low-pressure reverse osmosis membranes. By providing narrow spaced grooves with an aspect ratio of two to five along the tube, it is possible to have a membrane surface enlargement up to a factor two to five. If such grooves spiralise along the fibre/capillary Dean and/or Taylor vortices can be induced in the medium/liquid flowing along the grooves. This will reduce the build-up of a cake layer during filtration, which also contribute to the filtration permeate flow rates. The hollow fibre/capillary membranes according to the phase separation micro moulding method may also be bent (spiralised) during assembly in a filtration module to induce additional Dean vortices in filtration runs. Grooves may be provided on the membrane's bore surface, but can also be provided on the shell surface in order to reduce the flow resistance of the permeate through the module, while maintaining its supporting properties.

The micro structures introduced in the nascent polymeric film have to retain their original shape until phase separation solidifies and fixates the desired micro structure. This can be accomplished by using highly viscous solutions and/or defining systems that instantaneous demix after the micro-structure has been introduced.

That these micro-grooves can really be provided shows the following example:
25 wt % Polyimide (Matrimid® 5218, Ciba Geigy) is dissolved at 50 °C in 33.75 wt % N-Methyl pyrrolidone, and 22.50 wt % acetone. This solution is extruded through a tube-in-orifice spinneret/nozzle (needle ID = 0.2 mm, orifice OD = 0.5 mm) at a rate of 8 m/min at 50 °C. After passing through a temperature (50 °C) and humidity (< 10 %) controlled air gap of 200 mm the nascent fibre enters a coagulation bath filled with water of 20 °C. As bore liquid a mixture of 30 wt.% water and 70 wt % NMP was applied. After thorough rinsing with water the fibres were immersed in ethanol for 4 hours followed by immersion in hexane for another 4 hours, after which the fibres were dried in air. These hollow fibres have an outside diameter of 0.53 mm, an inside diameter of 0.41 mm, an oxygen permeability of 4.0 x 10-6 $cm^3.cm^{-2}.s^{-1}cm.Hg^{-1}$ and a high oxygen/nitrogen selectivity of 6.8. The calculated dense skin layer thickness is around 0.4 micrometer. Looking at the shell surface (skin layer) imprinted lines or micro-grooves in the direction of the fibre length can clearly be observed (see Fig. 29). Since the fibre is perfectly gas selective, the micro-grooves have not distorted the skin layer.

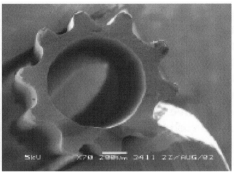

Fig. 29 SEM photographs showing a cross-section of a micro structured gasseparation membrane, Left: Microstructured hollow fibre membrane, fibre geometry resembles spinneret geometry well due to rapid structure fixation by immediate coagulation, Right: Microstructures are reduced in size and wall thickness is increased due to reflow of polymer solution before final structure fixation due to retarded coagulation. (Courtesy Aquamarijn Research/Membrane Technology Group, University of Twente).

Another advantage of creating micro-grooves of a certain shape might be the inter-locking effect the grooves can have when co-extrusion of two polymeric layers is performed. An example of such an inter-lock structure **14, 15** is given in Fig. 28b. Applying a triple layer spinneret a first polymeric solution can be extruded through the first orifice **3**, while a second polymeric solution can be extruded through the second orifice **5**. This way a composite hollow fibre structure can be produced. The production of such a composite structure would be desirable for example when a costly functional polymer has to applied. A thin layer of this material is often sufficient to create the desired structure with its functionality (e.g. certain pore size, certain porosity, certain selectivity for gasses, liquids, vapours, etc.) the second layers only acts as a mechanical support layer. When two materials are chosen, which are very different from each other, like an elastomeric material (e.g. PDMS, EPDM) and a glassy material (e.g. PSF, PES, Nylon, PI) or a hydrophobic material (e.g. PSF, PES, PI) and a hydrophilic material (e.g. sulfonated PES, carboxylated PES, sulfonated PEEK, sulfonated PSF) the adhesion between the two layers can be very problematic. Not only will these materials behave differently in the membrane forming process, which can cause delamination. Furthermore these materials might show different swelling or expansion in certain environments (liquids, vapours, temperature, etc.), which can results in delamination of the two layers as well.

An example of the production of such a laminated co-extruded hollow fibre is decribed below:

A first polymer solution consists of 17 wt % polysulfone (Udel P3500, Amoco) dissolved in 52.4 wt % N-Methyl pyrrolidone and 30.6 wt % diethylene glycol. A second polymer solution consists of 30 wt. % sulfonated polyethersulfone, 35 wt % N-Methyl pyrrolidone and 35 wt % acetone. The first solution is extruded through the first orifice of a triple layer spinneret (needle OD = 0.6 mm, 1st orifice OD = 1.0 mm, 2nd orifice ID = 1.1) and the second polymer solution is extruded simultaneously through the second orifice both at a rate of 6 m/min at 50 °C. The second polymer solution was extruded at a rate of 0.43 ml/min. After passing through an air gap of 30 mm the nascent fibre enters a coagulation bath filled with water of 20 °C. As bore liquid a mixture of 25-wt % water and 75 wt % NMP was applied. After thorough rinsing with water the fibres were immersed in a 10 wt % glycerol solution for 48 hours, after which the fibres were dried in air. The polyethersulfone layer was porous and had a thickness of approximately 120 micrometer, while the sulfonated polyethersulfone layer was rather dense with a thickness of 2 micrometer.

Beside hollow fibres/capillaries also solid fibres/capillaries can be produced with microstructures on the shell side of these fibres/capillaries, similar to the production of hollow fibres/capillaries. The only difference in the production process is that the needle is not present and a bore liquid is not applied or a tube-in-orifice spinneret/nozzle is used and instead of a bore liquid a polymer solution (either the same polymer or a different polymer) is extruded parallel to the polymer solution coming from the first orifice. Applying the triple layer spinneret/nozzle even a third polymer solution can be extruded in parallel. A certain microstructure on the shell of solid fibres/capillaries might be desirable to improve adhesion of certain species like pigments, biological cells, etc.

Creating micro-structures on the bore surface of the fibres/capillaries can be achieved by structuring the needle **2** of the extrusion nozzle or spinneret either at the needle's inside **12** or outside **11, 13** surface (see Fig. 28). Creating microstructures at the shell side of the fibres/capillaries can be achieved by microstructuring the surface of the tube that forms the outside border of the orifice **10, 16**. Structuring the needle **2** (either inside or outside or both) and the tube forming the outside border of the orifice of a spinneret/nozzle can also be applied to create micro-structures on both bore side and shell side of the fibres/capillaries. In order to create spiralised structures, the needle or orifice containing the texture should be able to rotate (with or without forward and backwards motion) or the polymer dope should be leaving the nozzle/spinneret in a spiralised flow direction. At a high level of spiralisation nearly orthogonal structures can be made with respect to the length of the capillary. Once the microstructure is imprinted in the polymer dope solution this structure should be maintained until the structure solidifies by phase separation. This can be achieved by using a highly viscous polymer dope and/or by instantaneous precipitation of the polymer dope after leaving the nozzle/spinneret.

The efore mentioned methods and experiments of creating microstructures in hollow fibre and capillary polymeric surfaces are not restricted to fibres or capillaries used as filtration membranes only. Microstructures can also be applied for instance to, capillaries for capillary electrophoresis, electro-chromotography and also biomedical devices.

REFERENCES

[1] C.J.M. van Rijn, Membrane filter as well as a method of manufacturing the same, PCT patent application 95/1386026.
[2] W. Nijdam, Aquamarijn, Masters Thesis, University of Twente, 1995.
[3] S. Kuiper, Ph.D-Thesis, University of Twente, 2000.
[4] J.H. Brauker et. al., Porus polymeric membrane structure, Baxter Int.,US5,807,407.
[5] J.D. Jacobson, Microporous filter membrane, Baxter Int., EP1194216.
[6] Tests made in 1995 at Baxter, Nivelles, Belgium.
[7] Keller, Berkeley University, US patent 5,660,680.
[8] H.V. Janssen, Thesis, University of Twente.
[9] J.Elders, C.J.M. van Rijn, Mould as well as a method to produce such a mould, NL 1001220, 17-091995.
[10] Arch Chemicals N.V., Keetberglaan 1A, Havennummer 1061, B-2070 Zwijndrecht, Belgium.
[11] S.Y. Chou, P.R. Kraus, W. Zhang, L. Guo and L. Zhuang, Sub-10 nm imprint lithography and applications, J. Vac. Technol. B 15(6) (1997) 2897-2904.
[12] H.V. Jansen, M.J. de Boer, H. Wensink, B. Kloeck and M.C. Elwenspoek, The Black Silicon Method VIII: A study of the performance of etching silicon using SF6/O2-based chemistry with cryogenical wafer cooling and a high density ICP source Microsystem Technologies, Vol.6 No.4 (2000).
[13] G.J. Burger, Thesis, University of Twente, 1995.
[14] J. Brugger, J.W. Berenschot, S. Kuiper, W. Nijdam, B. Otter and M. Elwenspoek, Resistless patterning of sub-micron structures by evaporation through nanostencils, Microelectronic Engineering 53, 403 (2000).
[15] H.V. Jansen, J.G.E Gardeniers, J.Elders, H.A.C. Tilmans and M.C. Elwenspoek, Applications of fluorocarbon polymers in micromechanics and micromachining, Sensors and Actuators A, 41-42 (1994) 136-140.

[16] R.J. Jaszewski, H. Schift, B. Schnyder, A. Schneuwly and P. Gröning, The deposition of anti-adhesive ultra-thin teflon-like films and their interaction with polymers during hot embossing, Applied Surface Science 143 (1999) 301-308.

[17] M.T. Gale, Replication techniques for diffractive optical elements, Microelectronic Engineering 34 (1997) (321-339).

[18] Millipore, Whatmann.

[19] General Electric Plastics, Distributor: B.V. Snij-Unie HiFi, Zoutketen 23, 1061 EX, Enkhuizen, Nederland.

[20] All masks and etch recipes provided by Aquamarijn.

[21] S. Kuiper, Thesis 2000 Chapter 9, University of Eindhoven.

[22] General Electric Plastics, Distributor: B.V. Snij-Unie HiFi, Zoutketen 23, 1061 EX, Enkhuizen, Nederland.

[23] S.Y. Chou, P.R. Krauss and P.J. Renstrom, Imprint of sub-25 nm vias and trenches in polymers, Appl. Phys. Lett. 67 (1995) 3114.

[24] L. Mashev and S. Tonchev, Formation of holographic diffraction gratings in photo-lacquer, Appl. Phys. A 26 (1981) 143-149.

[25] H. Tan, A. Gilbertson and S.Y. Chou, Roller nanoimprint lithography, J. Vac. Sci. Technol. B 16(6) (1998) 3926-3928.

[26] M. Wessling, C.J.M. van Rijn, J.N.Barsema, W. Nijdam, NL1016779, 2 december 2000, Mould, method for fabrication of precision products, in particular microsieves and filtration membranes.

[27] C.J.M. van Rijn, L. Vogelaar, W. Nijdam, J.N.Barsema, M. Wessling, WO0243937 Publication date: 6 June 2002, Method of making a product with a micro or nano size structure and product.

[28] Later trials for a period of several months during the winter all failed. Praying for success Cees van Rijn gave at one day two blasts with his breath before and after the casting process and discovered together with Laura Vogelaar the missing link in the production process: the relative humidity of the air. (Note, in the late summer the relative humidity of air is relatively high).

[29] C.J.M. van Rijn, W. Nijdam, L.A.V.G. van der Stappen, O.J.A. Raspe, L. Broens and S. van Hoof, Innovation in yeast cell filtration: Cost saving technology with high flux membranes, Proc. EBC Congress, Maastricht (1997) 501-507.

Chapter 7

Microfiltration

1. INTRODUCTION

The ultimate goal in membrane microfiltration is to achieve a low flow resistance, a high chemical resistance and a well controlled pore size distribution of the membrane in order to obtain a high operational flux, long standing times and good separation behaviour. Cross-flow microfiltration is a well-known technique for the removal of microparticles and microorganisms from fluid streams, and has therefore many (potential) industrial applications in food and bioprocess engineering [1].

Conventional microfiltration processes are mostly based on the principle of size exclusion. Larger particles cannot enter the pores in the membrane, and accumulate on the surface of the membrane. Blocking of the filtration membrane is prevented by the application of a cross-flow, see Fig. 1. Still, accumulation of retained material on the membrane surface and membrane fouling are the limiting factors in this process, to such an extend, that in many applications, people have taken refuge to using ultrafiltration membranes with much smaller pore sizes (< 0.1 micron) instead of microfiltration membranes. However, ultra-filtration gives unwanted retention of macromolecular material, and demands an energy input which is an order of magnitude higher than for microfiltration processes.

Fig. 1 Schematic process of cross-flow: part of the fluid with small particles permeates through the membrane filter, part of the fluid with large particles is retained and flows parallel to the membrane filter.

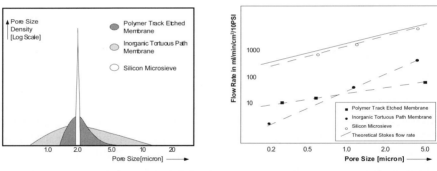

Fig. 2 Left: pore size distribution and Right: clear water flux of various microfiltration membrane filters, courtesy Aquamarijn Research.

Microsieve filters consist of a thin membrane with well-defined uniform pores, and for most applications the membrane layer is reinforced by a support structure. Inorganic membranes and in particular ceramic membranes [2] have a number of advantages over polymeric membranes, such as high temperature stability, relative inertness to chemicals, applicable at high pressures, easy to sterilise and recyclable. However, they are not used extensively because of the high costs and relatively poor control in pore size distribution (see Fig. 2, Fig. 3). Also the effective membrane layer is very thick in comparison to the mean pore size (typically 50-1000 times), which results in a reduced flow rate.

A microsieve having a relatively thin filtration or sieving layer with a high pore density and a narrow pore size distribution on a macroporous support will show better separation behaviour at high flow rates (see Fig. 2). The support structure contributes to the mechanical strength of the total microsieve membrane surface. The openings in the support should be made as large and numerous as possible in order to maintain the flow rate of the membrane layer and to reduce interaction of the support with the fluid.

Fig. 3 SEM micrographs of membrane filters. Left: Organic phase separated membrane filter, Middle: Ceramic sintered filter, Right: polymeric track etched membrane filter, courtesy Aquamarijn Research.

The use of microsieves with very thin membrane layers will result in a more energy- and cost-saving micro filtration separation technology for present and future innovative applications, such as micro liquid handling, modular fluidic systems or micro total analysis systems [3].

Fig. 4 SEM micrographs. Left: detail of microsieve surface for microfiltration, Middle: overview of microsieve surface, Right: support side of a microsieve structure for microfiltration [4].

Compared to other microfiltration membranes, microsieves have an extremely small flow resistance [5]. An accumulation of retained particles in front of the sieve (cake layer formation) will –more than for other membranes– greatly increase the flow resistance. It is therefore important to keep the surface free of particles during filtration. This is usually done by applying a cross-flow in which larger particles will be removed from the membrane surface. Permeate flow reversal (back pulsing) is a more advanced method to remove smaller particles from the surface of the membrane filter. The required cross-flow velocity to remove larger particles is dependent on several variables such as the ratio between the particle and the pore size, the transmembrane pressure and cross-flow channel dimensions. In order to be able to filter under optimum conditions with a properly designed filtration module the influence of these variables on the accumulation of particles on the microsieve surface has to be investigated, which is described in detail in section 6.2. In section 6.3 the effect of shear-induced diffusion of large and small particles between adjacent laminar fluid layers is discussed.

When using a microfiltration membrane with a very low flow resistance, it is paramount to prevent the formation of a cake layer, because the resistance of even a thin cake layer might already severely reduce the optimum operational flux. An additional advantage of a clean membrane surface is that the retention characteristics of the filtration process is only determined by the membrane layer itself and not by the additional permeation characteristics of the cake layer.

2. CROSS FLOW MICROFILTRATION OF LARGE PARTICLES[*][6]

In case the particles in the fluid are relatively large compared to the pore size, a cross-flow of the fluid may prevent the accumulation of these particles at the surface of the microsieve. This effect was first observed in 1995 [7,8] with a small cross flow filtration module (channel height 100 micron) with a microsieve with a pore size of 2 micron and a baker's yeast solution with a mean particle size of 5 micron at transmembrane pressures up to 0.05 bar.

In 1997 a PhD together with two students [9] was asked to study this effect more in detail and to derive a theoretical model for the observed phenomenon. A description of this model is now being given together with an empirical verification.

In 1986 Fischer and Raasch [10] introduced a single-particle model for pore blocking in cross-flow microfiltration. They assumed that a critical ratio of transmembrane pressure to cross-flow-drag force exists below which the filter medium remains clean. This ratio was determined from experiments with woven metal sieves with mesh sizes of 25, 42 and 80 µm. Lu et al. [11] made an attempt to calculate the critical ratio by analysing the forces acting on a trapped particle. But whereas Fischer and Raasch used a sieve with a well-defined mesh size, Lu et al. regarded a filter medium with an unknown surface roughness and therefore had to use fitting parameters. De Balmann et al. [12] continued to develop Fischer and Raasch' model by estimating the two forces acting on a

---

[*] Partial reprinted with permission of J. of Membrane Science (2000), Elsevier Science.

spherical particle trapped on a circular pore. For cross flow microfiltration with a microsieve this critical ratio can be derived in a fully analytical way.

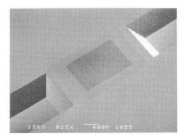
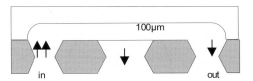

Fig. 5 Small cross flow module with a cross flow inlet (in), a cross flow outlet (out), a permeate outlet and a glass cap to facilitate microscopic examination of the microsieve surface, courtesy Aquamarijn Research [13].

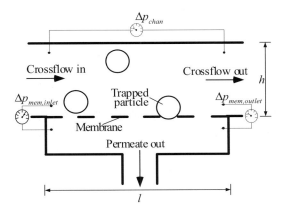

Fig. 6 Spherical particles in a cross-flow-microfiltration module.

Fig. 6 shows a simplified situation with relevant cross-flow variables: h is the height of the cross-flow channel above the sieve and l the length, Δp_{chan} denotes the pressure drop across the channel and $\Delta p_{mem,inlet}$ and $\Delta p_{mem,outlet}$ are the transmembrane pressures at the inlet and the outlet of the cross-flow channel, respectively. Fig. 7 shows how the pressure and drag force act on a trapped particle. The pressure force keeps the particle trapped while the cross-flow-drag force tries to release the particle. The ratio of their moments towards pivot position A determines the release of the particle.

Magnitude of the forces acting on a trapped spherical particle
Besides the pressure force and the cross-flow-drag force, several other forces are acting on a trapped particle. The most relevant are the lift force, adhesion forces and the permeate-drag force. The magnitude of these five forces will be calculated in order to determine their influence on a trapped particle.

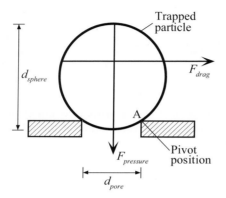

Fig. 7 Pressure and drag force acting on a trapped particle in a cross-flow module.

Pressure force
The pressure force $F_{pressure}$ is caused by the pressure difference Δp_{mem} across the membrane (transmembrane pressure). Its magnitude is equal to the pore area A_{pore} times the pressure difference:

$$F_{pressure} = \Delta p_{mem} A_{pore} = \Delta p_{mem} \tfrac{1}{4}\pi d_{pore}^2 \tag{1}$$

with d_{pore} the diameter of the circular pore. The moment $M_{pressure,A}$ of this force towards pivot position A is given by:

$$M_{pressure,A} = F_{pressure} \tfrac{1}{2} d_{pore} = \Delta p_{mem} \tfrac{1}{8}\pi d_{pore}^3 \tag{2}$$

Drag force
The cross-flow exerts a tangential drag force F_{drag} on the trapped particle. This force is dependent on the cross-flow velocity near the particle. In case of a laminar flow the velocity profile will be parabolic, as shown in Fig. 8.
Integration of the parabolic profile leads to the velocity distribution $v(y)$. For a rectangular channel one obtains:

$$v(y) = \tfrac{3}{2}<v>\left\{1-\left(\frac{y}{\frac{1}{2}h}\right)^{2}\right\}$$

(3)

where $<v>$ is de average cross-flow velocity, h the height of the rectangular channel and y the distance to the channel centre.

The velocity v_c near the centre of the sphere can be found by substituting $y = \tfrac{1}{2}h - \tfrac{1}{2}d_{sphere}$:

$$v_{c} = \tfrac{3}{2}<v>\left\{2\frac{d_{sphere}}{h} - (\frac{d_{sphere}}{h})^{2}\right\} \cong 3\frac{<v>d_{sphere}}{h}$$

(4)

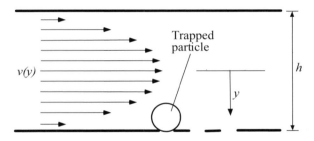

Fig. 8 Velocity profile of a laminar flow in a cross-flow channel.

The last step in Eq.(4) is valid if $d_{sphere} \ll h$.

If a sphere is subject to a uniform creeping flow (Re$\ll$1, based on the sphere) without the influence of a wall, the drag force is given by the linear Stokes equation:

$$F_{drag,Stokes} = 3\pi\eta v d_{sphere}$$

(5)

with η the dynamic viscosity of the fluid.

If the sphere is fixed to a wall and subject to a linear creeping shear flow (which usually applies to a microscopic particle in the boundary layer of a cross-flow channel), the Stokes equation has to be multiplied by 1.70 while substituting [14,15] $v = v_c$:

$$F_{drag} = 1.70 \cdot F_{drag,Stokes} = 5.10\pi\eta v_c d_{sphere} \qquad (6)$$

This equation is only valid for a sphere that remains at the entrance of the pore. If the sphere is so small that it sinks deeply into the pore, the actual drag force will be significantly smaller than calculated with Eq.(6).

In order to calculate the moment of the drag force towards pivot position A, the point of impact of this force on the sphere has to be known.

As the upper part of the sphere is subject to higher velocities, it is obvious that the point of impact of the drag force will be above the centre C of the sphere. The distance a from the force line to the centre can be calculated using an expression for the torque T towards C exerted on the sphere [14]:

$$T = 0.944\pi\eta v_c d_{sphere}^2 \qquad (7)$$

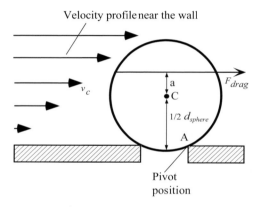

Fig. 9 Impact of the drag force on a sphere in a linear creeping shear flow with wall influence.

Now a can be calculated:

$$a = \frac{T}{F_{drag}} = \frac{0.944\pi\eta v_c d_{sphere}^2}{5.10\pi\eta v_c d_{sphere}} = 0.185 d_{sphere} \qquad (8)$$

For the moment of the drag force $M_{drag,A}$ towards pivot position A follows:

$$M_{drag,A} = F_{drag}\left(\tfrac{1}{2}d_{sphere} + a\right) = F_{drag} 0.685 d_{sphere} = 3.50\pi\eta v_c d_{sphere}^2 \qquad (9)$$

With Eq.(4) one obtains the moment of the drag force as a function of the average velocity in the cross-flow channel:

$$M_{drag,A} = 10.5 \frac{\pi \eta <v> d_{sphere}^3}{h} \tag{10}$$

As De Balmann et al. [12] neither used the correction factor of 1.70 for the drag force (see Eq.(6) nor the exact point of impact (they used $0.5d_{sphere}$ instead of $0.685d_{sphere}$), their obtained value for $M_{drag,\ A}$ is a factor of 2.33 smaller than that given in Eq.(9).

Lift force

As the cross-flow velocity near the upper part of the sphere is higher, a lift force perpendicular to the cross-flow will arise. For a resting particle in a laminar boundary layer the magnitude of this lift force can be estimated with [16]:

$$F_{lift} = 2.15\eta^{0.5} \rho^{0.5} v_c^{1.5} d_{sphere}^{1.5} \tag{11}$$

with ρ the density of the filtrate. In order to compare the magnitude of the lift force and the drag force their ratio can be calculated:

$$\frac{F_{drag}}{F_{lift}} = \frac{5.10\pi \eta v_c d_{sphere}}{2.15\eta^{0.5} \rho^{0.5} v_c^{1.5} d_{sphere}^{1.5}} = 7.45 \left(\frac{\eta}{\rho v_c d_{sphere}} \right)^{0.5} = \frac{7.45}{d_{sphere}} \left(\frac{\eta h}{3\rho <v>} \right)^{0.5} \tag{12}$$

If some values are substituted for a dilute yeast cell suspension ($\eta = 10^{-3}$ Pa·s, $d_{sphere} = 5$ μm, $\rho = 10^3$ kgm^{-3}) and a fairly high shear velocity ($<v>$=2.5 m/s and h=0.5 mm), one obtains:

$$\frac{F_{drag}}{F_{lift}} = \frac{80 \cdot 10^{-6}}{d_{sphere}} = 16 \tag{13}$$

If this result is combined with the fact that the arm of the drag force ($0.685d_{sphere}$) is at least a few times larger than the arm of the lift force ($0.5d_{pore}$), it is clear that one can neglect the influence of the lift force on a trapped sphere with the size of a yeast cell or smaller.

Adhesion forces

On a microscopic scale adhesion forces like Vanderwaals forces and electrostatic forces can significantly influence the behaviour of a particle near a membrane. However, in certain cases they are so small that they can be neglected. In the description of the experiments further down this it is shown how to suppress the adhesion forces to a negligible level with respect to the drag and pressure force.

Permeate-drag force

The conditions that are needed to clean a blocked membrane by using a cross-flow is a relevant parameter. A microsieve with a pitch (pore to pore distance) larger than the sphere diameter can get completely blocked and will therefore show a negligible permeate flow. If the pitch is smaller than the sphere diameter, the spheres can block only a limited number of pores. In this case the permeate flow might exert a significant drag force on a trapped particle in the same direction as the pressure force.

If one assumes that the permeate-drag force can be estimated with the linear Stokes equation (Eq.(5)), one can compare the permeate-drag force with the pressure force:

$$\frac{F_{pressure}}{F_{drag,permeate}} = \frac{\frac{1}{4}\pi d^2_{pore}\Delta p_{mem}}{3\pi\eta v_{permeate}d_{sphere}} = \frac{d^2_{pore}\Delta p_{mem}}{12\eta v_{permeate}d_{sphere}} \tag{14}$$

In earlier work [17] it is shown that the permeate velocity through a microsieve with a porosity of 20% (i.e. the pitch is twice the pore size) can be calculated with:

$$v_{permeate} = \frac{4.3\cdot 10^3 d_{pore}\Delta p_{mem}}{\eta} \tag{15}$$

Insertion of this equation into Eq.(14) yields:

$$\frac{F_{pressure}}{F_{drag,permeate}} = 19\frac{d_{pore}}{d_{sphere}} \tag{16}$$

This equation shows that for small particle-pore ratio's (but $d_{sphere}/d_{pore} > 1$) the permeate-drag force plays a minor role in keeping particles trapped on a pore. For membranes thicker than microsieves (e.g. track-etched membranes), the permeate-drag force becomes even smaller while the pressure force remains the same. In further calculations the permeate-drag force will be neglected, but keeping in mind that for large particle-pore ratio's or porosity's higher than 20% it can play a significant role.

The existence of a permeate flow has another influence on the flow conditions around a particle: it will affect the parabolic profile. This effect will not be incorporated.

Release of a trapped sphere

The two most relevant forces are the cross-flow-drag force and the pressure force. The ratio of their moments is (Eqs.(2) and (10)):

$$\frac{M_{drag,A}}{M_{pressure,A}} = \frac{10.5 \frac{\pi \eta <v> d^3_{sphere}}{h}}{\Delta p_{mem} \frac{1}{8} \pi d^3_{pore}} = 83.9 \frac{\eta <v>}{h \Delta p_{mem}} \left(\frac{d_{sphere}}{d_{pore}} \right)^3 \tag{17}$$

On using an expression for the pressure drop Δp_{chan} across the rectangular cross-flow channel [18],

$$<v> = \frac{\Delta p_{chan} h^2}{12 l \eta} \tag{18}$$

Eq.(17) can be written as:

$$\frac{M_{drag,A}}{M_{pressure,A}} = 7.0 \frac{h}{l} \frac{\Delta p_{chan}}{\Delta p_{mem}} \left(\frac{d_{sphere}}{d_{pore}} \right)^3 \tag{19}$$

A trapped sphere will be released from the entrance of a pore if the moment of the drag force exceeds the moment of the pressure force. This leads to the 'particle-release condition' for a pore:

$$\boxed{\frac{\Delta p_{mem}}{\Delta p_{chan}} < 7.0 \frac{h}{l} \left(\frac{d_{sphere}}{d_{pore}} \right)^3} \tag{20}$$

The pores on the inlet of the cross-flow channel are subject to the highest transmembrane pressure ($\Delta p_{mem,inlet}$). If they are kept clean by the cross-flow, the whole surface will be kept clean. Therefore the particle-release condition for the entire microsieve becomes:

$$\boxed{\frac{\Delta p_{mem,inlet}}{\Delta p_{chan}} < 7.0\frac{h}{l}\left(\frac{d_{sphere}}{d_{pore}}\right)^3} \tag{21}$$

This remarkably simple equation shows at what transmembrane pressures the pores can be kept free.

In order to prevent back flow of the permeate through the membrane near the outlet of the channel, $\Delta p_{mem,inlet}$ on the inlet should be larger than Δp_{chan}. If this condition is added to Eq.(21) one obtains:

$$1 < \frac{\Delta p_{mem,inlet}}{\Delta p_{chan}} < 7.0\frac{h}{l}\left(\frac{d_{sphere}}{d_{pore}}\right)^3 \tag{22}$$

This equation is visualised in Fig. 10 for a constant ratio of d_{sphere} to d_{pore}. The graph shows that cake layer free filtration without permeate backpuls is only possible if the ratio of h to l is larger than a certain critical value, no matter what cross-flow or transmembrane pressure is applied! From Eq.(22). it is deducted that filtration without pore blocking can only occur if:

$$\boxed{\frac{l}{h} < 7.0\left(\frac{d_{sphere}}{d_{pore}}\right)^3} \tag{23}$$

This equation can be used as a rule of thumb for the design of a cross-flow module for cake layer free filtration, provided that certain conditions are met.

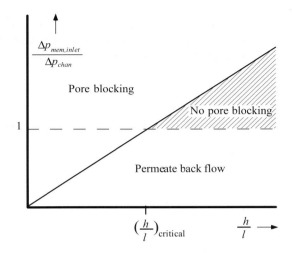

Fig. 10 Visualisation of Eq. (22). The shaded region is the only region where filtration without cake layer formation is possible.

These conditions are: laminar cross-flow, rigid spherical particles, circular pores smaller than but of the order of the particle diameter, dilute suspensions and a membrane with microsieve-like properties (flat, smooth, narrow pore-size distribution and cylindrical pores).

Tubular membranes

The equations derived are only applicable for flat sheet membranes. It is interesting to know whether the particle-release conditions for tubular membranes can be described by similar equations.

The velocity distribution for a tube is [18]:

$$v_{tube}(y) = 2 < v_{tube} > \left\{ 1 - \left(\frac{y}{\frac{1}{2}D} \right)^2 \right\}$$ (24)

where D is the diameter of the tube. In the same way as for rectangular channels (Eqs. (3) and (4)) one obtains the cross-flow velocity $v_{c,tube}$ near the centre of a spherical particle on the surface of a tubular membrane:

$$v_{c,tube} = 4 \frac{< v_{tube} > d_{sphere}}{D}$$ (25)

The factor 3 for rectangular channels has changed to a factor 4 for tubular channels. The average velocity $<v_{tube}>$ is given by [18]:

$$<v_{tube}> = \frac{\Delta p_{tube} D^2}{32 \eta l} \tag{26}$$

Combination Eqs. (25) and (26) gives:

$$v_{c,tube} = \frac{\Delta p_{tube} d_{sphere} D}{8 \eta l} \tag{27}$$

Eqs.(4) and (18) give a similar result for the velocity $v_{c,chan}$ near the centre of a particle on the wall of a rectangular channel:

$$v_{c,chan} = \frac{\Delta p_{chan} d_{sphere} h}{4 l \eta} \tag{28}$$

Comparison of Eqs.(27) and (28) shows that for $D=h$ and equal channel pressure drops the velocity near a trapped particle is 2 times higher for rectangular modules than for tubular modules. Therefore the particle-release condition for tubular modules becomes (compare with Eq.(21):

$$\boxed{\frac{\Delta p_{mem,inlet}}{\Delta p_{chan}} < 3.5 \frac{D}{l} \left(\frac{d_{sphere}}{d_{pore}} \right)^3} \tag{29}$$

and the module geometry demand for tubular membranes (compare with Eq.(22)):

$$\boxed{\frac{l}{D} < 3.5 \left(\frac{d_{sphere}}{d_{pore}} \right)^3} \tag{30}$$

In order to verify Eq.(20) two different approaches are followed. The first approach consists in checking pore blocking by observing the surface of a small microsieve with different pore sizes through a microscope. In the second approach one tries to detect pore blocking on a larger sieve with uniform pores by measuring permeate fluxes.

List of symbols

All symbols are in SI-units

Δp_{mem}	TMP for a certain pore
$\Delta p_{mem,inlet}$	TMP at the begin of the membrane
$\Delta p_{mem,outlet}$	TMP at the end of the membrane
Δp_{chan}	Pressure drop across a rectangular cross-flow channel
Δp_{tube}	Pressure drop across a tubular cross-flow channel
F_{drag}	Cross-flow-drag force
$F_{drag,Stokes}$	Drag force according to the linearised Stokes equation
$F_{drag,permeate}$	Permeate-drag force
$F_{pressure}$	Force of the TMP on a particle
F_{lift}	Lift force on a particle
$M_{pressure,A}$	Moment of $F_{pressure}$ towards pivot position A
$M_{drag,A}$	Moment of F_{drag} towards pivot position A
T	Torque
a	Distance of the force line of F_{drag} to the centre of a sphere
C	Centre of a sphere
A	Pivot position
A'	Shifted pivot position
d_{pore}	Pore diameter
d_{sphere}	Sphere diameter
A_{pore}	Surface area of a pore
l	Channel length
h	Height of a rectangular channel
w	Width of a rectangular channel
D	Tube diameter
y	Distance to channel centre
v	Cross-flow velocity
v_c or $v_{c,chan}$	Cross-flow velocity near the sphere centre for a rectangular channel
$v_{c,tube}$	Cross-flow velocity near the sphere centre for a tubular channel
$<v>$	Average cross-flow velocity
$v_{permeate}$	Permeate velocity
η	Dynamic viscosity
ρ	Density

Experimental set-up for microscope observations

A microsieve is made with a 1μm thin membrane containing circular pores in the range of 1.5 to 7.0 μm. The pores are arranged into 5 groups of each 100 identical pores. The pitch is set at 20 μm to prevent trapped particles to touch each other. This means that all pores can get blocked and therefore the influence of the permeate flow on the trapped particles will be minimal. In a next section microsieves with much higher porosities will be used in order to determine the effect of a large permeate flow. Fig. 11 shows a schematic top view of the test microsieve in a cross-flow channel.

A cross-flow channel is created by gluing a hollow glass plate on top of the microsieve. Fig. 12 shows a schematic view of the fabrication process.

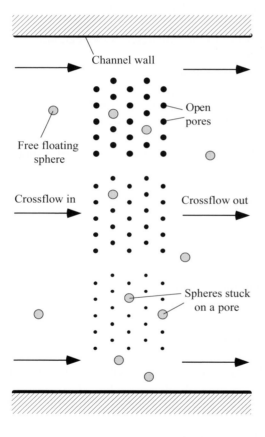

Fig. 11 Schematic top view of the test sieve in a cross-flow channel. The actual sieve contains 5 groups of each 100 pores.

3 modules were constructed with different channel heights (135, 260 and 350 μm). The length and width are 39 and 10 mm, respectively. The channels are low enough to ensure that the vertical parabolic cross-flow profile is fully developed near the perforations. The large width of the channels ensures that the horizontal profile is hardly developed. Therefore all the perforations will be subject to a uniform flow, especially because they are placed in the middle of the channel. As test particles polystyrene spheres with a diameter of 6.3 and 10.6 μm were used. The results of the experiments with these spheres were compared with the results obtained with cells of baker's yeast.

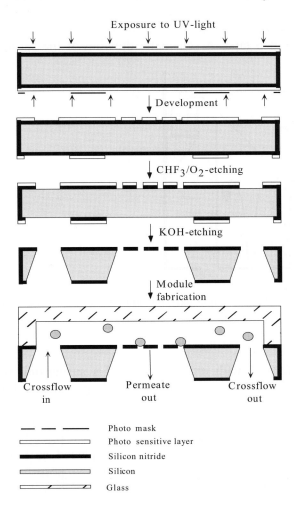

Fig. 12 Schematic view of the fabrication process of a test module.

As Eq.(20) does not require the measurement of any cross-flow or permeate flux, one can keep the set-up fairly simple. With a suspension of known spheres in a module with known dimensions, only Δp_{mem} and Δp_{chan} have to be measured. The suspension flows from a highly placed vessel through the module to a lower placed vessel, so Δp_{chan} is fully determined by the difference in height between the vessels (the flow resistances of the tubes are negligible). If the system is in equilibrium, the water level in the permeate tube denotes the zero position for Δp_{mem}. Δp_{mem} can be increased by lowering the permeate tube and its value is then simply the distance of the tube end to the zero position.

The level in the top vessel is kept constant by a peristaltic circulation pump. The surface of the membrane can be checked in-line with a microscope and images are recorded on video tape. A schematic drawing of the set-up is shown in Fig. 13.

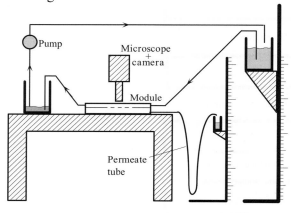

Fig. 13 Set-up for verification of the single-particle model through microscope observations

Each experiment is performed with one module and a suspension with spheres of one diameter, which means that h, l and d_{sphere} are kept constant. For a certain difference in height between the two cross-flow vessels the equilibrium water level in the permeate tube is determined (this is usually exactly in the middle of the two cross-flow vessels, because the lowest vessel is placed at the same level as the module). Then, the permeate vessel is placed as low as possible to create such a transmembrane pressure that all the pores in the membrane are blocked by the spheres. Next the permeate vessel is raised slowly until the spheres on the group with the smallest pores are released (this release takes place in a fairly small pressure interval). The height of the permeate tube end is determined and subsequently the permeate vessel is raised again, until the spheres on the next group of pores are released. This procedure is continued until all the groups are free. Each release pressure is determined three times. After this the upper cross-flow vessel is raised to increase the cross-flow and the whole procedure is repeated. The experiment is repeated with the other two modules and the other two suspensions.

Results and discussion of microscope observations

If the polystyrene spheres are suspended in water, they tend to stick strongly to the membrane and the rest of the system. The addition of a small amount of detergent (0.1% Teepol) turns out the be the solution to the problem: the particles completely lose their tendency to stick. The detergent may have changed surface properties such as charge of the membrane and particles.

From Eq.(20) a linear relation is expected between the transmembrane pressure at which particles are released and the pressure drop across the channel.

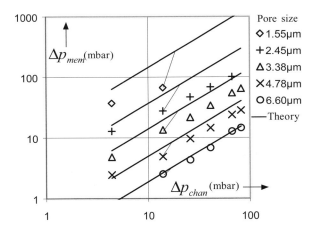

Fig. 14 Transmembrane pressure at which 10.6 μm polystyrene spheres are released from a pore as a function of the pressure drop across a 260 μm high cross-flow channel. The straight lines are drawn according to Eq.(20).

Fig. 14 shows the results of an experiment where particle-release pressures were determined. The results of the other experiments are similar. It appears that the relation between Δp_{mem} and Δp_{chan} is quite linear as expected from Eq.(20), although for higher pressures the data show a digressive behaviour. The data that belong to the largest pores (6.6 μm) are in perfect agreement with the theory. For smaller pores the particles tend to be more difficult to release than predicted.

In order to obtain a general verification of Eq.(20) the results of all the experiments are gathered with the different modules and particles and are plotted together in one graph (see Fig. 14). The graph shows that for a particle-pore ratio around 2, Eq.(20) gives a fairly good prediction of the release pressure. For larger ratios the particles are more difficult to remove than expected. This deviation might be caused by a shift in the pivot position of the particle on the pore. If the pore is much smaller than the sphere, the tangent of the sphere in the point of contact A is almost parallel to the membrane surface. A small deformation of the sphere caused by the transmembrane pressure will result in a shift of point A to a point A' away from the pore (see Fig. 16). Besides a larger contact area (larger adhesion force), this shift in pivot position will give the pressure force a larger arm. If the arm increases by a factor of 2, the release pressure will decrease by a factor 2. This deformation effect will be stronger for larger transmembrane pressures, which is in agreement with the digressive behaviour in Fig. 14.

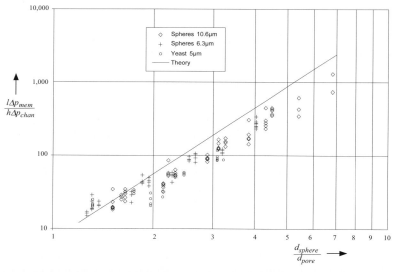

Fig. 15 Results of all microscope measurements. The straight line is drawn according to Eq.(20).

The egg-shaped yeast cells turn out to follow the theory reasonably well, although they tend to stay on the pores a little longer than the polystyrene spheres. This may probably be due to the fact that the cells are regarded as spheres with a diameter of 5 μm, which is only a rough estimation. Furthermore, they are easier to deform than polystyrene spheres. The size distribution of the trapped cells may be much smaller than that of the suspended cells, because the cells are 'caught' by slowly decreasing the cross-flow. In this way the smallest cells block the pores first and the larger cells will remain in suspension.

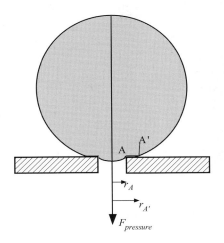

Fig. 16 Shift in pivot position for a slightly deformed sphere with a particle-pore ratio of 5. r_A is the original arm of the pressure force, $r_{A'}$ is the arm in case the pivot position has shifted to point A'.

Flux measurements

If, for a certain cross-flow, the transmembrane pressure is increased beyond a certain critical value, trapped particles will no longer be released. For a low-porosity microsieve with a pitch larger than the particle diameter all the pores can get blocked and thus the permeate flux will probably decrease for pressures larger than this critical value. So by measuring the flux it should be possible to register pore blocking and verify Eq.(20) in a more realistic filtration situation than with the model membrane in our microscopic observation experiments.

Set-up for flux measurements

Diluted suspensions of yeast cells (0.2-4.0 g/l are made in a ¼ strength Ringers solution at 5 °C) and a membrane with a pore size of 2.50 µm and a pitch of 7.50 µm. An SEM picture of the membrane is given in Fig. 17.

As Eq.(20) shows that the smallest yeast cells (diameter about 5 µm) block the pores easier than the larger cells, one expects a blocked membrane to be covered mainly with 5 µm cells. These small cells can not touch each other and should therefore be able to block the membrane completely. This blocking will lead to a strong decrease in flux and should therefore be easy to detect.

The microsieve is placed in a module with a continuously adjustable channel height. The channel length is 39 mm and the width 25 mm. Fig. 18

shows a schematic image of the constant-pressure cross-flow system that is used. The transmembrane pressure can be adjusted by altering the height of the permeate and/or cross-flow vessel. Permeate fluxes are measured as follows.

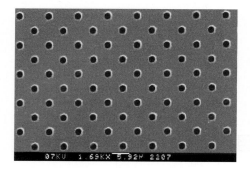

Fig. 17 The low-porosity sieve that is used for the flux measurements, courtesy Aquamarijn Research.

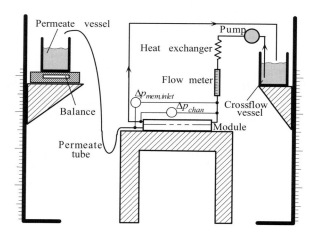

Fig. 18 Set-up for verification of the single-particle model through flux measurements.

For a pressure above the particle-release pressure the volume of permeate was determined that has to pass through the microsieve to obtain a more or less constant flux. Each flux measurement starts with passing this volume through the sieve. This is done to give the yeast cells the time to block the pores. Finally, the time is measured for 25 ml of permeate to pass through the sieve. For these 25 ml the permeate flux is calculated.

Results and discussion of the flux measurements

The graph in Fig. 19 shows that for a low-porosity microsieve with uniform pores the flux reaches a maximum value for a certain critical transmembrane pressure. For pressures larger than this critical pressure the trapped yeast cells can not be released by the cross-flow. The value of the release pressure can be calculated with Eq.(21).

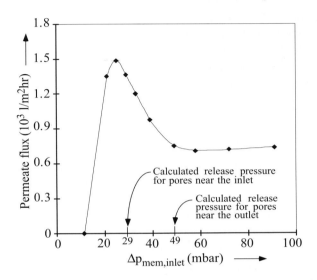

Fig. 19 Permeate fluxes for a yeast suspension filtered through the microsieve shown in Fig. 17.

The applied cross-flow in Fig. 19 was 1.6 l/min and the channel height 1.0 mm. The measured pressure drop across the cross-flow channel was 20 mbar. With a pore size of 2.50 μm and yeast size of 5 μm all the necessary information is obtained to calculate the release pressure. Insertion of the variables in Eq.(20) leads to a release pressure at the cross-flow inlet of 29 mbar. This value corresponds quite well to the experimentally determined value of the critical pressure. Notice that the reason why the flux is zero at about 10 mbar (half the pressure drop across the channel) is that the inlet transmembrane pressure is plotted, not the average pressure.

The calculated release pressure predicts when the inlet of the channel starts to get blocked. The pores downstream are still open then, for they are subject to a smaller transmembrane pressure caused by the pressure drop across the channel. Hence in Fig. 19 the transmembrane pressure for pores near the outlet of the channel is 20 mbar (the pressure drop across the channel) lower than for pores near the inlet. The outlet pores will get blocked for an inlet pressure of

29+20=49 mbar. For higher pressures the flux should stop decreasing, because all the pores are blocked. This can indeed be observed in Fig. 19. The fact that the flux does not become zero indicates that the yeast cells do not perfectly block the pores.

High-porosity microsieves

For high-porosity microsieves the pitch of the pores is often smaller than the particle size. Now the pores can not all be blocked at the same time, because a trapped particle will prevent surrounding pores to get blocked. Therefore a maximum value for the flux like in Fig. 19 will probably not be observed. Furthermore, the permeate flow through the open pores will influence the release conditions of the yeast cells by altering the parabolic profile and giving rise to a permeate-drag force.

Results

The low-porosity microsieve used in Fig. 19 was replaced by a high-porosity microsieve (pore size 1.94 µm and pitch 3.00 µm, porosity 33%). For a channel height of 1.1 mm, a cross-flow of 1.0 l/min and a pressure drop across the channel of 6 mbar, the graph in Fig. 20 was obtained. As expected, the flux graph does not show a maximum. However, a kink in the graph is observed that indicates an increase in the microsieve resistance due to trapping of cells. Insertion of the filtration conditions into Eq.(20) leads to a particle-release pressure of 20 mbar. This value roughly corresponds to the observed kink in the graph, which confirms the assumed minor influence of the permeate-drag force on the trapped cells. Eq.(16) showed that this force can be neglected for particles of the order of the pore size. In practice one often uses pores that are an order of magnitude smaller than the particles. For microsieves this is not necessary, as the small pore size distribution ensures that all particles are retained, even if the pores are only a fraction smaller than the particles. And in case the suspension consists of a broad range of particles, mainly the small particles will be trapped. These small particles will probably be of the order of the pore size and can therefore be described by the model.

In experiments where the release pressures were higher (for instance for larger channel heights or larger cross-flows) it was observed that both for low and high-porosity sieves the deviation from the theory increased. This is in agreement with the results in Fig. 14 and Fig. 15.

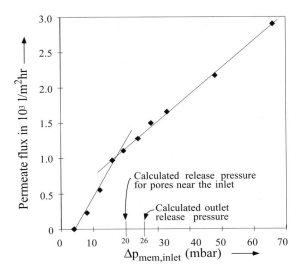

Fig. 20 Permeate fluxes for a yeast suspension filtered through a high-porosity microsieve.

Critical flux or critical pressure?

Prior to all yeast-filtration experiments the flux of water without yeast cells was measured. When comparing permeate flux with water flux a remarkable effect is observed. Although the first part of the graph in Fig. 20 seems to show no pore blocking, the permeate flux is about 4 times smaller than the water flux. Observation of the microsieve surface through a microscope during filtration shows that for pressures below the release pressure, particles get trapped on a pore, but their residence time is only a fraction of a second. Apparently the permeate flow drags the cells towards the sieve surface, but the transmembrane pressure is not large enough to keep the cells trapped. For higher yeast concentrations a lower flux has been measured, but the position of the kink remained the same. Probably a higher concentration causes more cells per second to be trapped and released, thus keeping the pores blocked for a longer time. The occurrence of this phenomenon, called dynamic fouling, was first expounded by Le and Howell [19] and further calculated by De Balmann et al. [12]. It is now confirmed that it indeed occurs as they described.

For the low-porosity microsieve (used for the graph in Fig. 19) the permeate flux is almost as high as the water flux. Possibly, in this case the permeate-drag force is not large enough to push the cells onto the pores. A moving particle experiences two forces perpendicular to the membrane: a permeate-drag force towards the membrane and a lift force in opposite direction.

The ratio of the two determines the direction of movement and was extensively described by Altena and Belfort [20]. Suppose the ratio is smaller than 1, which means that the particle moves away from the membrane. If the membrane porosity is increased (without altering the pore size), the flux and thus the permeate-drag force will increase and the ratio can become larger than 1. Now the particle moves towards the membrane and will –temporarily or permanently– block a pore. Apparently a certain 'critical flux' has to be exceeded in order to drag the particle towards a pore. Whether it stays at the pore depends (among other variables) on the transmembrane pressure, which is independent on the porosity. Above a certain 'critical pressure' the trapped particle will not be released by the cross-flow-drag force. The fact that the flux is dependent on porosity whereas the transmembrane pressure is not, implies that –depending on the degree of porosity– the filtration may be governed by a critical flux or by a critical pressure. It is not only the porosity that determines which of the two governs the filtration, but membrane thickness, cross-flow velocity, pore size and particle size are also important factors. For example, for a very thick membrane the flux may be too low to drag particles to the membrane, but once they arrive at a pore (e.g. by mutual collisions or by a pressure peak caused by a closing or opening valve) they will remain trapped. It has actually be seen this valve effect occurring after each backpulse during yeast-cell filtration of lager beer [20]. Using Eq.(20) and an expression for the ratio of permeate-drag force to lift force [19] it should be possible to calculate under which conditions filtration with a certain sieve is governed by a critical flux or by a critical pressure. The result may determine which operation strategy should be applied: filtration at constant flux or at constant pressure. Of coarse this analysis is based on the here described conditions. For filtration of e.g. protein solutions with pore sizes not much larger than the size of the protein molecules, a different analysis should be made.

Conclusions

The forces acting on a trapped particle were calculated as a function of several cross-flow variables. It was made plausible that the cross-flow-drag force and the transmembrane-pressure force are the major forces. Comparison of these forces leads to a very simple expression that describes under what conditions a trapped particle is released from a pore. The model was developed for microsieves, but may also be applied for other smooth membranes with a wider pore-size distribution. In that case the largest pore in the membrane determines which cross-flow conditions have to be applied. The model is valid for filtration of dilute suspensions of rigid spherical particles through membranes with circular pores under laminar cross-flow. The forces are calculated under neglect of inertia forces, which –for the boundary layer of a laminar cross-flow– is usually the case for particles of the order of the size of a yeast cell or smaller. Furthermore the pores should be smaller than, but of the order of the particle size.

Special microsieves with different pore sizes were fabricated with silicon micromachining technology to verify the theory. The experimental results are in fairly good agreement with the theory, although for larger particle-pore ratio's (>2) the particles (polystyrene spheres and yeast cells) tend to be more difficult to release than expected. Higher pressures (>0.05 bar) also cause a relatively difficult release of the particles. Both effects are possibly caused by deformation of the particles.

Experiments with a high-porosity sieve show that for pressures below the release pressure the permeate flux is significantly lower than the water flux, whereas a low-porosity sieve does not show this effect. This lower flux appears to be caused by temporarily captured particles. The effect leads us to the hypothesis that particle capture and release may be governed by a critical flux or by a critical pressure, depending on the filtration conditions.

The final overall conclusion is that microsieves appear to be excellent membranes for the experimental verification of filtration models. The well-defined pores and uniform distribution together with the flat and smooth surface give low particle adhesion and reproducible results. Moreover, pore size, shape and distribution can be customised, which increases the possibilities for model verification.

3. CROSS FLOW MICROFILTRATION OF POLYDISPERSE SUSPENSIONS, CONCENTRATION POLARISATION AND SHEAR INDUCED DIFFUSION*[21]

Most fluidic media subjected to a microfiltration process are polydisperse suspensions like e.g. beer and milk. The fouling layer is not only a build up of particles larger than the pore size of the microsieve but also of particles smaller than the pore size. Concentration polarisation is the gradually build up of all these particles above the microsieve surface. A detailed study on concentration polarisation and interdiffusion processes during microfiltration has been studied here in order to improve a better insight on cake layer formation.

Concentration polarisation is reduced by several back-transport mechanisms, however, for particles in the size range 0.5-30 μm shear-induced diffusion is the predominant one. Shear-induced diffusion is caused by the hydrodynamic interactions between particles in shear flow, which are strongly dependent on the particle size. Because of this, the polydispersity of suspensions is expected to have a large effect on cake layer formation. Although this aspect is of considerable practical relevance, it has only received minor attention in literature.

The effect of particle size distribution on microfiltration was investigated by studying the steady-state flux of bidisperse suspensions with a ratio of particle radii of 3. The flux appeared to be completely determined by the smaller particles; it remained almost constant between a volume fraction of 0.2 and 1 of small particles. This trend in the steady-state flux was in good agreement with calculations based on a shear-induced particle migration model for polydisperse suspensions. Since polydispersity is a common property of suspensions this may contribute to optimisation of flux and selectivity in microfiltration processes, which depend largely on adequate control of cake layer formation. This can be achieved by implementation of the developed model in process control routines.

In microfiltration processes, concentration polarisation mostly leads to the formation of a cake layer at the membrane surface, which has a large negative effect on flux and selectivity. Therefore, concentration polarisation can be considered a key factor in the control of microfiltration processes. Concentration polarisation is governed by two simultaneous processes: (1) convective transport of particles towards the membrane along with the permeate flow and (2) back-transport of particles from the concentrated layer into the bulk phase. In case of an excess particle flux towards the membrane, a cake layer builds up. When convective transport balances back-transport, the situation can be considered a

---

*Contribution J. Kromkamp et al., Desalination (2002). Partially reprinted with permission of Elsevier Science.

steady-state. In this situation only minor variations in cake layer resistance and flux are expected.

Shear-induced diffusion can be considered the predominant back-transport mechanism during microfiltration of suspensions with particles in the range of 0.5-30 μm [22]. This mechanism comprises the diffusive movements of particles in a shear field that are caused by hydrodynamic interactions. In the presence of a concentration or shear rate gradient, shear-induced diffusion leads to a net migration of particles in the direction of a lower concentration or shear rate, therewith effectively reducing concentration polarisation.

Using the fundamental knowledge about shear-induced diffusion, the microfiltration behaviour of monodisperse suspensions can, in general be predicted well. However, practically relevant suspensions are seldom monodisperse. Polydispersity, that is known to influence permeate flux and flow resistance of cake layers, seems to be the rule. Incorporation of effects of polydispersity in shear-induced diffusion models for microfiltration would therefore be an important step toward a better control of concentration polarisation during microfiltration of practically relevant suspensions.

The effect of the particle size distribution of a bidisperse suspension on the steady-state flux is investigted and results are compared with predictions obtained with a particle migration model for polydisperse suspensions [23].

Shear-induced migration model for polydisperse suspensions

For the quantitative description of shear-induced diffusion, most microfiltration models use the research of Eckstein and co-workers [24] and Leighton [25] and Acrivos [26]. The research of these authors was directed at self-diffusion of monodisperse particles in a suspension under shear. A frequently applied equation for shear-induced diffusion, reported by Leighton and Acrivos, is:

$$D_s = 0.33 \, \gamma \, a^2 \phi^2 \left(1 + 0.5e^{8.8\phi}\right) \tag{1}$$

All symbols are explained in the nomenclature list. For the steady-state permeate flux, Davis and Sherwood [27] derived the following equation, using an exact similarity solution with shear-induced diffusion as the dominant mechanism for particle back-transport

$$\langle J \rangle = 0.072 \, \gamma_0 \left(\phi_w \, a^4 \, / \, \phi_b \, L\right)^{1/3} \tag{2}$$

Shear-induced self-diffusion of particles in polydisperse suspensions has so far not been reported.

Phillips and co-workers [28] proposed a phenomenological model for shear-induced particle migration. The model is based on particle migration potentials. Migration is induced by two mechanisms, a non-uniform frequency of particle interactions in a shear field and a non-uniform viscosity field. Shauly and co-workers [23] extended this model to polydisperse suspensions with various particle sizes. For the system investigated here, the total flux of particles of species i can be described with

$$j_i = -k \gamma \bar{a} \, a_i \, \phi \phi_i \left[\nabla \ln (\gamma \phi_i) + \frac{\bar{a}}{a_i} \nabla \ln \mu^\lambda \right] \tag{3}$$

The term outside the brackets equals the shear-induced diffusion coefficient for the ith particle.

The migration potential in the shear-induced particle migration model consists of an interaction and a viscosity term. For the viscosity of monodisperse suspensions, Krieger [29] found:

$$\mu = \left(1 - \frac{\phi}{\phi_m} \right)^{-m} \tag{4}$$

where m = 1.82 and the maximum packing density for a monodisperse suspension (ϕ_{m0}) is assumed to be 0.68. This expression was extended by Probstein and co-workers [30] to describe the effective viscosity of a bidisperse suspension. In the approach of Probstein and co-workers the maximum packing density is not taken as a constant but is calculated as a function of the species local concentrations ϕ_1 and ϕ_2, as is shown in equation below:

$$\frac{\phi_m}{\phi_{m0}} = \left[1 + \frac{3}{2} |b|^{3/2} \left(\frac{\phi_1}{\phi} \right)^{3/2} \left(\frac{\phi_2}{\phi} \right) \right] \tag{5}$$

with $b = (a_1 - a_2) / (a_1 + a_2)$.

Nomenclature

a	particle radius (m)
$\bar{a}$	average particle radius (m)
b	dimensionless function
D_s	shear-induced diffusion coefficient (m²/s)
$\langle J \rangle$	length-averaged permeate flux (m/s)
j_i	migration flux of particle i (kg/m².s)
k	dimensionless particle migration parameter
L	filter length (m)
P_{tm}	transmembrane pressure (bar)

Greek symbols:

ϕ	particle volume fraction
ϕ_b	particle volume fraction in the bulk suspension
ϕ_m	maximum packing density
ϕ_{m0}	maximum packing density for a monodisperse suspension
ϕ_w	particle volume fraction at the channel wall
$\dot{\gamma}$	shear rate (s⁻¹)
$\dot{\gamma}_0$	nominal shear rate at the membrane surface (s⁻¹)
λ	dimensionless particle migration coefficient in Eq.(3)
μ	viscosity (kg/m.s)

Materials and methods

Microfiltration set-up

Microfiltration experiments were performed with microsieves with a membrane surface area of 30.2 mm² and slit-shaped pores with a length and width of 3 and 0.7 µm respectively. The thickness of the membrane was only 1 µm.

The feed for microfiltration consisted of suspensions of polystyrene spherical particles (surfactant-free, sulphate polystyrene, Interfacial Dynamics Corporation, USA) in a neutrally buoyant mixture of demineralised water and glycerol (density 1.055 g/ml at 20°C).

The experiments were carried out in a parallel plate device with a channel height of 0.25 mm. The temperature of the feed was kept constant at 25°C. The pressure was monitored with two pressure sensors in the parallel plate device, one before and one after the microsieve. The pressure at the permeate side of the microsieve was atmospheric. The feed flow rate was monitored with a rotameter and the permeate flux was measured with an analytical balance. Since the microsieve was divided into porous and non-porous fields, the permeate flux was calculated based on the surface area of the porous fields only.

Fig. 21 SEM micrograph of a microsieve with slit shaped pores.

Results and discussion

Predicted steady-state fluxes

The diffusion coefficients for the separate particles in a bidisperse suspension is calculated as a function of the suspension composition and related these values to the diffusion coefficients in case of a monodisperse suspension (Fig. 22). The ratio of the particle radii a_2/a_1 was 3 (equal to that in the experiments). The total volume fraction of particles ϕ in the concentration polarisation layer was always 0.6.

Starting from a monodisperse suspension of particle 1 ($\phi_1/\phi=1$), the diffusion coefficient first increases with increasing amount of particle 2. This is due to the increase of the mean particle radius (Eq.(33): $\bar{a}$) in the suspension. From a volume ratio ϕ_1/ϕ of 0.75 down to 0, the decreasing volume fraction of particle 1 (ϕ_1) in the suspension causes the diffusion coefficient to decrease. In contrast to particle 1, particle 2 shows a monotonous decrease of the diffusion coefficient with decreasing volume ratio. This is because the mean particle radius and the concentration of particle 2 both decrease. Apparently, the diffusion coefficient of the smaller particle in the bidisperse suspension has a different dependency on suspension composition as compared to that of the larger particle.

But what does this imply for the steady-state flux during microfiltration To analyse this effect, the steady-state flux using Eq.(32) was calculated, but adjusted for the relative change of the diffusion coefficients of both particles 1 and 2, as compared to the monodisperse case. With these model calculations two relations are obtained for the steady-state flux as a function of suspension composition (Fig. 23). One can however expect that the lowest value for the steady-state flux will actually be measured during microfiltration, because this

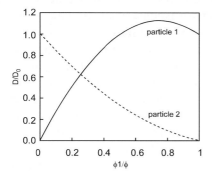

Fig. 22 Relative diffusion coefficient for the separate particles in a bidisperse suspension as a function of suspension composition for a total volume fraction of 0.6 and $a_2/a_1 = 3$, as calculated on basis of the particle migration model.

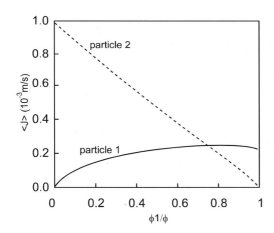

Fig. 23 Length-averaged steady-state permeate flux for cross-flow microfiltration, calculated for the separate particles (a_1=0.8 μm; a_2=2.4 μm) in a bidisperse suspension based on the particle migration model. The bold lines indicate the predicted actual steady-state flux. Microfiltration conditions: $\gamma_0 = 2.0*10^3 \text{ s}^{-1}; \phi_w = 0.6; \phi_b = 1.5*10^{-5}; L = 4.1*10^{-3} \text{ m}$

represents the situation in which neither for particle 1 nor for particle 2 an excess particle flux exists.

According to Fig. 23, this would imply that for a volume ratio ϕ_1/ϕ lower than 0.75, the back-transport flux of particle 1 determines the steady-state flux, while for a volume ratio ϕ_1/ϕ higher than 0.75 particle 2 does (except for a volume ratio ϕ_1/ϕ of exactly 0 or 1). From this, it follows that the flux is comparable to the flux of the monodisperse suspension of the smaller particle for a volume ratio from 0.25 up to 0.75. For volume ratios ϕ_1/ϕ near 0 or 1, the experimentally determined flux values should be considered with care, because measured fluxes may easily be influenced by particle depletion in the concentration polarisation layer or by very slow cake layer formation.

Please note, in the calculation only the concentration difference between concentration polarisation layer and bulk phase for the migration potential is taken into account. The assumption is made that the shear rate and the viscosity do not change the migration potential significantly from that for the monodisperse case. This seems a reasonable assumption. Using the model of Probstein and co-workers (Eq.(5), one can calculate that for the bidisperse suspension, the volume fraction in the concentration polarisation layer would need to rise from $\phi = 0.60$ to 0.66 (in the most extreme case) in order to achieve the same viscosity as for the monodisperse suspension. This increase in volume fraction gives an increase in steady-state flux with only 3%.

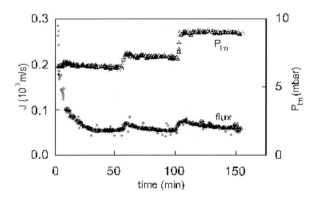

Fig. 24 Flux and transmembrane pressure during microfiltration of a bidisperse suspension of polystyrene particles (a_1=0.8 µm, ϕ_1/ϕ=0.8 and a_1=2.4 µm, ϕ_2/ϕ=0.2).

The transmembrane pressure was step-wise increased during the run. Microfiltration conditions: $\gamma_0 = 2.0*10^3$ s^{-1}; $\phi_b = 1.5*10^{-5}$; $T = 298$ K

Experimental results

Steady-state flux levels were determined for monodisperse suspensions with particle radii of 0.8 and 2.4 µm and for bidisperse suspensions, consisting of mixtures of these two particles. This was done by step-wise increasing the transmembrane pressure during a filtration run and measuring the steady-state flux (an example is shown in Fig. 24). Due to the low membrane resistance of microsieves, already at an extremely low transmembrane pressure a cake layer is formed and a steady-state flux level is achieved. In a steady-state situation, an increase of the transmembrane pressure only resulted in a temporary increase of the permeate flux. The subsequent decrease of the flux can be explained with extra accumulation of particles on the membrane surface, resulting in a thicker cake layer.

From calculations on bidisperse suspensions, it is expected that the particle with the largest excess particle flux would determine the steady-state flux level. As can be seen from Fig. 25, for all bidisperse mixtures the steady-state flux was comparable to the flux of the monodisperse suspension with the smaller particles. The variation in flux between the mixtures was relatively small. Apparently, in the bidisperse mixtures, the smaller particle had a larger excess particle flux and determined the steady-state flux level. This is in agreement with the presented model predictions.

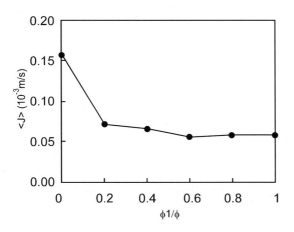

Fig. 25 Measured steady-state flux for microfiltration of a bidisperse suspension of polystyrene particles (a_1=0.8 μm; a_2=2.4 μm) as a function of suspension composition. Microfiltration conditions: $\gamma_0 = 2.0 * 10^3 \text{ s}^{-1}; \phi_b = 1.5 * 10^{-5}; T = 298 \text{ K}$

Comparison of the measured steady-state flux levels with values that are predicted by the shear-induced diffusion model reveals that the measured fluxes are about 5 times lower. This might be due to an overestimation of the predicted steady-state flux for short filter lengths (L). According to Eq.(32), the flux of monodisperse suspensions increases with the diameter of the particles with a power-dependence of 1.33. For the particles investigated here (a_2/a_1=3), a flux ratio (J_2/J_1) of 4.3 is expected, while a flux ratio of 2.7 was measured. This effect will be subject of further research.

Conclusions

The steady-state flux for microfiltration of a bidisperse suspension with a ratio of particle radii of 3 has a level that is comparable to that of a monodisperse suspension of the smallest particle (in the bidisperse suspension). The trend in the observed behaviour could be described accurately with a particle migration model for polydisperse suspensions. This model takes the effect of the particle size distribution on the shear-induced migration of individual particles into account.

Measures to improve filtration performance

The flux and separation characteristics of a membrane filtration process will change during operation. For high flux membranes, such as microsieves, a fast accumulation and high concentration of non-permeating species near the membrane surface (boundary layer) may seriously hamper the permeation and separation performance of the membrane and should be avoided. The specific process conditions, types of membranes and modules, all influence what changes in flux and specificity are likely to occur and also what can be done to mitigate their effects by inducing a suitable flow pattern near the membrane, which causes back-mixing between the liquid near the membrane and the bulk liquid. This can be effected by e.g. turbulent cross flow, mixing elements, spacers, fluid vibrations, frequent back washing and/or periodic back pulsing. Operating cycles, which can include operations such as flow reversals, back pulses, and periodic cleaning regimes, are therefore commonly employed. Optimising how to do these is an emerging discipline in its own right.

Sophisticated theoretical, design modelling approaches can be applied to a membrane filtration process. These models often start with the full conservation of mass and momentum equations and are numerically integrated throughout the module. Such an approach will more accurately represent the real flow patterns and mass transfer resistances and driving forces that exist.

An important mass transfer resistance is formed by the hydrodynamic mass transfer boundary layer between the bulk solution and the first interface where separation can occur (analogous to the concentration (boundary) layer mentioned previously). The next mass transfer resistance may be formed by an adsorption layer from reversible equilibrium partitioning onto the surface of the membrane surface itself. This mass transfer resistance is sometimes extended with another term contributing for a fouling layer from irreversible adsorption or direct deposition of material on the membrane surface. Next the mass transfer properties through the membrane itself have to taken into account and finally the hydrodynamic mass transfer resistance between the permeate interface and the bulk permeate fluid has to be included.

4. MICROFILTRATION OF LAGER BEER

The Grolsche Bierbrouwerij Nederland and Aquamarijn Micro Filtration started collaboration since 1995 with the purpose to investigate the possibilities of replacing diatomaceous earth (kieselguhr) filtration of lager beer with microsieve filtration [31].

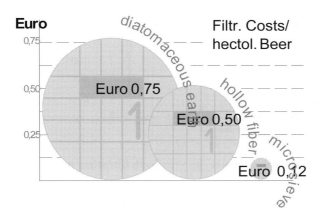

Fig. 26 Global cost comparison of yeast cell filtration methods.

Nowadays tubular ceramic or polymeric membrane systems are being developed as an alternative for diatomaceous earth filtration. Complications in using these type of filtration membranes are yeast cell clogging and protein adsorption leading to a fast flux decline and subsequent elaborate in-line cleaning procedures. A pilot plant was built in which the performance of the microsieves was tested. Clarification of lager beer is an important operation during the brewing process. Rough beer is filtered in order to eliminate yeast cells and colloidal particles responsible for haze. Common beer-filtration systems are based on kieselguhr. However, the exploitation costs of these systems are rather high.

Cross-flow microfiltration with polymeric or ceramic membranes may be an alternative. Several studies have been carried out, but often problems like poor permeate quality (i.e. high turbidity or protein and aroma retention) or insufficient fluxes are encountered [32,33]. Experiments at the Grolsch breweries by the author in 1995-1996 in which a microsieve was used for the filtration of beer [34] showed a permeate flux of $4 \cdot 10^3$ l/m^2/hr during a period of at least 5 hours without any increase in transmembrane pressure (see Fig. 27).

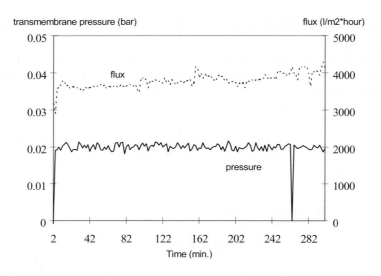

Fig. 27 Behaviour of flux and pressure in yeast-cell filtration of lager beer with a microsieve.

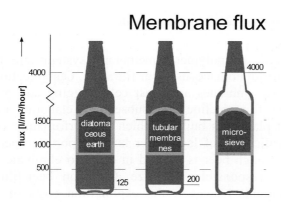

Fig. 28 Flux comparison of yeast cell filtration methods.

This flux is one to two orders of magnitude higher than typical fluxes obtained with diatomaceous earth or other membranes. In the experiments the formation of a cake layer was diminished by using a cross-flow configuration in combination with backpulse techniques and a transmembrane pressure of only 20 cm H_2O (0.02 bar). As the microsieve is made of an inert material it may be cleaned with aggressive chemicals or by steam sterilization. After cleaning the new permeate still has the tendency to foam. Many other membranes may give problems on this point, as it is difficult to remove cleaning agents from the large inner surface and from dead-end pores. An additional advantage is the absoluteness of the filtration: the uniform pores do not permit a single yeast cell to pass through the sieve.

Moreover, extensive cleaning procedures are required, as beer turns out to cause severe fouling [35,36,37]. Ceramic membranes have an advantage over polymeric membranes regarding fouling, as they can withstand harsh cleaning methods. However, the obtained fluxes are usually significantly lower. Ceramic membranes with a small flow resistance would therefore be highly desirable for beer filtration. Microsieves made with silicon micromachining consist of a thin micro-perforated silicon nitride membrane attached to a macro-perforated silicon support. The membrane thickness is of the order of the pore size, thus allowing high fluxes and relatively simple cleaning procedures. Moreover, the membrane is optically flat and smooth (surface roughness typically below 10 nm), which hampers adsorption of foulants. Furthermore, the pores are uniform in size and distribution, which may be important for the quality control.

Membrane fouling during filtration of lager beer with microsieves can well be studied through in-line microscopic observations. PhD student Kuiper [38] was willingly asked in 2000 by Aquamarijn to perform a number of experiments with lager beer from "The Grolsche Bierbrouwerij' at Enschede. All microsieves with a poresize less than 1.2 µm for this study were manufactured by Aquamarijn with thanks for the photolithography to Dimes at Delft.

It was observed that the main fouling was caused by micrometre-sized particles, presumably aggregated proteins and/or polysaccharides/cellulose's residues. These particles formed flocks covering parts of the membrane surface. Most of the flocks could be removed by a strong temporary increase in cross-flow. Underneath the flocks a permanent fouling layer was formed inside the pores. This made frequent removal of the flocks crucial in delaying the process of permanent in-pore fouling.

Besides the fouling process the influence of pore size on permeate flux and turbidity was investigated. Centrifuged beer appeared to give a significantly clearer permeate than rough beer. For centrifuged beer and a microsieve with a pore diameter of 0.55 µm a haze of 0.23 EBC was obtained during 10.5 hours of filtration at an average flux of $2.21 \cdot 10^3$ l/m^{2}hr. For a sieve with slit-shaped perforations of 0.70×3.0 µm^2 a haze of 0.46 EBC was obtained during 9 hours of filtration at an average flux of $1.43 \cdot 10^4$ l/m^{2}hr. This flux is about two orders of magnitude higher than is commonly obtained with membrane-filtration of lager beer. Concentration of the beer by a factor of 12 in a 3 hour run hardly influenced the magnitude of the flux.

After several hours of filtration an irreversible fouling layer prevented further filtration. Examination of the sieves showed a remarkable form of fouling. Carpet-like structures covered certain areas of the membrane surface. The origin of these structures was not known, but it was clear that they completely blocked the pores. The SEM micrograph in Fig. 29 shows such a local fouling layer.

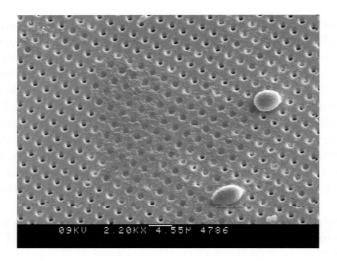

Fig. 29 Carpet-like fouling [39] of a microsieve observed after filtration of rough lager beer (Courtesy Aquamarijn Research).

Had the carpets slowly grown or were they formed in the system (for instance on the tube walls or in the centrifugal pump) and subsequently dropped on the sieve surface? The best way to investigate the origin of this peculiar form of fouling was in-line observation of the microsieve surface through a microscope. A set-up was build for such observations and also the dependence of flux and permeate haze on the pore size was investigated.

Experimental set-up [*][38]

Rig set-up

In order to prevent CO_2 from escaping, beer is normally filtered under pressurised conditions. However, to keep the set-up flexible, a rig was build that can only be used under atmospheric pressure. The escape of CO_2 will change the pH of the beer and may therewith influence the fouling process.

The rig mainly consists of silicone tubing and was designed for a constant-pressure filtration under cross-flow circulation of the feed. This constant

---

[*] Partially reprinted with permission of J. of Membrane Science.

pressure is obtained by a difference in height between the feed vessel (volume 2 l) and the permeate outlet. A schematic illustration of the rig is presented in Fig. 30.

Three pressure transducers (PT) are used to monitor the pressure drop over the cross-flow channel (ΔP_{chan}) and the transmembrane pressure (ΔP_{mem}). Furthermore a flow meter (FM) and a thermometer (TM) are used to monitor the cross-flow conditions. The upper part of the filter module consists of a glass plate with a 0.17 mm thickness, which allows for observation of the microsieve

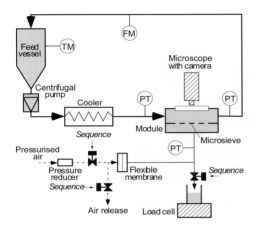

Fig. 30 Schematic illustration of the cross-flow rig used for the filtration of lager beer.

surface through a microscope. The microscope (Leica) is equipped with adjustable objectives, in order to correct for the distance that the light has to travel through the glass and beer. Backpulses are obtained by periodically pressurising the permeate via a dense flexible membrane. In this way no external liquid is added to the permeate. The three valves are actuated all at the same moment with the same signal. The permeate valve and the air-release valve are in a normally open position and the valve in the pressurised-air tube is normally closed. The amount of permeate is measured with a load cell connected to a computer.

Microsieves

For the experiments several small microsieves were fabricated with an area of 5.5×5.5 mm². The membranes contain circular pores with diameters of 0.55 µm, 0.80 µm and 1.5 µm. Furthermore two membranes with slits were used with a slit length/width ratio of 5. The widths of the slits are 0.70 µm and 1.5 µm. An SEM micrograph of such a microsieve with slits is shown in Fig. 31. The membrane thickness is 1.0 μm for the 1.5 μm pores and 0.8 μm for the other pores. The channel height (space between sieve and glass) is 1.0 mm, the channel width 10 mm and the length 9.0 mm.

Lager beer

Lager beer was obtained direct from the brewery (Grolsche Bierbrouwerij Enschede). The beer was taken from two different stages of the brewing process: just before and just after centrifugation. During centrifuging the yeast content decreases by several orders of magnitude and also some aggregated proteins are removed.

Fig. 31 SEM micrograph of the sieve with the largest slit-shaped perforations (1.5×7.5μm^2).

Experiments

The experiments can roughly be divided into three subjects: membrane fouling, permeate turbidity and permeate flux.

Fouling was studied by observing the sieve surface through the microscope, while varying the filtration conditions. For conditions where cake-layer formation could largely be prevented by the cross-flow and backpulses, permeate samples were collected and the turbidity was analysed at the brewery.

Finally, the microsieves that produced the clearest permeates were used for flux measurements in long-run experiments.

Unless otherwise specified all filtrations were performed at a temperature of 5°C, a cross-flow of 50 l/h (with a resulting pressure drop across the module of 0.030 bar) and an average transmembrane pressure of 0.15 bar. The backpulse pressure was -0.05 bar and the pulse duration 0.05 s. The pulse interval was varied, depending on the rate of pore obstruction, but was usually of the order of seconds.

Microscope observations: results and discussion

Yeast cells

For rough beer and a transmembrane pressure of 0.15 bar all microsieves were covered by a monolayer of yeast cells within a fraction of a second. This monolayer caused a flux decline by approximately one order of magnitude. During a backpulse it was observed that all yeast cells were removed from the surface.

When lowering the transmembrane pressure, the rate of pore obstruction declined fast (faster than the decline in pressure). Whereas initially the yeast cells arrived randomly at the surface, they showed a remarkable obstruction mechanism at lower pressures. Once a yeast cell was trapped, other cells were trapped in the upstream direction of this cell. An avalanche effect occurred an a monolayer of yeast cells grew in upstream direction. Fig. 32 shows this effect in a series of frames captured from videotape recorded during filtration.

On the open areas it was observed that yeast cells were trapped on a pore, but dragged away by the cross-flow a fraction of a second after arrival. Apparently the transmembrane pressure was not large enough to keep the cells trapped. This phenomenon has been theoretically described by De Balmann et al.[40]. However, beer also contains other –smaller– particles like protein aggregates and cell fragments. These particles were often not dragged away and formed an obstruction for the yeast cells so that the avalanche effect could start. For even smaller transmembrane pressures the avalanche effect no longer occurred.

Formation of flocks

When centrifuged beer was used it could be observed that many small particles (of the order of the pore size) were trapped on the pores. Like the yeast cells these particles could be removed with a backpulse. However, very few particles were not removed. They seemed to be attached to the membrane by invisible 'wires' with a length of approximately 1 µm.

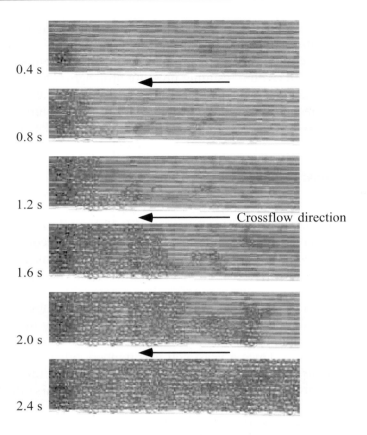

0.4 s

0.8 s

1.2 s

Crossflow direction

1.6 s

2.0 s

2.4 s

Fig. 32 Different stages of the pore-obstruction process. The numbers indicated denote the time that has passed since the last backpulse. The cross-flow direction is from the right to the left.

The wires were stuck in the pores or on the surface between the pores. The particles obstructed the pores, but during a backpulse they were lifted off the surface. After the pulse they immediately obstructed the same pores again. Such particles appeared to be able to catch other particles and after a certain time (which varied from minutes to hours) the stuck particles had gathered a flock-like structure around them that was largely lifted off the surface during each backpulse. Especially along the edges of the membrane fields the flocks were numerous. They were usually attached at only a few points of the membrane surface. Fane [41] reported a similar fouling phenomenon during filtration with a Whatman Anopore membrane. He used 'DOTM' (Direct Observation Through the Membrane) and observed that the flocks grow by accumulating other flocks. In cross-flow the flocks reached critical size and then detached due to increased axial drag. Here a similar behaviour was observed, although the detachment is more an exception than a rule. Banplain et al. [42] did not use a direct observation method in their study of fouling mechanisms, but nevertheless arrived at similar conclusions. Comparing permeate fluxes with classical filtration models, they concluded that the main phenomenon limiting the filtration of beer is the formation of aggregates of colloids that can form bridges over the pores by a mechanism of dendrite build-up. Such a dendrite build-up is confirmed by our observations of 'invisible wires' connecting particles to the surface. Interesting is the fact that they filtered a clarified (kieselguhr-filtered) beer, but still found that on-pore fouling has a stronger influence on flow resistance than in-pore fouling.

The formation of flocks occurred for centrifuged beer as well as for rough beer. Fig. 33 shows a close-up photograph of a flock on a membrane with 1.5×7.5 μm^2 slits during filtration of rough beer. The pictures were captured from videotape and represent the situation just before and just after a backpulse. The yeast cells are all removed, but the flock (it is hanging loosely over the unperforated area) remains. The flocks may be composed of chill-haze proteins, which are large proteinaceous colloids formed at low temperature through the aggregation of hydrophilic proteins with a phenolic substance as the coagulating agent [43]. Protein aggregates may also be formed under influence of shear stress in the pump.

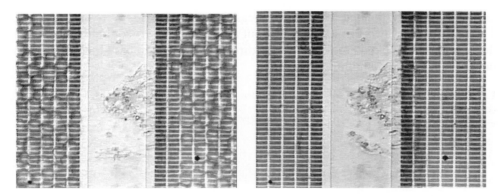

Fig. 33 Loosely-attached flock on a membrane with 1.5×7.5 μm^2 slits during filtration of rough lager beer. The picture on the left shows the situation during filtration and the picture on the right immediately after a backpulse. Cross-flow direction is from the right to the left.

Xu-Jiang et al.[44] show that the type of pump plays an important role in the aggregation of protein. They suggest that high shear stress in the pumps causes denaturation of proteins, which can subsequently form aggregates. This might be an explanation for the observation that during the run the concentration of particles increased stronger than might be expected from concentrating the feed.

As detachment of the flocks sometimes occurred under influence of the cross-flow-drag force, it should be possible to exploit this effect by applying a stronger cross-flow. A successful method turned out to be a short (several seconds) 'cross-flow boost' to 130 l/h (compared to 50 l/h under normal filtration conditions). In combination with gas sparging this method gave even better results. Nearly all flocks could be removed. The method only works if the permeate flow is temporarily stopped, so that the flocks are not pushed onto the membrane by the transmembrane pressure. Periodic stopping of the permeate flow in combination with air bubbling was earlier described by Tanaka et al. [45]. Using this method they found a significant increase in flux for a suspension of baker's yeast.

In-pore fouling

As mentioned before, the microsieves consist of several rectangular membrane fields. Microsieves were constructed with half of these fields placed perpendicular and the other half parallel to the cross-flow. Besides the formation of flocks on top of the pores, both kinds of fields showed an irreversible fouling that began on the downstream side of each field. The fouling layer slowly grew in the upstream direction. A consequence was that the perpendicular fields were largely blocked at the end of a run, whereas the parallel fields were largely open. The fouling layer was not well visible, but it could be observed that it was inside the pores. The blocked areas could be indirectly observed in case there was a gas bubble behind the membrane field.

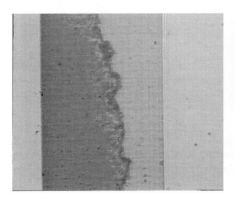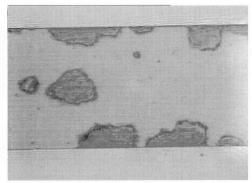

Fig. 34 Permanently blocked pores (the dark regions) made visible by a gas bubble underneath the membrane during a backpulse. The cross-flow direction is from right to left.

During a backpulse the clean areas lightened up as the bubble touched the membrane, whereas the fouled areas remained dark because the permeate could not be pushed through the pores and hence the bubble could not touch the membrane. Another indirect way to see the fouled areas was the capturing of particles. The fouled areas did not capture any particles. Fig. 34 shows two video frames recorded during a backpulse. The dark spots indicate the place of the fouling layer. During filtration these spots were hardly visible.

On some spots also the fields parallel to the cross-flow suffered from the in-pore fouling. These were exactly the spots where flocks were observed earlier. Apparently, underneath these flocks a permanent fouling layer can grow and therefore attention should be paid to detach them frequently. The process of a permanently fouled area that slowly grows in up-stream direction appeared for all microsieves, regardless of pore size and shape.

Protein precipitation

During filtration the unperforated areas of the microsieve remained clean. The beer components did not show a visible tendency to adhere to the surface. However, when the feed was cooled down from 5°C to –1°C a severe precipitation of presumably proteins was observed. Small transparent particles (smaller than the pore size) precipitated in the pores and on the surface, herewith completely clogging the microsieve.

Heating up to the original temperature of 5°C made the layer disappear again. Precipitation could be prevented by leading the feed through a bypass along the sieve during cooling down, while closing the module. After the end temperature had been reached, the module was opened and precipitation on the microsieve was not observed.

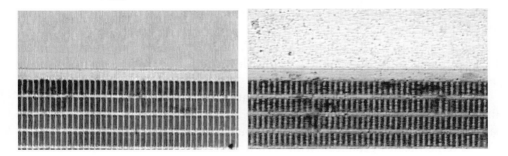

Fig. 35 Precipitation on the surface after cooling down from 5°C to –1°C.

Table 1

Haze of permeate and feed for rough beer and centrifuged beer.

	Rough beer *Haze (EBC)*	Centrifuged beer *Haze (EBC)*
Feed	27.6	1.24
1,5 µm slits	1.40	1.05
1.5 µm circles	1.13	0.89
0.80 µm circles	0.76	0.58
0.70 µm slits	0.71	0.47
0.55 µm circles	–	0.28

Permeate haze: results and discussion

Flocculation in the permeate

Permeate samples collected during the first experiments showed a severe flocculation within a day, which made the haze results dependent on the time passed between collection and measurement. In the brewery sometimes PVPP (Polyvinylpolypyrrolidone) is added prior to kieselguhr filtration to remove the polyphenols. Polyphenols are known to form aggregates with protein. Addition of PVPP (0.15 g/l) prior to the experiments appeared to solve the problem: aggregation of protein in the permeate was no longer observed. In all next runs PVPP was therefore added to the feed.

Haze values

For two batches of beer (rough beer and centrifuged beer) the permeate haze was determined. Samples of the feed collected before filtration were analysed as well.

The haze values of the feed show that centrifuging removes a large part of the particles responsible for haze. After filtration the centrifuged beer gives significantly lower haze values than the rough beer.

The permeates produced with 0.70 µm slits and 0.55 µm circles are below the haze limit demanded by the brewery (0.50 EBC). The value of 0.28 EBC for the 0.55 µm pores is even comparable to what the brewer commonly obtains after kieselguhr filtration. An SEM micrograph of this microsieve is given in Fig. 36.

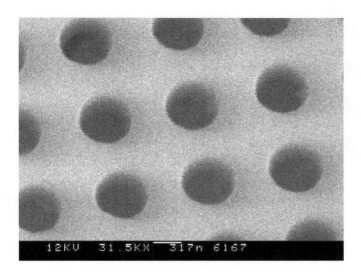

12KU   31.5KX   317n 6167

Fig. 36 SEM micrograph [46] of a microsieve with a 0.55 μm pore diameter (courtesy Aquamarijn Research).

Similar investigations for ceramic membranes (Ceramem Corporation) on the effect of pore size on permeate turbidity were reported by Burrell and Reed [47]. They filtered two commercial rough beers, using pore sizes of 0.5, 1.0 and 1.3 μm. The 0.5 μm membrane resulted in 'exceptionally bright filtrates', typically 0.4 EBC and never above 0.65 EBC. The 1 μm membrane resulted in a haze between 0.55 and 0.75 EBC and the 1.3 μm membrane between 0.6 and 3 EBC.

Permeate flux

Experiments

Under the microscope it could be observed that yeast cells quickly obstructed the pores, thus causing a fast flux decline. In order to prevent this, low transmembrane pressures, high cross-flow velocities or high backpulse frequencies are necessary. As the permeate of the centrifuged beer was significantly clearer than that of the rough beer, and as the permeate flux of centrifuged beer will be significantly higher, it was decided to use centrifuged beer for the flux measurements. The microsieves with 1.5 μm pores were no

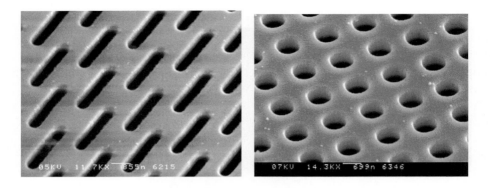

Fig. 37 SEM micrographs of a microsieve with 0.70×3.0 μm² slits and a sieve with 0.80 μm circles [46].

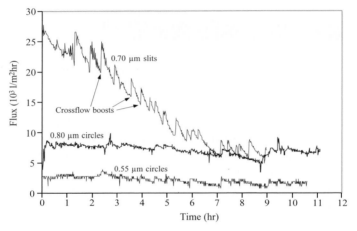

Fig. 38 Flux behaviour for filtration of centrifuged lager beer with three different microsieves.

longer used, as the produced permeates were not much clearer than the feed.

The experimental conditions were chosen as specified in former paragraph, and the backpulse interval was set at 1.0 s. During the filtration runs several cross-flow boosts were carried out in order to remove the formed flocks.

Results and discussion

Fig. 38 shows the flux results of three different sieves over a period of approximately 10 hours.

The graph shows that the differences in fluxes between the three microsieves are quite large. The lowest flux was obtained with the 0.55 μm pores: $2.21 \cdot 10^3$ l/m²hr over a period of 10.5 hours. This is still more than an order of magnitude larger than is commonly obtained for membrane filtration of lager beer. The microsieve with slits produced the highest flux, but the rate of fouling was significantly larger than for the microsieves with circular pores. For the microsieve with slits it was difficult to remove the flocks with a cross-flow boost: the peaks in the graph show that the flux after a cross-flow boost did not reach the flux after the previous boost. The flocks appeared to be strongly attached to the membrane. For the 0.80 μm pores it was easier to detach the flocks and for 0.55 μm only half of the cross-flow boost was sufficient to remove the flocks.

The feed-vessel volume of 2 L was not sufficient for the microsieves with 0.70 and 0.80 μm pores. When the vessel was nearly empty, fresh beer was added during the experiments. A significant change in flux was not observed.

In order to be able to make a fair comparison between the different microsieves regarding flux decline, the fluxes should be plotted as a function of permeate volume rather than time.

The flux decline of the microsieve with slits looks less severe in comparison with the other microsieves. The horizontal axis represents the volume of beer that passed through the microsieve, which is a better measure for the fouling probability than time.

Besides permeate flux, the water flux of the sieves was measured before each filtration. The results are listed in Table 2, together with some other relevant results of the three long-run experiments.

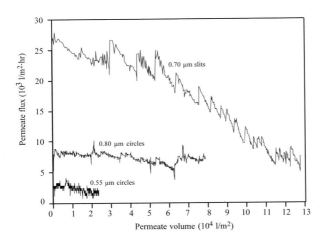

Fig. 39 Permeate flux plotted against accumulative permeate volume per square metre of sieve area.

For the 0.70 μm slits and 0.80 μm circles it appears that the initial beer flux is approximately a factor of 10 smaller than the water flux. This large difference has several causes. The viscosity of the beer at 5°C is typically 3 Pa·s, which is 3 times larger than the viscosity of water at 20°C. Another cause for the large difference is due to the backpulses that occur every second (water fluxes were measured without back pulses). Finally, in-between two backpulses the membrane is partially blocked with particles like aggregated proteins. This is one of the reasons for the even larger difference (a factor of 26) between water and beer fluxes for the microsieve with 0.55 µm pores.

Table 2

Results of three long-run experiments.

	Porosity (%)	Average beer flux ($l/m^2 \cdot hr$)	Initial beer flux ($l/m^2 \cdot bar \cdot hr$)	Water flux at 20°C ($l/m^2 \cdot bar \cdot hr$)	Permeate haze (EBC)
0.70 µm slits	31	$14.3 \cdot 10^3$	$18 \cdot 10^4$	$18 \cdot 10^5$	0.46
0.80 µm circles	22	$7.24 \cdot 10^3$	$5.6 \cdot 10^4$	$5.8 \cdot 10^5$	0.58
0.55 µm circles	24	$2.21 \cdot 10^3$	$1.8 \cdot 10^4$	$4.7 \cdot 10^5$	0.23

The smaller pores retain more particles and will thus cause a faster flux decline in-between two backpulses. It may therefore be effective to increase the backpulse frequency for the sieve with 0.55 µm pores.

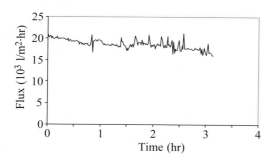

Fig. 40 Concentration of lager beer with 0.70 µm slits. The concentration factor at the end of the run was 12.

Concentration factor

During the runs the permeate was not recycled. This implies that the beer was concentrated during filtration. Due to the large dead volume of the set-up the maximum concentration factor was approximately 4. In order to investigate the flux behaviour as a result of increasing concentration, the tubes and heat exchanger were replaced by smaller ones, thus obtaining a smaller dead volume.

A concentration experiment was carried out with the 0.70 μm slits for almost identical conditions as before. The only adjustment was (besides the smaller dead volume) the backpulse period. It was decreased by a factor of 2 to 0.5 s in order to anticipate the expected increase in pore-blocking rate.

After concentration by a factor of 12 the decrease in flux was only 13 %. The average flux over a period of 3.5 hour was $18.3 \cdot 10^3$ l/m^2 hr. Combined with the cross-flow of 50 l/h this results in an average ratio of permeate flux over cross-flow of 1.1%.

Some conclusions

A cross-flow microfiltration rig was built in order to study fouling of microsieves through in-line microscope observations. The fouling process started with the formation of loosely attached flocks on the surface, gradually followed by in-pore fouling underneath these flocks. Strong attachment of the flocks to the sieve surface was prevented by applying a periodic backpulse. Most of the flocks could be removed by a strong temporary increase of the cross-flow, if necessary in combination with gas sparging. Using this method, filtration intervals of approximately 10 hours were achieved with average fluxes of more than two orders of magnitude higher than is commonly obtained with membrane filtration. Using a microsieve with a 0.55 μm pore size a permeate haze of 0.23 EBC was obtained during 10.5 hours of filtration at an average flux of $2.21 \cdot 10^3$ l/m^{2}hr. A microsieve with slits of 0.70×3.0 μm^2 produced a less clear permeate (0.46 EBC), but the average flux over 9 hours was huge: 14.3×10^3 l/m^{2}hr. In another run over 3 hours the feed was concentrated by a factor of 12, while the permeate flux decreased by only 13%.

A good temperature control appeared to be an important factor in keeping the sieves clean. Cooling down of the beer in the rig should be avoided, as this led to precipitation of presumably protein on the surface and inside the pores.

The experiments were performed on small (0.3 cm^2) microsieves. Scaling up will lead to larger pressure drops across the cross-flow channel. This problem may be avoided by dividing the channel in several short channels with a spacer.

Scaling up

It has been shown that on a 0.3 cm^2 microsieve the fouling process may largely be controlled by periodic backpulses and cross-flow boosts. During such boosts the pressure drop across the module rises from 0.03 bar to approximately 0.3 bar. For scaling up it is likely that 6 inch wafers [48] will be used for the microsieve production. The channel length will then increase by a factor of 17 compared to the sieves that were used in this work. For such long channels a cross-flow boost would create a pressure drop of approximately 5 bar, which will cause a high transmembrane pressure at the inlet of the channel, thus hindering detachment of the flocks. It may therefore be necessary to divide the cross-flow channel into several short parallel channels with the help of a spacer that is placed above the microsieve. In this way the channel height can remain low.

Chemical and enzymatic cleaning

After filtration part of the fouling layer could be removed with warm water. Addition of standard enzymes membrane-cleaning agents removed even more of the layer, although the results varied quite strongly. With harsh chemical-cleaning methods it was possible to restore the original water flux, but the microsieves had to be removed from the rig fur such cleaning in order to protect the rig.

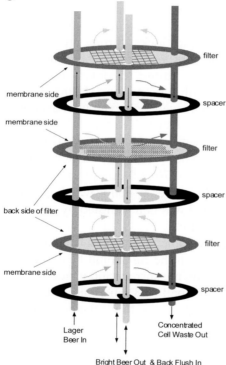

Fig. 41 Filtration stack of a number of microsieves and spacers [49].

5. MICRO FILTRATION AND FRACTIONATION OF MILK

Introduction[*] [50]

Milk is used as a starting material for many dairy products. The functionality of the various components in milk (e.g. serum proteins, casein and fat) could be utilized more effectively if they were available as separate components. Therefore, fractionation of milk is of interest, not only for improvement of product quality but also for economic reasons.

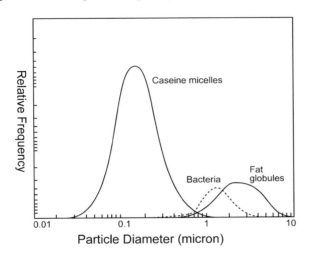

Fig. 42 Particle size distribution in an averaged quantity of milk [52].

For the production of a milk product with a good shelf-life and a good taste, it is necessary to seek a balance between the desire to provide a product with the most favourable possible bacteriological quality which is at the same time not adversely affected by necessary heat treatment [51].

Microfiltration is often applied to sterilize liquids, for instance in the pharmaceutical industry. Filters are generally applied for this purpose with a pore size smaller than or equal to 0.22 - 0.45 µm. Casein micelles are on average about six times smaller than micro-organisms, but for both components there is a size distribution of considerable width, a number of large casein micelles are larger than the smallest bacteria. Maubois [52] has done experiments showing the occurrence of the different particles in milk in accordance with size.

The types of membrane applied for microfiltering of milk can generally be characterised as membranes with a determined (broad) pore size distribution, e.g. ceramic sintered membranes and a thickness that is a multiple of and up to

---

[*] Contribution of Gerben Brans, Food Process Technology, University of Wageningen (2002).

even hundreds of times larger than the average pore diameter. There is no direct relation between the bacterial retention and the average pore size of the membrane. During microfiltration of milk a high degree (log 2-3 reduction) of removal of bacteria can be achieved with a ceramic membrane with a mean pore size of 1.4 µm while it is known that barely more than half the bacteria are smaller than 1.4 µm. The retention of the bacteria greatly depends, among other things, on the process of the build up of a cake layer (milk proteins, micelles, fat globules etc.) forming the effective (dynamic) membrane layer covering the ceramic membrane layer itself. In addition, the retention of the bacteria of coarse also depends on the morphology of the membrane filters, such as the tortuousity of the pores. Said factors determine the degree of contamination of the filter during performing of the process.

Table 3

Typical composition of Dutch cow milk [53].

	% of whole milk	% of dry matter
Water	86	0
Total protein	3.5	25
Casein	2.6	17
Serum proteins	0.8	5
Fat	4.0	29
Lactose	4.7	34
Ash/salt	0.8	6
Rest	1.8	

This makes the application of microfiltration for removing bacteria complex, because it is necessary to take into account this dependence on the bacteria retention in the choice of the membrane filters and the filtering process conditions [54]. In order to realise sufficient removal of bacteria, operations are usually carried out, in respect of the relatively low selectivity, under conditions wherein a part of the casein is retained by the microfilter. Because a balance has to be found between product optimisation (the fewest possible bacteria) and cost minimisation (the highest possible flux of the liquid flow), the permeate yield of the microfiltration process is usually limited in this respect to less than or equal to 95%. In this process the fat-rich fraction of the milk is moreover pre-separated from the milk and treated separately. In the described situation microfiltration of milk with the fat would only provide a worse result through rapid contamination of the filter.

A process for preparing sterile milk with membranes with a pore size between 0.05 and 0.2 µm may also be obtained [55] by keeping the largest part of the casein in the retentate. This fat-rich and casein-rich retentate is subjected to a heating at high temperature and then added to the permeate. A great drawback of this process is that, in addition to the fat, the largest part of the casein in milk is also retained by the microfilter. This has the result that a large part of the milk still has to be sterilised by means of another sterilisation technique, usually high heating, which causes impairment of the taste. In addition, the relatively low flux results in considerably higher costs.

Membranes seem to be an obvious choice for the fractionation of milk [56]. However, the current selectivity and efficiency of membranes are not sufficient to fractionate milk on a commercial scale. The development of membranes with low resistance and defined uniform pore size (microsieves) is an improvement that could make the fractionation of milk feasible.

Milk can be regarded as a diluted emulsion of fat globules in milkplasma. The milkplasma consists of suspended and dissolved particles and water. The typical composition of Dutch cow milk is outlined in
Table 3.

The composition of milk is subject to change and varies per cow (race, age, stadium of lactation), the season, climate and feed. Milk cannot be regarded as a static system; it consists of many equilibriums and also irreversible changes take place. The temperature influences the concentration of dissolved ions, the pH, and the size of casein micelles.

Fat globules

The fat consisting of triacyl glycerol's is mainly located in the fat globules. The fatty acid composition is very diverse regarding chain length and degree of saturation. This gives milk fat its specific flavour and mouth feel. At room temperature, part of the fat is liquid, while the other part is solid. A membrane that contains protein, phosphatides, and glycerides surround the fat globules. The membrane stabilizes the globule and resembles the cell membrane of the milk-producing cell. Glycoproteins in the membrane cause steric hindrance and prevent cluttering of the globules. In experiments it is important to know the storage history of the milk, because cooling, heating, agitation and aeration can affect the physical properties of the fat and the milk.

The diameter of the fat globules in raw milk is between 0.1 and 15 µm, with an average of 3.4 µm. After homogenisation of the milk the average fat globule size is 0.6 µm and has a narrower distribution. Casein and serum proteins are thought to help the stabilization of the enlarged surface area of the homogenized fat globules.

Casein micelles

The protein in the micelles accounts for 2.6% of the milk. The complexes consist of sub-micelles containing casein protein, calcium, phosphate and water. On the outside, they have a "hairy" surface, caused by the tails of the κ casein.

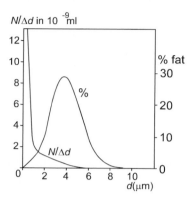

Fig. 43 The size distribution of milk fat globules in untreated milk

The diameter of the micelles is between 20 and 300 nm with an average between 110-120 nm. The size of the micelles depends on the calcium concentration and temperature.

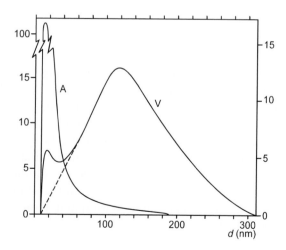

Fig. 44 The size distribution of the casein micelles in untreated milk.

Casein micelles contain α_{s1} α_{s2}, β, and κ casein. During the cheese-manufacturing process, the negatively charged κ casein is removed and the micelles will coagulate (the renneting process). The molar mass of the individual casein molecules is 20-25 kD.

Serum proteins and other N containing compounds

The serum proteins can be divided in: α lactalbumin and β lactoglobulin, BSA, proteose-pepton, immunoglobulines, lactoferrine, transferrine, and traces of enzymes (e.g. lactoperoxidase, katalase, lipase). The group of other N containing compounds includes peptides, amino acids, urea, and ammonia.

The fraction organic acids mainly consist of citrate, formiate, acetate, and lactate. Also, some free fatty acids are present in milk.

Lactose and salts

Lactose is the major sugar in milk. Other sugars (e.g. glucose) are only available in trace amounts. Various ions are present in milk: Ca^{2+}, Mg^{2+}, Na^+, K^+, Cl^-, PO_4^{3-}, SO_4^{2-}, CO_3^{2-}. The ion strength in milk is approximately 0.075 M.

The rest fraction consists of vitamins, carotenoids, dissolved air, sterols. Furthermore, some undesired components can be present in milk, like somatic cells, micro-organisms or antibiotics.

Table 4

Milk serum proteins.

Serum protein	Mol weight kD	Diameter (nm)	Amino acids	Conc in milk (g/l)
α lactalbumin	14	3.8	123	1.2
β lactoglobulin	18	5.2	162	3.2
BSA	66	6.4	582	0.4
immunoglobulines	150-900	10-30		0.8
Proteose pepton	4-40			0.8
lactoferrine	86	6.8		0.1
transferrine	76			0.1

Membranes in the dairy industry

Daufin and co-workers [57] recently review application of membranes in the food and dairy industry. Membranes can be used in the following processes:

Separation of fat globules, from whole milk or cream (microfiltration)
Reduction of bacteria and spores in skim milk (microfiltration)

Concentration of casein micelles prior to cheese making (ultrafiltration)
Production of native casein protein isolates (ultrafiltration)
Fat removal from cheese whey (ultrafiltration)
Purification of serum (whey) proteins by ultrafiltration and dialysis
Recovery of lactose by nanofiltration
Brine purification in cheese processing (cleaning step in the brine circulation)

Separation of fat globules

Goudedranche et al. described the separation of milk fat globules from raw milk. They used a 2 μm ceramic membrane to separate the fat phase. The fat fractions with a globule size above and below 2 μm were evaluated on texture and sensory properties in different consumer products against the reference cream. Application of small fat globules led to products with finer textural characteristics compared to products with the large size fat globules and the reference cream.

Removal of bacteria from skim milk ("cold pasteurisation")

Microfiltration can reduce the amount of viable bacteria at lower energy costs than a heat treatment without affecting the taste of the milk due to heating. A commercial process, called the bactocatch system is already in use. To reduce fouling high cross flow velocities are used (typically 6 to 8 m/s). It is possible to filter for at least 6 hours at a permeate flux of $1.4*10^{-4}$ m/s at a concentration factor of 10. By use of a reversed asymmetric ceramic membranes (0.87 μm) and optimising the back-pulsing system, Guerra and co-workers [58] could achieve the same performance at lower cross flow velocities (0.5-1 m/s). Their experiments resulted in a bacterial spore reduction by a factor of 10^4-10^5 and a 100% casein transmission.

Concentration of casein micelles

The native calcium-casein-phosphate complex can be concentrated by a membrane with smaller pore size (typically in the range 0.1-0.2 μm). Using an 0.2 μm ceramic membrane, Vadi [59] found a maximum flux of $1.5*10^{-5}$ m/s at a cross flow velocity of 7.1 m/s and concentration factor of 10. At a cross flow velocity of 6.9 m/s and concentration factor of 3, Pouliot et al. [60]found a flux of $2.5*10^{-5}$ m/s through a 0.22 μm membrane. This is in agreement with the results of Vadi who found fluxes of $2.4*10^{-5}$ and $2.0*10^{-5}$ m/s at concentration factors of 2 and 4. Le Berre and co-workers [61] experimented with a 0.1 μm ceramic membrane and found an optimum flux of $2.1*10^{-5}$ m/s at a concentration factor of 2. Recombination of milk fat and casein for the production of cheese will give some increased yield for cheese, and will improve the rennetting process compared to 'regular' milk.

Recovery of serum proteins

Whey is a side-product of the cheese industry containing water, serum proteins, lactose and salts. Whey can be concentrated with reverse osmosis as an alternative to evaporation. To purify the proteins, whey can be treated with ultrafiltration and dialysis to obtain whey protein concentrate and isolate. Milk serum protein is a highly nutritional protein source for humans, that can be used in infant foods, beverages and other food products. The serum proteins have useful physio-chemical functional properties like emulsification, foaming and gelling [62,57].

Fractionation of milk

The separation of milk components with multiple membrane steps or a filtration stack is not often described in literature. Kelly and co-workers [63] described two membrane steps to separate the calcium-casein-phosphate complexes from skim milk (0.1 μm) and the whey proteins from whey with a 5 kD MWCO membrane, but did not run the separations simultaneous. A three-step separation of skim milk was accomplished by Surowka [64]. They used polysulfone hollow fibre modules with molecular weight cut off's (MWCO's) of 10, 30 and 100 kDalton (kD), respectively. The major part of the proteins was retained by the 100 kD membrane (casein micelles), whilst the serum proteins were recovered by the 10 and 30 kD membranes. In general, membrane processes for milk have a rather low capacity due to either flux decline by fouling [65] or unfavourable process conditions that may prevent fouling.

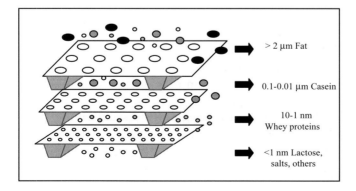

Fig. 45 Microfractionating scheme of milk with microsieves [66].

Such conditions are high cross-flow velocity and a low transmembrane pressure. The fractionation of milk with multiple membrane steps or a filtration stack is hardly ever described in literature.

Membrane theory:

When considering filtration processes, two parameters are of importance, namely the flux and the retention coefficient. The flux is a measure for the amount of product that can be obtained and determines the capacity of a membrane rig. The retention coefficient is a measure for the selectivity of the process. Many models for membrane processes are available in literature and the most important are outlined here.

Membrane retention (exclusion principle):

$$R = \begin{cases} \left(\dfrac{r}{r_p}\right)^2 & \textbf{When } r < r_p \\[2ex] 1 & \textbf{When } r > r_p \end{cases} \tag{6}$$

R: retention
r: particle radius
r_p: the pore radius

$R=1$ means that all particles are retained, while $R=0$ means that all particles pass through the membrane. This relation is valid for non-deformable rigid spheres. Deformable particles will generally have a lower retention. In practice, the membrane will not have one pore size, but a pore size distribution. If the membrane has a wide pore size distribution, the separation of particles is less sharp and the specificity is expected to be low.

Flux equations:

$$J = \frac{\Delta P - \Delta \Pi}{\eta (R_m + R_c)} \qquad \textbf{Darcy's law} \tag{7}$$

R_m and R_c are the resistance of the membrane and the cake layer, respectively
ΔP: pressure drop across the membrane
η: viscosity
J: permeate flux
$\Delta \Pi$: osmotic pressure difference

In micro filtration, the osmotic pressure difference is often negligibly small compared to ΔP.

If R_c is negligible, the flux can be described with the Hagen-Poiseuille equation for flow through uniform cylindrical pores:

$$J = \frac{n_p \pi r_p^4 \Delta P}{8 \eta l}$$
(8)

n_p: number of pores per unit area
r_p: pore radius
l: membrane thickness

The Hagen Poiseuille equation can be used to predict the clean water flux through the membrane. For microsieves, the orifice effects and the co-operative effect of pores in the neighbourhood must be taken into account. In practice, cake formation and concentration polarization will limit the flux. Two models, a cake model and a gel layer model are often used to describe these effects.

The cake resistance R_c can be calculated, if the cake consists of a packed bed of uniform rigid spherical particles with a certain height.

$$R_c = \frac{5(1-\varepsilon_c)^2 S_c^2 \delta}{\varepsilon_c^{\,3}}$$ **Kozeny-Carman equation**   (9)

ε_c: porosity of the cake (void volume/total volume)
δ: cake layer thickness
S_c: cake solids surface area/cake solids volume (for rigid spheres this is 3/r)

An extra factor can be introduced to account for compressibility of the cake. Although the cake layer model has been applied often, it is mostly difficult to use, because the cake height cannot be measured or accurately estimated.

Another approach is the classical gel layer model. It describes the flux as a function of the concentration, by assuming the build up of a concentration polarization layer. For a concentration below the gel concentration, the flux can be estimated:

$$J = k_m \ln\left(\frac{C_g}{C_b}\right)$$
(10)

and

$$Sh = a \, Re^b \, Sc^c$$

$Sh = k_m d_h / D$ = Sherwood number

Re= $\rho v d_h/\eta$ = Reynolds number
Sc= $\eta/\rho D$ = Schmidt number
a, b, c :   constants depending on the flow regime
k_m: mass transfer coefficient (=D/δ)
C_g: gel concentration
C_b: bulk concentration
D: diffusion coefficient (can be calculated with Stokes-Einstein equation)
δ: thickness of concentration polarization layer
d_h: hydraulic diameter
v: cross flow velocity

Stokes-Einstein relation for the diffusion coefficient:

$$D = \frac{k_b T}{6\pi\eta r} \tag{11}$$

k_b :   Boltzmann constant
T :   absolute temperature
r :   particle radius

Albeit helpful for a first evaluation, the cake layer model and the gel layer model are not suited to describe some effects that were observed in practice. The observed flux is generally higher than predicted by the models. This is caused by back transport mechanisms that reduce the formation of a cake or gel layer. Belfort et al. [67] describe several back transport mechanisms. The main back transport mechanism is Brownian diffusion for particles up to 1 µm, shear-induced-diffusion for particles between 1 and 10 µm and inertial lift for larger particles. This is valid for particles that are non-deformable rigid spheres.

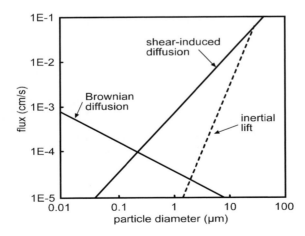

Fig. 46 Estimated fluxes for different particle sizes with different back-transport mechanisms.

Brownian diffusion is the diffusion of the particle itself. Shear-induced diffusion is caused by interactions of particles that undergo random displacements from the streamlines in a shear field. Inertial lift arises from non-linear interactions of a particle with the surrounding flow field.

For the various back transport mechanisms, the following equations are valid. Brownian diffusion:

$$\langle J \rangle = 0.81 \left(\dot{\gamma_0} D^2 / L \right)^{1/3} \ln(\Phi_w / \Phi_b) \tag{12}$$

Shear-induced-diffusion:

$$\langle J \rangle = 0.078 \dot{\gamma_0} \left(r^4 / L \right)^{1/3} \ln(\Phi_w / \Phi_b) \tag{13}$$

With shear induced diffusivity (cf. Eq.(31)) of spherical particles in sheared suspensions:

$$D = 0.33 \dot{\gamma} r^2 \Phi^2 \left(1 + 0.5 e^{8.8\Phi} \right) \tag{14}$$

Predicted flux with inertial lift:

$$\langle J \rangle = 0.036 r^3 \rho \dot{\gamma_0}^2 / \eta \tag{15}$$

$\langle J \rangle$:   averaged flux over membrane length (L)

$\dot{\gamma}$, $\dot{\gamma_0}$:   local shear rate, and overall shear rate, respectively

Φ, Φ_w, Φ_b :   total volume fraction, volume fraction in the boundary layer (w), and volume fraction bulk phase (b),

ρ :   density

The overall shear rate can be obtained via the wall shear stress via Blasius' law (laminar-turbulent):

$$\dot{\gamma_0} = \tau_w / \eta \qquad\qquad \tau_w = f \rho v^2 / 2 \qquad\qquad f = 0.0792 \, \mathrm{Re}^{-0.25} \tag{16}$$

τ_w :   wall shear stress
f :   friction coefficient

And under laminar conditions via:

$\dot{\gamma_0} = 8v / d$ tubes with diameter d,

$$\dot{\gamma_0} = 6v/h \qquad \text{rectangular channels with height } h$$

For milk, most particles are in the shear-induced-diffusion region. The back transport during filtration of fat globules and casein micelles can be described with shear induced diffusion, while serum proteins and lactose lie in the region where back transport can be described with Brownian diffusion.

It is not clear if there are interactions between different particles in milk (e.g. between the fat globules and casein micelles). If present, these effects may hinder separation during microfiltration. Another uncertainty is the effect of poly-dispersity. It is difficult to predict how each particle size influences the flux and how the cake layer or gel layer will develop in time with different particle sizes present.

In experimental data from literature, the fluxes of skim milk seem rather low. However, the fluxes seem reasonable when compared to predicted fluxes by the shear-induced-diffusion model. Note that the predicted fluxes are independent of the transmembrane pressure and the membrane resistance. Changing the membrane is expected to have little or no effect on the flux, since the layers on top of the membrane determine the flux. Possibilities to obtain higher fluxes are the use of coatings to reduce layer formation and a module design with fluid instabilities. Winzeler et al. [68] review different types of fluid instabilities. Instabilities can be created by pro-turbulences, corrugated membrane surface, inserts, pulsed cross flow, backpulsing or vortices.

Microsieves and Milk Microfiltration Experiments [69]

Microsieves have an optical flat and inert surface, which reduces risk of particle adsorption and facilitates cleaning. The low membrane thickness and high porosity enables a clean water flux that is between one and two orders of magnitude higher than for conventional membranes with similar pore size. To obtain a higher porosity, the shape of the pores can be changed to slits. Because the membrane is so well defined, the microsieve is an ideal system to validate models for micro filtration regarding fluxes and selectivity and optimise the channel, module, and process design by CFD calculations. Furthermore, the flat surface and low transmembrane pressure create a low shear field that enables the purification of shear-prone components. In the case of milk, these are the fat globules and native casein micelles.

Example 1

Skim milk is treated at 50 °C with a smooth (surface roughness of less than 100 nm) microsieve having a membrane layer with a pore size of 750 nm (STD 5%) and a thickness of 1000 nm. At a pressure of 25 mbar a flux is measured of

263 l/m2h (litres per m2 membrane surface area per hour). At a pressure of 37 mbar the flux increases to 630 l/m2h.

Example 2

Simulated Milk Ultrafiltrate (SMUF) as described by R. Jenness and J. Koops in Neth. Milk Dairy J. 16, 153 (1962), inoculated with Bacillus subtilis bacteria is treated in a dead end configuration, wherein the liquid speed tangentially of the membrane is zero, with a smooth filter with a pore size of 500 nm (STD 5%) and a thickness of 1000 nm. A decimal reduction in bacteria of 6.6 is measured.

Example 3

Skim milk is treated at 50°C with a smooth membrane filter having a filter layer with a pore size of 1200 nm (STD 5%) and a thickness of 1000 nm. At a pressure of 10 mbar a flux is measured of 3500 l/m2h.

Example 4

Skim milk is treated at 50°C with a smooth membrane filter having a filter layer with a pore size of 750 nm (STD 5%) and with a pore size of 1200 nm (STD 5%) at a transmembrane pressure of about 20 mbar. In neither case is a protein retention measured; a Kjeldahl determination of nitrogen gives a retention of <0.01.

The use of a microsieve with a thin membrane layer and pores of a precisely determined size produces a reliable filtration result when applied to milk, with an unexpectedly high selectivity between bacteria and casein micelles. This is surprising since the bacteria and the casein micelles differ little from each other in dimensions, and the size distributions even partially overlap each other (Fig. 42). In addition, higher flux levels are found than those reported for other known microfiltration devices.

When microfiltration with the above stated membrane filter is applied, it is possible to achieve a distinct separation between casein micelles on the one hand and bacteria and bacterial spores on the other. As is apparent from the dimensions of milk particles, a microsieve with a pore diameter of about 0.5 - 1.2 μm is preferably applied, wherein the casein micelles enter the permeate flow and the bacteria together with the fat globules enter the retentate flow.

In respect of the small differences in size between casein micelles and bacteria, it is important for a successful progression of the microfiltration process to achieve the clearest possible separation between said components. The use of a microsieve with pores of a precisely determined size is of great importance here. If these microsieves then have a very thin filter layer, a very smooth membrane surface and pores with a very small tortuousity, the total microfiltration process then displays an unexpectedly great improvement. A significant advantage is that the bacteria retention is practically constant during the filtration process when a determined membrane filter is applied, and is therefore no longer dependent on process history or process conditions and the like, since the pores of the membrane filter become much less contaminated than when a conventional microfilter is applied. Flux levels are herein attained which are higher than reported for known microfiltration devices. Calculations demonstrate that the flux of casein micelles increases still further when there is a further decrease in the filter layer thickness to for instance 500 nm or 200 nm. The bacteria are already retained before reaching the membrane, while casein micelles pass through the membrane filter relatively quickly, so that no accumulation of components occurs just prior to the membrane such that this exerts a strong influence on the effective retention of bacteria.

Further, the milk microfiltration process with the described membrane filters can surprisingly be carried out at very low transmembrane pressure. A maximum flux is usually already reached at a transmembrane pressure of less than 200 mbar. Under the transmembrane pressures applied in the examples the flux is negligibly small in conventional microfiltration membranes.

When the process of the present method is applied in the removal of bacteria in milk, the largest part of the casein in milk passes through the membrane and enters the permeate. The permeate yield can hereby be increased compared to the application of conventional microfiltration techniques, and only a small part of the milk (retentate and/or the cream) still has to be sterilised in other manner. In this way it is possible to prepare sterile milk with a considerably improved taste compared to sterile milk and UHT milk prepared in conventional manner, wherein a heating process of at least 1 second at 135 °C is applied as specified in the Dutch Commodities Act, Foodstuffs (B-1.3.1, schedule III, published 12-12-1999). It is also possible to prepare (highly) pasteurised milk with an improved shelf-life compared to pasteurises milk prepared in conventional manner, wherein a high temperature is applied for s short time (e.g. a minimum of 71.7 °C for 15 seconds), see above stated Dutch Commodities Act.

It is even found no longer necessary to skim the milk when the thin membrane filter as described above is applied. The fat particles are retained together with the bacteria in the retentate flow without this having an adverse effect on the permeation of casein micelles through the filter. In this way the permeate yield of the process is optimised. The fat content of the permeate flow can of course be brought to the desired level by mixing it with a desired quantity of sterile cream. The retentate could be used for this purpose after sterilisation by for instance heating.

6. MICROFILTRATION OF BLOOD-CELL CONCENTRATES

Introduction

In systems where cells should be separated from each other (cell-cell separation) a very thin perforated plate (microsieve) supported by a grid has the advantage that cells will remain only a very short time in the pores leading to a low activation, sticking or rupture of cells [70,71].

Because of its smoothness and biocompatibility the microsieve might be used for the filtration of white blood cells (leukocytes) from blood-cell concentrates. Experiments of Aquamarijn in collaboration with the Central Laboratory of Blood Transfusion Technology of the Red Cross at Amsterdam in 1996 showed no significant activation of blood platelets [72] (thrombocytes) or haemolysis of the red blood (erythrocyte) cells.

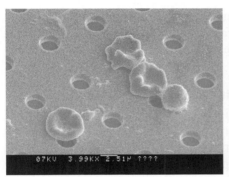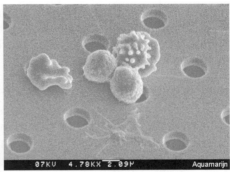

Fig. 47 SEM micrographs, red donut shaped blood cells and white blood cells retained on a microsieve. The samples were not prepared and obtained by drying to air, causing some deforms of the cells.

Table 5

Normal number values for different blood cells. Blood consists for 45% of blood cells, together called the hemacrit.

	Average size	Average number of cells/µl blood	Percentage of total white cells
Granulocytes: Neutrophils, Eosinophils, Basophils	8µm	5400 275 35	60 3 0.4
Monocytes	20µm	540	6
Lymphocytes	5-6µm	2750	31
Erythrocytes	Donutshape diameter, 8µm height, 2µm	$5 \cdot 10^{6}$	...
Platelets	2-4 µm	$3 \cdot 10^{5}$	...

The narrow pore-size distribution allows separation of cells that differ in size or deformability, for example erythrocytes and leukocytes. It is important that such a separation is done at very small transmembrane pressures, otherwise haemolysis will occur and/or leukocytes will be deformed and pushed through the pores.

Biocompatibility and surface smoothness

Depending on the specific application [73] bio or blood compatible materials should be used for the construction of the membrane. These materials can be thin films deposited on the used membrane surfaces. The thin films (like titanium, titaniumoxide, titanium nitride, siliconnitride etc.) result in very smooth surfaces (typical surface roughness <100nm) where no adhesion of proteins or rupture of cells will occur. Some of these materials are presently used in dental and chirurgical applications [74], e.g. as a non-corrosive dental coating or as a biocompatible film on pacemakers. The biocompatibility of these materials is strongly related to the low surface roughness [75] and their intrinsic hydrophilic nature, however for every different application the biocompatibility should still be tested.

Cell deformability, Retention and Transit Time of Blood Cells

Even at blood cell concentrations as high as 80% blood is still a liquid [71], due to the high deformability of the erythrocytes. Normal particle suspensions would have become solid at such a concentration. The high deformability of the erythrocyte is important for the cell to pass through narrow vessels in the circulatory. In fact the erythrocytes are selected in the spleen on deformability [76]. Here only the cells that can pass the narrow capillaries are returned to the circulatory. These capillaries have a diameter smaller than 1μm and a length of 3μm.

Because of this high deformability, the erythrocytes are pulled away from the wall when blood is moving in narrow vessels [71]. Thrombocytes being less, deformable, are pulled towards the wall when blood is moving. These properties are an advantage for their function. Erythrocytes should only be transported in the blood fluid, the transport being faster in the centre of the vessels. Thrombocytes should repair vessels, so they are best concentrated near the wall.

The erythrocyte deformability is mainly dependent on the viscoelastic properties of the membrane and on the difference in the viscosity of intra- and extracellular fluid. Blood cells (without cytoskeleton, such as the red blood cells and the thrombocytes) are like 'bags' filled with protoplasm. Due to the high deformability, sieving of the blood cells is not easy. However the cells can swell due to liquids like soap and afterwards the cell membrane can be fixed with glutaraldehyde. Cells treated with solutions of 0.011% to 4% glutaraldehyde show an increase in transit time [77]. An increase in transit time up to 100-fold has been found with a 2% solution. By this the cells will be less deformable, but they can't be put back in the body.

Many studies on the deformability of blood cells have been performed. Some illnesses like diabetes or anaemia give rise to a decrease of the deformability of the erythrocytes. The minimum pore diameter an erythrocyte can negotiate is just below 1μm. An erythrocyte will pass [78] through a 3μm pore with a pressure of only 1cm H_2O. Observed is also an influence of blood storage time on the deformability. This influence is dependent on temperature.

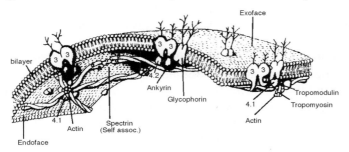

Fig. 48 A drawing of the lipid bilayer and the skeletal mesh work of a red blood cell membrane [79].

While at a storage temperature of 4°C only a small difference is observed, at a temperature of 25°C a more pronounced difference is found.

Generally the retention of cells is determined [77] by size and deformability. The size of the blood cells from large to small is given by:

Table 6

Comparison of the size and deformability of different blood cells.

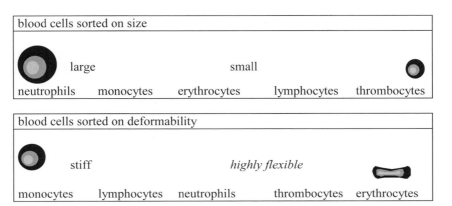

The transit time of polymorph nuclear leukocytes, like neutrophils, through cylindrical micropores with a diameter of 8μm and a length of 19μm has been measured [80] at a membrane pressure of 3cm H_2O. The mean transit time is at least 10-fold longer for the neutrophils than for the erythrocytes. A neutrophil will pass through a 5μm pore with 10cm H_2O with a transit time of seconds.

At lower pressures the transit time is even longer and the retention higher. At very low transmembrane pressure differences of 0-2.8cm H_2O retention of leukocytes in general will increase from 30% to 90%. A difference in filterability of granulocytes, monocytes and lymphocytes is observed.

Leukocyte depletion of erythrocyte and thrombocyte concentrates.

Nowadays advanced centrifugal techniques are used to separate whole blood in erythrocyte and thrombocyte concentrates. However both concentrates are still being contaminated with relatively large amounts of leukocytes.

These blood cell concentrates are normally being used for blood transfusion.

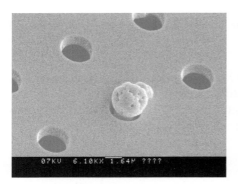 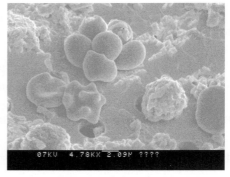

Fig. 49 SEM micrographs, Left: Passage of a granulocyte through a pore, Right: A number of lymphocytes accumulating before a pore entrance.

The person who receives the transfused blood often has a low response of the immune system so leukocytes containing viruses or immune antibodies are highly unwanted (especially lymphocytes with a cell nucleus). At present the remaining leukocytes in the concentrates are further depleted mainly with use of non woven filters. The mechanism of depletion is adsorption to the fibbers (depth filtration) through enhancing the activation of the leukocytes. Historically wool was being used, activating not only the white cells but also the thrombocytes. In advanced filters additional filtration is obtained by a kind of sieving through using compact sheets of very thin fibbers with a diameter of only a few microns. The additional filtration with non-woven filters however is not 100% and gives also rise to a relatively great loss of bloodcells (10% to 40%) due to the large dead volume of the filter pack.

A depletion mechanism solely based on sieving with use of thin perforated membrane plates would be a good alternative. Most existing membranes however are not suitable for cell-cell separation because of the low flow rate and the adsorption of cells to the membrane surface often with a high surface roughness comparable with the size of the cells.

Materials and methods

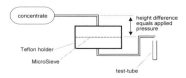

Fig. 50 Test set-up.

Preparation of the filtration membranes

Microsieves (Aquamarijn Microsieve® SiN1.3, SiN1.4, SiN1.5 and SiN1.6) with a siliconnitride membrane layer (thickness of 1µm and pore sizes of 3, 4, 5 and 6µm) have been used for performing leukocyte depletion experiments. Before trial the membranes have been cleaned in fuming and subsequently boiling nitric acid. The surface area of all microsieves was 12cm^2 with a pore density of $1 \cdot 10^6$ pores per cm^2.

Filtration studies

For each of the membrane evaluations, the microsieve was placed into a Teflon based holder and connected in series between the concentrates and test-tubes (see figure 4). The applied pressure was adjusted by changing the height difference between the concentrate and the filter holder. Before and after the experiments, samples have been taken for component counting.

Results

When filtrating the cell concentrates the pressure across the membrane slightly rises in the beginning with a few cm H$_2$O. Cautions have been taken to reduce this effect at very low transmembrane pressures during experiments. Therefore the addition of a small amount of plasma has been used for wetting the membrane surface with pore size 3µm, for larger pore sizes wetting is not necessary. The results of the experiments are shown in next Tables.

Table 7

Leukocyte depletion of erythrocyte concentrates using a microsieve with various pore sizes.

Filter Pore size [µm]	Volume [ml]	Volume loss after exp.	Hemato- cryt	Erythro- cyte *10^6/ml	Increase of Hemolysis	WBC before experiment * 10^3/ml	WBC depletion after experiment	Filtration time [sec.]	Membrane Pressure [cm H$_2$O]
4	200	1.5 %	55 %	6.5	0.01 %	11	85 %	100	50
6	200	2 %	64 %	7.5	0.15 %	6000	95 %	150	75

Table 8

Leukocyte depletion of thrombocyte concentrates using a microsieve with various pore sizes.

Filter pore size [µm]	Volume [ml]	Volume loss after experiment	Thrombocyte * 10^9 / ml	Thrombocyte activation after experiment	Leukocyte before experiment * 10^3 / ml	Leukocyte depletion after experiment	Membrane pressure [cm H$_2$O]
3	25	2 %	1.1	none	480	60 %	10
3	10	4 %	1.2	none	510	85 %	3
3	5	3 %	1.3	none	490	95 %	0-3
5	5	1 %	1.1	none	500	95 %	2
5	25	10 %	0.6	none	810	>99 %	0-3

Discussion

In a dead-end configuration, as has been used in these experiments, the leukocyte depletion of the concentrated cell suspensions is complicated due to the leukocytes that remain at the sieving surface. As a result, a second layer is formed by the retained leukocytes. At high leukocyte concentrations this layer may even act as a second filter for smaller cells or even proteins that could have passed through the pores of the sieve. This phenomenon is known as 'concentration polarization' and may be being circumvented by using a 'cross-flow' configuration through continuously transporting all cells from the surface larger than the pore size to the cross-flow exit.

Moreover the interaction time of the (polymorph nuclear) leukocytes with each individual pore will be reduced severely, avoiding the cells from slipping through. Pulsatile flow may be used to inhibit the slipping effect. The membrane surface may also be cleaned by pulsatile back-flushing during depletion.

In cross-flow configuration the height of the cross-flow channel above the membrane surface will be less than 50µm. A pre-filter with a pore size of 20µm should be used to avoid 'microcluts' when using a cross-flow channel with a very small height.

7. MICROSCREENS, A NEW TOOL FOR RAPID MICROBIOLOGICAL DETECTION

In very dilute suspensions it might be important to have a fast determination of the kind and concentration of particles (e.g. liquid contaminated with bacteria). The small flow resistance of the microsieve allows a large amount of liquid to be concentrated on a very small surface, herewith simplifying the analysis of the suspended particles.

In the production chain for the beverage, dairy, beer brewery and pharmaceutical industry, microbiological control is of vital importance to validate the quality of the beverages. The detection of micro-organisms (yeast cells and bacteria) as early as possible can prevent the occurrence of production losses.

Currently the 'plate count' method is being used in e.g. breweries. A well defined volume of beer with potential contaminants is being filtered through a paper filter plate that captures the contaminants. The filter is subsequently placed on a cultivating plate, that is selected for specific yeast and/or bacteria and is stored in an incubation stove. After 48-96 hours of incubation, the number of yeast/bacteria colonies are being count on the filter plate with the eye, which number is a measure for the amount of living micro-organisms present in the original beer sample. The plate count method is rather straightforward, however the main disadvantage is the long time needed between sampling and counting of the results. In the mean time the beer may already have been delivered at the customer.

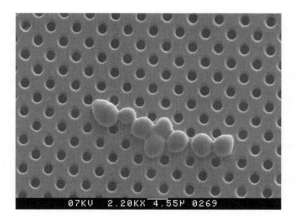

Fig. 51 SEM micrograph 1, surface of a prototype micro analysis screen with a pore size of 1.5 micron. On the screen surface a number of yeast cells can be observed.

Especially for those cases in which a fast count result (within half an hour) is required, the 'plate count' method cannot be used and an alternative rapid test method is highly wanted.

Fig. 52 Photograph, typical MAS test disc. The flow rate is about 20 ml/min (dependent on the viscosity of the liquid and the amount of contaminants), while the sieve surface area is less than 4.5x4.5 mm.

In collaboration with a few breweries Aquamarijn therefore has developed a new fast test method with a microsieve to count directly with use of a fluorescence microscope the number of dead and/or living micro-organisms in a given beer sample.

Microsieve as a Microscreen

For the first time microsieves with such small pores can be used for microscreen applications in which until today only conventional filtration membranes were used. With respect to the current filtration membranes the microsieve has a number of specific advantages that might favour the newly proposed microbiological detection method.

The microsieve is for example characterised by a very low flow resistance, a regular and precise pore geometry and a small surface sample area. With these features the sample liquid can be filtered through the microsieve using a vacuum in a relatively short time and all micro-organisms will be concentrated on this small surface area. Furthermore the microsieve is chemically inert, has no disadvantageous fluorescence back light scattering and is optical flat, which facilitates the staining and detection of the micro-organisms with a microscope in a relatively short time.

The Micro Analytical Screen (MAS)

In many laboratories and research institutes world-wide new test methods to screen yeast's and bacteria are being developed. In these developments emphasis is being put on alternatives for the incubation method in order to save time. Many new methods e.g. ATP will diminish the total screening time from three days to one day or at best 12 hours, but faster methods and especially more selective to the type of micro-organism under test are needed. Also the minimum detection limit of many new methods is rather high (PCR > 10-100 cells depending on the matrix).

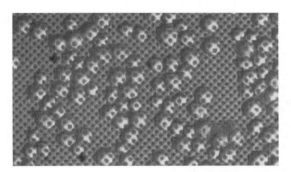

Fig. 53 Photograph, Yeast cells collected on a microsieve.

In view of more stringent regulations (new HACCP standard) it will be necessary to lower the detection limit. Because the total sieve surface of the MAS is rather small, it is easy to count individual micro-organisms.

For automated detection methods the image analysis has been rapidly evolved to a helpful tool for fast classification about number and shapes of micro-organisms. The recognition of these micro-organisms is simplified with good image preparation using proper staining of cells and arranging cells on a surface in a regular geometry. MAS fulfils in this respect these requirements.

Furthermore, with the MAS method it is possible to differentiate with selective staining agents between dead and living cells, something that is difficult with 'plate count' (living cells only), ATP (living cells only) and PCR (dead plus living cells) methods.

Detection of micro-organisms in beer

The research at the beer breweries aims at the filtration behaviour, throughput, application of different staining techniques (dead/alive, fluorescence) and the introduction of image analysis automation.

When new analysis techniques are being introduced it is important to validate the technique. In the experiment described here, the new MAS analysis method is being compared with the 'plate count' method.

In experiments samples with different yeast cell concentrations have been analysed using both methods. The result shows that with the MAS technique a higher number of yeast cells were counted than with the 'plate count' method. This can be explained by the fact that with the MAS technique all individual cells can be counted that form a colony, whereas in the 'plate count' method these cells together would count as one single cell after cultivation. In the MAS detection also less vivid cells are being counted that of course would not have been able to multiply on agar and thus are not found in the 'plate count' method.

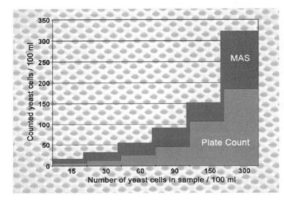

Fig. 54 Comparison of the MAS method with the 'plate count' method: The new rapid MAS microsieve test turns out to be more sensitive than the 'plate count' method.

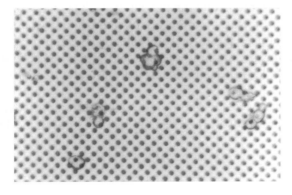

Fig. 55 Microscope picture of coloured yeast cells on a micro sieve surface.

A wide variety of colouring techniques can be used in order to improve the recognition of the micro-organisms on the microsieve surface. Microsieves are inert which makes it possible to use all present colouring agents and chemicals without colouring or attacking the microsieve surface.

Because of the absence of adsorption of the staining agent at the microsieve surface, the microsieve will vanish in a fluorescence picture, enabling it to interpret easily the microscope image of potentially present micro-organisms.

Bacteria

Much development is being initiated towards the detection of anaerobe beer spoilage bacteria using microsieves with a pore size of 0.45 micron. With the 'plate count' method it is difficult to work in an oxygen-free atmosphere for a long period. This measure is not necessary for the rapid MAS method, selective colouring agents can be used to locate quickly the bacteria on the microsieve surface.

Fig. 56 Microscope picture of yeast cells on a microsieve with a picture insert: Fluorescent image of yeast cells coloured with acridine orange on same micro sieve surface. The microsieve surface has vanished and the number of collected cells is easy to count.

Other microscreen applications

The presented microbiological detection method for a fast detection of micro-organisms may be used in applications other than beer screening.

The MAS method may also be applied for the quality control of drinking water on the presence of Cryptosporidium, Legionella and E-Coli. Furthermore in the biomedical field it may be applied to observe e.g. cell growth. For recognition of cell aberrations it is worthwhile to place the cells well arranged on a regular sieve grid.

Verification of new theories in membrane research is often complicated by variations in membrane properties. For example, the formation of a cake layer depends on the size and distribution of the pores and the flatness and smoothness of the membrane. The verification of a model that predicts such a cake layer formation would be more accurate using a microsieve with its well defined properties.

REFERENCES

[1] G. Daufin, J.-P. Escudier, H. Carrère, S. Bérot, L. Fillaudeau and M. Decloux, Recent and emerging applications of membrane processes in the food and dairy industry, Trans IChemE, 79 (2001) 89-102.
[2] K.K. Chan and A.M. Brownstein, Ceramic Membranes Growth Prospects and Opportunities, Ceramic Bulletin, Vol. 70 (1991) 703-707.
[3] C.A. Smolders, New Membrane Materials and Processes: A Survey of Work in The Netherlands, Membranes, Oxford & IBH Publishing, 1992, ISBN 81-204-0686-9.
[4] S. Sanchez, Onstream MST, Eindhoven, The Netherlands.
[5] S.Kuiper, C.J.M. van Rijn, W. Nijdam, M.C.Elwenspoek, Development and applications of very high flux microfiltration membranes. Journal of membrane science, 150 1-8 (1998) ISSN 0376-7388.
[6] S.Kuiper, C.J.M. van Rijn, W. Nijdam, et al., Determination of particle release conditions in microfiltration, J. membrane science, 180 15-28 (2000).
[7] E. van Gaasbeek and C.J.M. van Rijn, Cross flow filtration of Yeast Cells, Aquamarijn student report, 1995.
[8] J. Peters and D. Groot, A self cleaning micro filtration system, Aquamarijn student report, 1997.
[9] S.Kuiper, D. Groot and J. Peters University of Twente (1997).
[10]E. Fischer and J. Raasch, Model tests of the particle deposition at the filter medium in cross-flow filtration, Proceedings of the 4th World Congress on filtration, Belgium, 1986, Part II, 11.11.
[11]W. Lu and S. Ju, Selective particle deposition in cross-flow filtration, Sep. Sci. and Technol. 24(7&8) (1995) 517-540.
[12] H. de Balmann, P. Aimar and V. Sanchez, Membrane partition and mass transfer in ultrafiltration, Sep. sci. and Technol. 25(5) (1990) 507-534.
[13] Manufactured by dr. ir. N. Tas for Aquamarijn (1995).
[14] A.J. Goldman, Investigations in low Reynolds number fluid-particle dynamics: volume 1, steady settling motion of arbitrary particles at low Reynolds numbers, New York University, 1966.
[15] M.E. O'Neill, A sphere in contact with a plane wall in a slow linear shear flow, Chem. Eng. Sc., Vol. 23 (1968) 1293-1298.
[16] G. Rubin, Widerstands- und Auftriebswerte von ruhenden kugelformigen Partikeln in stationären wandnahen laminaren Grenzschichten, dissertation University of Karlsruhe, 1977.
[17] C.J.M. van Rijn and M. C. Elwenspoek, Micro filtration membrane Sieve with Silicon Micro Machining for Industrial and Biomedical Applications, IEEE proc. MEMS (1995) 83-87.

[18] R.B. Bird, W.E. Stewart and E.N. Lightfoot, Transport phenomena, Wiley International Edition, New York, 1960.

[19] M.S. Le and J.A. Howell, Alternative model for ultrafiltration, Chem. Eng. Res. Des. 62 (1984) 373-380

[20] F. W. Altena and G. Belfort, Lateral migration of spherical particles in porous flow channels: application to membrane filtration, Chem. Eng. Sci., 39 (2) (1984) 343-355.

[21] J. Kromkamp, M. van Domselaar, K. Schroën, R. van der Sman, R. Boom, "Shear-induced diffusion model for microfiltration of polydisperse suspensions", Desalination, Vol. 146 (2002) 63-68.

[22] G. Belfort, R.H. Davis and A.L. Zydney, The behavior of suspensions and macromolecular solutions in cross-flow microfiltration, J. Membr. Science, 96 (1994) 1-58.

[23] A. Shauly, A. Wachs and A. Nir, Shear-induced particle migration in a polydisperse concentrated suspension, J. Rheol., 42 (1998) 1329-1348.

[24] E.C. Eckstein, D.G. Bailey and A.H. Shapiro, Self-diffusion of particles in shear flow of a suspension, J. Fluid Mech., 79 (1977) 191-208.

[25] D.T. Leighton and A.A. Acrivos, Measurement of the shear-induced coefficient of self-diffusion in concentrated dispersions of spheres, J. Fluid Mech., 177 (1987) 109-131.

[26] D.T. Leighton and A.A. Acrivos, The shear-induced migration of particles in concentrated suspensions, J. Fluid Mech., 181 (1987) 415-439.

[27] R.H. Davis and J.D. Sherwood, A similarity solution for steady-state cross-flow microfiltration, Chem. Eng. Science, 45 (1990) 3203-3209.

[28] R.J. Phillips, R.C. Armstrong, R.A. Brown, A.L. Graham and J.R. Abbott, A constitutive equation for concentrated suspensions that accounts for shear-induced particle migration, Phys. Fluids A, 4 (1992) 30-40.

[29] I.M. Krieger, Rheology of monodisperse latices, Adv. in Colloid and Interface Science 3 (1972) 111-136.

[30] R.F. Probstein, M.Z. Sengun and T.-C. Tseng, Bimodal model of concentrated suspension viscosity for distributed particle sizes, J. Rheol 38 (1994) 811-829.

[31] C.J.M. van Rijn, W. Nijdam, L.A.V.G. van der Stappen, O.J.A. Raspe, L. Broens and S. van Hoof, Innovation in yeast cell filtration: Cost saving technology with high flux membranes, Proc. EBC Congress, Maastricht (1997) 501-507

[32] D.S. Ryder, C.R. Davis, D. Anderson and F.M. Glancy, G.N. Power, Brewing experience with cross-flow filtration, MBAA Tech Quart 25(2) (1988) 67-79.

[33] Q. Gan, R.W. Field, M.R. Bird, R. England, J.A. Howell, M.T. Mckechnie and C.L. O'shaughnessy, Beer clarification by cross-flow microfiltration: fouling mechanisms and flux enhancement, Trans. Inst. Chem. Eng 75, part A, January (1997).

[34] P.J.I. Janssen, Replacement of kieselguhr filtration by high flux microsieves, Report of research at the Grolsch Brewery, Groenlo, 1997.

[35] Q. Gan, R.W. Field, M.R. Bird, R. England, J.A. Howell, M.T. Mckechnie and C.L. O'shaughnessy, Beer clarification by cross-flow microfiltration: fouling mechanisms and flux enhancement, Trans. Inst. Chem. Eng 75, part A, January (1997).

[36] P. Banplain-Avet, N. Doubrovine, C. Lafforgue and M. Lalande, The effect of oscillatory flow on cross-flow microfiltration of beer in a tubular mineral membrane system -

Membrane fouling resistance and energetic considerations, J. Mem Sci. 152 (1999) 151-174.

[37] B. Czech, Cross-flow filtration of beer – experience within the brewery, The Brewer (1995) 103-110.

[38] S. Kuiper, Thesis 2000 Chapter 9, University of Twente.

[39] Internal Aquamarijn research report 1997.

[40] H. de Balmann, P. Aimar and V. Sanchez, Membrane partition and mass transfer in Ultrafiltration, Sep. Sci. and Technol. 25(5) (1990) 507-534.

[41] A.G. Fane, New insights from non-invasive observations of membrane processes, Proceedings (Vol. 1) World Filtration Congress 8, Brighton (2000) 11-17.

[42] P. Banplain, J. hermia and M. Lenoël, Mechanisms governing permeate flux and protein rejection in the microfiltration of beer with a Cyclopore membrane, J. Mem. Sci. 84 (1993) 37-51.

[43] Q. Gan, J.A. Howell, R.W. Field, R. England, M.R. Bird and M.T. McKechinie, Synergetic cleaning procedure for a ceramic membrane fouled by beer microfiltration, J. Mem. Sci. 155 (1999) 277-289.

[44] Y. Xu-Jiang, J. Dodds, D. Leclerc and M. Lenoël, A technique for the study of the fouling of microfiltration membranes using two membranes in series, J. Mem. Sci. 105 (1995) 23-30.

[45] T. Tanaka, H. Itoh, K. Itoh, K. Nakanishi, T. Kume and R. Matsuno, Cross-flow filtration of baker's yeast with periodic stopping of permeation flow and bubbling, Biotechnol. and Bioeng. 47 (1995) 401-404.

[46] Microsieve manufactured by Aquamarijn with photolithographic steps from DIMES, Delft.

[47] K.J. Burrell and R.J.R. Reed, Cross-flow microfiltration of beer: laboratory-scale studies on the effect of pore size, Filtration and Separation (1994) 399-405.

[48] Onstream MST, Eindhoven The Netherlands.

[49] C.J.M. van Rijn, O.Raspe, and S.C.J.van Hoof, Device for filtering a fermented liquid, EP19981125, Grolsche Bierbrouwerij BV.

[50] G. Brans, Food Process Group, University of Wageningen.

[51] J. Kromkamp, FCDF & Food process Group, University of Wageningen.

[52] J.-L. Maubois, Bulletin of the IDF, 320, 37 (1997).

[53] Wouters J.T.M., Melkkunde: een inleiding in samenstelling, structuur en eigenschappen van melk, University of Wageningen, 1999.

[54] WO 96/36238.

[55] WO 97/49295.

[56] FCDF research centre, Deventer.

[57] G. Daufin, J.-P. Escudier, H. Carrère, S. Bérot, L. Fillaudeau and M. Decloux, Recent and emerging applications of membrane processes in the food and dairy industry, Trans IChemE, 79 (2001) 89-102.

[58] A.Guerra, G. Jonsson, A. Rasmussen, Waagner, E. Nielsen and D. Edelsten, Low cross flow velocity microfiltration of skim milk for removal of bacterial spores, Int Dairy Journal 7 (1997) 849-861.

[59] P.K. Vadi and S.S.H. Rizvi, Experimental evaluation of a uniform transmembrane pressure cross-flow microfiltration unit for the concentration of micellar casein from skim milk, Journal of Membrane Science 189 (2001) 69-82.

[60] M.Pouliot, Y. Pouliot and M. Britten, On the conventional cross-flow microfiltration of skim milk for the production of native phosphocaseinate, Int Dairy Journal 6 (1996) 105-111.

[61] O.Le Berre and G. Daufin, Microfiltration (0.1 µm) of milk: effect of protein size and charge, Journal of Dairy Research 65 (1998) 443-455.

[62] A.L.Zydney, Protein separations using membrane filtration: new opportunities for whey fractionation, Int. Dairy Journal 8 (1998) 243-250.

[63] P.M.Kelly, J. Kelly, R. Mehra, D. Oldfield, E. Raggett and B.T. O'Kennedy, Implementation of integrated membrane processes for pilot scale development of fractionated milk components, Lait 80 (2000) 139-153.

[64] K.Surowka, A. Celej, Partitioning of protein and some trace elements during 3-step ultrafiltration of skim milk through membranes of decreasing permeability, Milchwissenschaft 51 (1996) 426-431.

[65] A.Makardij, X.D.Chen and M.M. Farid, Microfiltration and ultrafiltration of milk: some aspects of fouling and cleaning, Trans IChemE 77 (1999) 107-113.

[66] M. Girones, Z.Borneman and M. Wessling, Membrane Technology Group, University of Twente.

[67] G.Belfort, R.H. Davis and A.L. Zydney, The behavior of suspensions and macromolecular solutions in cross-flow microfiltration, Journal of Membrane Science 96, (1994) 1-58.

[68] H.B.Winzeler and G. Belfort, Enhanced performance for pressure-driven membrane processes: the argument for fluid instabilities, Journal of Membrane Science 80 (1993) 35-47.

[69] J.Kromkamp, C.J.M. van Rijn, Method for filtering milk, WO/0209527.

[70] P.Wilding, et. al., Manipulation and Flow of Biological Fluids in Straight Channels Micromachined in Silicon, Clin. Chem., Vol. 40 (1994) 43-47.

[71] W.Uijttewaal, On the motion of particles in bounded flows - applications in hemo-rheology, Ph.D-Thesis, University of Utrecht, 1993.

[72] C.J.M. van Rijn, W. Nijdam and M.C. Elwenspoek, High flow rate microsieve for biomedical applications, Proceedings ASME Dynamic Systems and Control Division, San Fransisco (1995) 995-10.

[73] P.D. Nguyen, E.A. O'Rear and B.M. Fung, Dynamic evolution of clotting phe-no-mena in vitro and perfluorochemical oxygen transport across a membrane-bound thrombus model, Biomaterials, Art. Cells, Art. Organs 17 (1989) 245-262.

[74] H.Brauner, Corrosion resistance and biocompatibility of physical vapour deposition coatings for dental applications, Surface and Coatings Technology, Vol. 62 (1993) 618-625.

[75] P.F.Williams, Fundamental aspects of biocompatibility, Boca Raton: CRC Press Inc., 1981.

[76] E.Ogura et al., Measurement of Human Red Blood Cell Deformability Using a Single Micropore on a Thin Si3N4 Film, IEEE Transactions on Biomedical Engineering, Vol. 38 (1991) 721-725.

[77] G.P.Downey, D.E. Doherty, B. Schwab, E.L. Elson, P.M. Henson and G.S. Worthen, Retention of leuko-cytes in capillaries: role of cell size and deformability, The American Physiological Society 69 (1990) 1767-1778.

[78] G.B. Nash, , Filterability of blood cells: methods and clinical applications, Biorheology 27 (1990) 873-882.

[79] S.Chien, Blood Cell Deformability and Interactions: From molecules to Micromechanics and Micro-circulation, Microvascular Research, Vol. 44 (1992) 243-254.

[80] G.Moessmer and H.J. Meiselman, A new micropore filtration approach to the analysis of white cell rheology, Biorheology 27 (1990) 829-848.

Chapter 8

Lab and Fab-on-a-Chip

"Small is beautiful."

[E.F. Schumacher]

1. INTRODUCTION

A microfabricated lab-on-a-chip combines micro fluidic sample handling and analysis steps into a single tool. Generally, the lab-on-a-chip concept [1] aims to replace in a number of cases conventional laboratory equipment and methods, including measuring and dispensing equipment, as well as more complex analytical equipment. As an analytical tool for combinatorial chemistry, high-throughput screening and process optimisation the "lab-on-a-chip" format offers unique opportunities due to the possibilities of parallel processing, integrated sensing and automation. As a production tool, the "fab-on-a-chip" offers the advantages of enhanced heat and mass transport, leading to higher safety levels, higher product selectivities and better defined reaction schemes. Also here small-sized reaction devices such as microreactors provide exciting ways and many advantages to achieve these goals.

A lab-on a-chip consists of a micro fluidic board (fluid channel substrate) with a number of small modular fluidic components such as pumps [2], valves, channels, reservoirs cavities, reaction chambers [3], mixers [4], fluidic interconnects [5], diffusers, nozzles, microsieves, porous media, etc attached or sealed on the fluidic board. Furthermore the fluidic board is provided with an electrical wiring system to enable an interface from e.g. a laptop computer to electrical components such as actuator/motors [6,7], valves [8], heaters, pressure-, flow [9]- and temperature sensors. Such an approach is commonly known as micro total analysis system (TAS) [10,11,12,13] or modular assembled total analysis systems (MATAS) [14]. TAS is ideal for continuous monitoring of chemical concentration in industrial, chemical and biochemical processes [15]. As such, the TAS concept has many potential applications in biotechnology, process control, and the environmental and medical sciences.

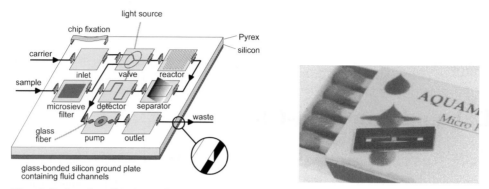

Fig. 1 Left: visualisation of a Lab-on-a-Chip, Right: Microsieve modular component (Courtesy Aquamarijn Research).

Devices, which include an optical or visual detection element, e.g., for use in fluorescence or calorimetric based assays, will generally be fabricated, at least in part, from a transparent polymeric material to facilitate detection. For example, the devices may include reaction chambers for performing synthesis, PCR amplification or extension reactions. The devices may also include sample analysis elements for the detection of a particular characteristic of the sample, such as a capillary channel for performing electrophoresis on the sample or an optical wave guide structure. Additionally, device materials are also selected for their inertness to critical components of an analysis or synthesis to be carried out by the device, e.g., proteins, nucleic acids and the like.

TAS fluidic boards themselves with micro-sized features are relatively difficult and elaborate to fabricate. Fabrication methods currently involve the use of photolithography to either generate a fluidic board substrate by etching trenches into the substrate, such as a silicon wafer or a glass [16] plate. Sometimes this substrate is covered by a silicon wafer or a glass plate to close the channels and chambers. Such structures are, however, rather expensive to produce. Inexpensive substrates that can be produced in large quantities, e.g. in the form of a roll good, would be welcomed. Moulding of such fluidic board products may be proceed by applying a mouldable material on a rigid support and bringing a mould in line contact with the mouldable material. The net result is a two-layer structure in which a microfluid processing architecture-bearing layer is integrally bonded to the polymeric product. A second product may be bonded to the moulded article to form a cover layer overlying the microfluid processing architecture. Most of the known products have utilized clamp-shell type devices, which have problems with sealing and leakage, particularly when products are being formed from polymers.

Also there is a need for polymer-based products that can be produced efficiently in commercial-scale quantities, e. g., in the form of a roll good, and that can be subsequently selectively tailored to perform a variety of functions including heating, mixing, cell sorting, capillary array electrophoresis, combinatorial chemistry, electro-chromatography and adaptations for the use of kinetic inhibition assays, competition immunoassays, enzyme assays and nucleic acid hybridisation assays.

2. LAB-ON-A-CHIP FOR PARTICLE FRACTIONATION[*]

To illustrate the complexity of lab-on-a-chip devices a presentation is given of one of Netherlands (Twente) earliest examples as designed by Aquamarijn [17] in 1995 and manufactured in a FUSE application experiment [18] in the period 1997-1999.

The lab-on-a-chip entails a silicon fluidic board with silicon based modular components for the fractionation of particles on size. A cascade of miniature cross-flow microsieve modules were developed for fractionating at relatively low pressures preventing rupture of biological cells, irreversible fouling etc. In order to offer a fully functional fluid handling system with a small dead volume for fractionation of small sample volumes, also other functional components were made with silicon micro machining techniques. For the assembly of the components on the fluidic board a special hot melt plastic technique was introduced.

The objective of the FUSE application experiment was to develop a micro fractionating system with three different microsieve filter units:

- the system is capable of separating samples (cells, bacteria, particles) on size in different fractions (<2μm, 2 - 5μm, 5 - 10μm and >10μm).
- system contains 3 microsieve cross flow units, pore size 2, 5 and 10μm.
- monitoring of transmembrane pressure of each microsieve.
- monitoring of flow in system.
- ability to control transmembrane pressure of each microsieve.
- ability to control cross flow (<2ml/min).
- sample size less than 0.1 ml.
- system volume <2ml.

---

[*] This section has been co-written with W. Nijdam, Aquamarijn Research.

The micro fractionating system consists of a fluid channel board (like a printed circuit board) with a number of small microfiltration membranes (microsieves), valves, fluid resistors, and pressure- and flow sensors for the handling of the liquid flows. A computer will control the system with a graphical software application for analysis and demonstration purposes.

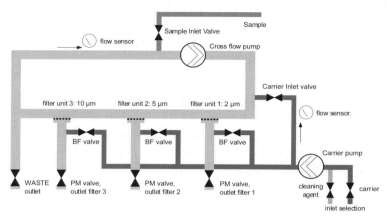

Fig. 2 Schematic diagram of the micro fractionating system.

The main advantage of the micro fractionating system are: (i) overall small dimensions (therefore it is possible to analyse small sample volumes), (ii) faster response times due to the small volumes less than 1 ml and (iii) the possibility of combining different domains (electronics, fluidics, mechanics) on a single board with the same technology resulting in a hand held device. Another advantage over larger systems is the ability to measure pressure differences and flow characteristics directly at the microsieves, which gives a possibility to control these parameters, and thereby the separation process (it is important to handle the cells carefully). Using miniature cross flow microsieve modules, the separation can be performed using low pressure, preventing cell damage.

Fluid handling components:
The components pressure sensor, temperature sensor, valves, fluid resistance sensors and microsieves were all fabricated with silicon micromachining technology. The pressure sensors and temperature sensors are commercially available as Silicon Micro System Technology components. An active valve based was not available but has been designed and produced [19]. Miniaturised valves are important for further downscaling. In the filtration analysis system eight valves are needed for optimal control of the flow handling.

The fluid resistors and the microsieves are also fabricated using MST to achieve a substantial down scaling.

Fig. 3 Photograph of the diced fluid channel board (credit card sized) with electrical wiring, microsieves, pressure sensors and fluid resistors.

Tools

Simulations in strength and flow behaviour have been carried out using the finite element method (Ansys). Calculations (verified with experiments) have been performed using standard computing tools like MathCad and Excel. With the layout editor CleWin, layouts were made for the production of the lithography masks.

Test methods

All fluidchannel board components were tested individually before assembly. This has proven to increase the total yield of working assembled systems. During and after the complete assembly of the prototype micro fractionating system the individual components and also the finished assembled system was tested on functioning. These tests include, electrical testing using a probe station (all sensors and fluid channel board), electrical testing using test software (finished system), leakage tests (all glued bonds) and microscopic inspection of all filtration membranes (microsieves). Normally filtration membranes are tested by means of a bubble point pressure test, however it turned out that microscopic inspection of the relatively small sieve areas is faster and more economical. A further functional check was done to test separately the performance of the valves, pump and the electronic circuitry.

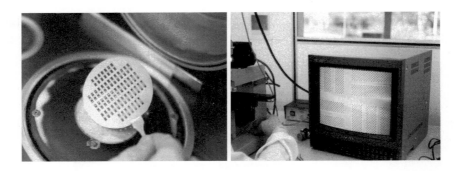

Fig. 4 Left: Photograph of a wafer with microsieves after Reactive Ion Etching, Right: optical inspection of the microsieves.

According to these specifications, several configurations have been tested to perform the fractionating function. Normally not all particles of one size range will be recovered in the permeate in a single pass along the membrane (for this a number of passes is required).

Table 1

Parameters of the micro fractionating system.

Functional specification	
Application:	separation of particles/cells/bacteria in sample in fractions with equal particle size in range: 0-2μm, 2-5μm, 5-10μm, and > 10μm .
Microsieves:	with pore size: 2, 5 and 10μm
	cross flow filtration modules (channel height = 100μm)
	filtration surface > 2mm^2
Cross flow speed:	adjustable, max. 1-2 ml/min
Transmembrane pressure:	adjustable, max. 15 cm H$_2$O
	overall pressure in system max. 2m H$_2$O
Sensors:	
Flow sensors:	in working range of system
	measurements of cross flow and carrier inlet flow
	flow sensor exists of two pressure sensors and a fluid restriction
Pressure sensors:	in working range of system
	measurements of transmembrane pressure over all three
	microsieve cross flow modules
Fluid channel board:	size: credit card
Material:	all components must be biocompatible

Another important constraint on the system's functioning was, that it should not happen that particles of a wrong size slip through the filtration membrane. It was found that one should also be able to fine-tune or lower the transmembrane pressure, because cells may be deformable and therefore may pass a membrane with a smaller pore size than the actual cell size, especially when the transmembrane pressure is high. The efore mentioned requirements have led to a configuration with two pumps, a number of valves and microsieves positioned in a closed cross flow loop.

Biocompatible micro filtration membranes have been used with pore sizes 2.0 μm, 5.0 μm and 10 μm suitable for the fractionation of suspension with particles of different size.

The initial prototype valves made with silicon micromachining realised by a subcontractor were not operating adequately. Other commercial available ones (not made with silicon micro machining) have replaced these valves. However, the performance of the system relied crucially on the valves. Finally a working prototype for the micro filtration fractionating system was obtained.

Pressure sensors

For the pressure sensors two options were available:

Honeywell, type 24PC, available in package only, but the 'die' can be taken out of the chip very easily. SensorTechnics, type P382, available as a 'die'. These chips were better because they have not to be disassembled. Chip sizes of both sensors are equal. The main difference is sensitivity, but for the operation of the fractioning system both sensors were adeaquately functioning.

Table 2

Comparison of pressure sensors of Honeywell and SensorTechnics.

Pressure Range	Full Scale Output	
	Sensor Technics (P382)	Honeywell (24PC)
0-1 psi (0-70 cm H_2O)	20 mV	45 mV
0-5 psi (0-350 cm H_2O)	75 mV	115 mV

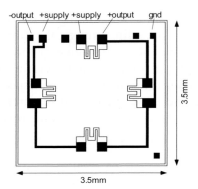

Fig. 5 Chip geometry of the pressure sensor (SensorTechnics).

The power supply of the pressure sensor has to be a constant current source in order to decrease the temperature sensitivity of the sensor (according to an application note of SensorTechnics).

Adequate information about electronic components, circuit design and software development for the control unit has been obtained by Syntens as well as 3T B.V, both represented in the Netherlands.

Temperature Sensor

In the used method for flow measurement with a pressure drop across a fluid restriction, it is important to know the viscosity of the fluid, which is dependent on the temperature. Also for a possible temperature controller of the system, the temperature of the fluid channel board should be known. For the temperature measurement a small, 'smart', temperature sensor will be used. The type is Smartec SMT-160-30-HE, available in a 'die'. The supply voltage is 5V and the output signal is a square wave with a duty cycle dependent on the temperature (duty cycle=$0.320+0.00470*t$ with t in °C).

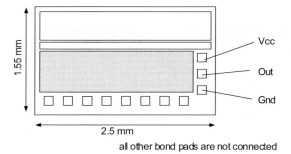

all other bond pads are not connected

Fig. 6 Chip geometry of the Smartec SMT-160-30-HE temperature sensor.

Valves

Valves for liquid applications in micro system technology are not commercially available. The valve specially developed for this project consists of a membrane (material: silicone rubber) bonded (vulcanised) between two silicon parts. In the bottom part, two openings are etched for fluid in- and output with a boss remaining between the in- and output. The silicone rubber membrane can be lifted using a magnetic actuator, thereby creating a fluid path under the rubber membrane. If the magnet is pushing the membrane downwards, the valve is closed.

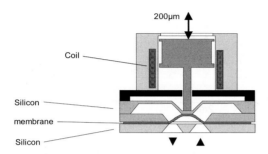

Fig. 7 Valve layout (drawn in open-state).

Besides the MST-components, also a control unit has been designed. Because the number of valves and sensors in the system is large, this unit consists of a laptop computer and interface electronics. The interface electronics has two parts: a data acquisition card (National Instruments, DaqPAD1200) and preamplifiers and power drivers designed in this experiment from and to the fractionating system.

Prototyping

An important phase in the experiment was the prototyping phase. Now the new technology could be put in practice. The system consists of a fluid channel board with on top a number of mounted components (micro-sieves, pressure sensors, micro valves and fluid restrictions). These components were manufactured in a clean room, a research/production facility where the air treatment is very important. It is easy to understand that particles in the air (dust and particulates, all coming from the human) will disturb the processing of the components, as the size of these particles is of the same order of magnitude as the features of the chip components.

Microsieve module

The microsieve module consists of a silicon wafer with a microsieve and on top a glass wafer with a channel structure. The channel height is 200μm and the distance from the module inlet to the module outlet is 11mm. The modules (as well as the fluid resistor modules) have not been separated by dice sawing but with a modified glass/silicon cutting technique without the disadvantage of particle contamination.

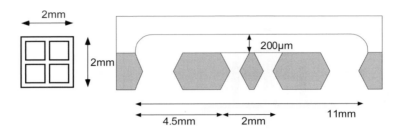

Fig. 8 Drawing of microsieve module. The module consists of a microsieve with a glass cover plate, a cross flow channel with a height of 200 micron and fluid ports.

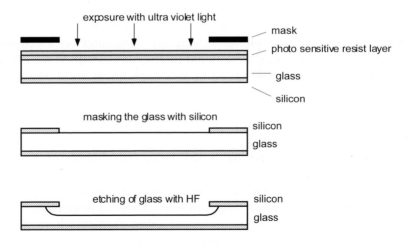

Fig. 9 Process steps for etching the glass wafer (top of microsieve module).

An overview of the silicon micromachining process steps for the module is given in Fig. 9. The processing of 1 wafer (3") will lead to 24 separate modules with equal pore size.

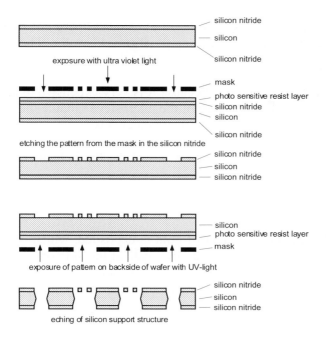

Fig. 10 Process steps of the silicon wafer with the microsieve containing extra openings for the microsieve module.

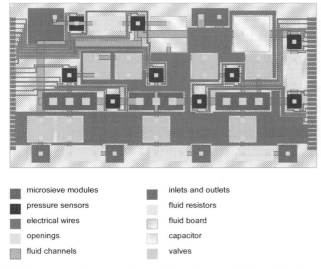

microsieve modules	inlets and outlets
pressure sensors	fluid resistors
electrical wires	fluid board
openings	capacitor
fluid channels	valves

Fig. 11 Overview of the masks and the layout of the fluid channel board.

Fluid channel board

In Fig. 11 the layout and the masks of the fluid channel board are shown. The fluid channel board (85 x 47 mm) fits on 1 silicon 4" wafer. The fluid channel board consists of a glass wafer bonded on a silicon wafer containing channels and openings.

The glass wafer contains no patterns, because it is very expensive to apply micro machining to a glass substrate. In case of the microsieve module the costs per module for a micro machined glass substrate are much lower.

The components (microsieve modules, pressure sensors, valves and fluid in- and outlets) are fixed on top of the silicon wafer of the fluid channel board with a plastic hot melt technique.

Electrical connections

On the silicon wafer a patterned layer of gold is deposited, which is used for wiring of the pressure sensors. Bond pads are made at the side of the fluid channel board and at each pressure sensor. The pressure sensors are connected with a wire bond to the fluid channel board and the board is connected to a flexible PCB to the electronics.

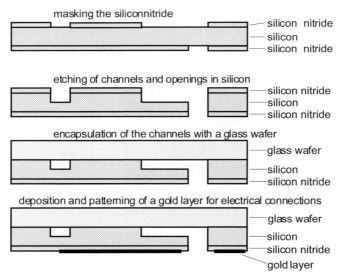

Fig. 12 Schematic overview of the process steps of the fluid channel board (note: the fluid channel board is up side down).

Future developments

The use of current printed circuits boards for modular assembly of fluidic devices will be the focus of Twentes activities in the future [20].

3. FAB-ON-A-CHIP FOR DE-HYDROGENATION PROCESS

Introduction

A lab-on-a-chip is primarily designed for analytical applications, whereas a fab-on-a-chip is ment for processing (cf. process-on-a-chip) of fine chemicals, nutrients, pharmaceuticals etc. Microfabricated reaction chambers [21,22,23] are considered to offer a number of advantages in processing of these fine materials over conventional batch-scale production or synthesis in stirred vessels. Improved mass and heat transfer properties enable the use of better-controlled reaction conditions giving higher yields than traditional bulk reactors. Also tuning of residence time and adequate heat management in microchannel reactors can avoid secondary reactions, hot spots and drifting conditions. This will lead to a higher selectivity and thus to a higher quality product while a more sustainable process with less energy input is needed.

Diffusion times of the reactants are very short and, in case of bubble and slug flow in microchannels, high gas-to-liquid and liquid-to-solid (catalyst) mass transfer rates can be reached because of the large gas/liquid/solid interfacial areas and high shear rates. It is believed that, for these reasons, microreactors will outperform conventional reactors, like stirred tanks and bubble columns, in which mixing is imperfect, residence time distribution causes poor selectivity, and reaction rates are low due to poor mass transfer.

In some cases microreactors can also be an attractive alternative for on-site, small scale energy generation, for instance hydrogen production at place of use in fuel cell applications, see Fig. 13 [17,24].

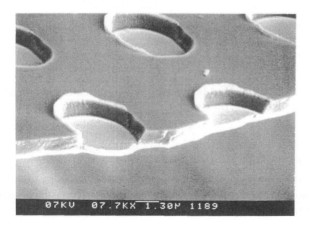

Fig. 13 Pd membranes for hydrogen purification with a thickness of 200 nm and diameter 4 μm hanging in the pores of a 1 μm thick silicon nitride membrane [17], courtesy Aquamarijn Research.

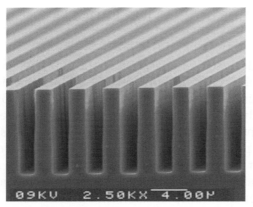

Fig. 14 SEM Photograph of a deep reactive ion etched silicon wafer, which can subsequently be transformed into a porous silicon base structure with an enhanced surface area, see Fig. 15.

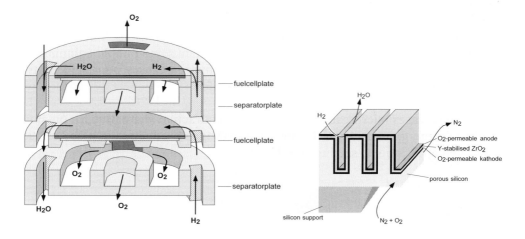

Fig. 15 Schematic outlook of a microfabricated fuelcell and a magnification of an oxygen permeable fuelcell plate, courtesy Aquamarijn Research (1998).

Enlarged surface area's can be achieved by anisotropic etching techniques to obtain enhanced mass transfer rates (see Fig. 14 and Fig. 15) for a range of applications including surface catalyzed reactions and membrane permeation. Application and integration of sensors and actuators, such as active micro valves and pressure, flow and temperature sensors, is necessary to control the proper process conditions. If necessary, the reactors may also be fitted with static micro mixers for appropriate (gas-liquid) mixing and dosing. Computational fluid dynamics simulations can provide valuable insight into a number of liquid phase processes, such as mixing, fluid transport, and separation. Although microreactors may be characterized by very high surface-to-volume ratios, the absolute geometric surface area is far too low to host sufficient catalytic sites, except for exceedingly fast reactions. Therefore, in most cases, porous layers will have to be used. These layers need to be designed to maximize heat and mass transfer. In this way excellent control over the conditions (concentrations and temperature) at the active sites can be achieved. In many cases this will result in significant improvement of selectivity in e.g. selective oxidations and selective hydrogenations.

Microfabricated Hydrogen Separation Reactor [*][25]

The University of Eindhoven and Aquamarijn have initiated in 2000 a PhD-project on the use of very thin Pd membranes to achieve high flux hydrogen separation at elevated temperatures between 300 and 600 °C.

Palladium is a very ductile material [26] with a relatively low yield strength, especially at elevated temperatures. Calculations have shown that the hydrogen permeation is diffusion limited in membrane layers with a thickness larger than 200 nm. Pure palladium also suffers from embrittlement at low temperatures and high hydrogen concentrations. Therefore in this project palladium-silver alloys are being studied, which suffer less from this effect. A brief review is given about the current work showing the potential of this technology. Results on palladium membranes hanging in a perforated membrane support structure (see Fig. 13) will be published elsewhere.

Experimental

A palladium-silver alloy layer (Pd-Ag) was prepared by co-sputtering on a flat side of a etched <110> silicon wafer. Anodic bonding of thick glass to silicon was used to package the membrane and create a very robust module. This membrane has the potential to be used for hydrogen purification and other applications [27,28]. The presented fabrication method allows the development of a module for industrial applications that consists of a stack with a large number of glass/membrane plates.

The increased demand for hydrogen in recent years in many sectors such as petrochemical and semi-conductor processing and fuel cell applications has led to a revival in methods for separation and purification of hydrogen from gas mixtures [29]. However, conventional technology is limited by the high cost of

---

[*] Partially reprinted with permission from J. of MEMS (2003).

palladium through use of fabrication methods resulting in thick palladium films ($>>$ 1 µm). For instance, palladium/silver metal multi-tubular assemblies with a membrane thickness of 50 µm or more have been used for many years for laboratory purification of hydrogen. The relative high wall thickness of these tubes incorporated in such assemblies reduces the hydrogen flux, which inhibits the application for larger-scale chemical production, apart from the investment cost of the precious metal.

In the last decade, a substantial research effort has been focused on achieving higher fluxes by depositing thin Pd or Pd-Ag membranes on porous support materials. However, the perm selectivity is often poor, because the pores of the support are not well covered by the Pd (Ag) layer during the deposition of the Pd (Ag) layer [30]. This is probably due to the fact that the Pd (Ag) has a tendency to deposit on top of the support material or on top of the previously deposited layer, and not to enter the pores, resulting in pin-holes.

Microtechnology is able to produce very thin, pin-holes free Pd membranes, therewith dramatically increasing the flux efficiency as well as the perm selectivity of the membrane, and possibly decreasing production costs. So far, only chip-diced Pd membranes have been reported [31,47].

Fabrication of Pd-membrane

A 3 inch, <110>, double side polished silicon substrate is coated with 1 µm of wet-thermally oxidized SiO_2. The SiO_2 coating serves as a protective layer during subsequent etching steps. First of all, fan-shaped structures are imprinted on the silicon wafer by standard photolithography, followed by removal of SiO_2 in a buffered oxide etch (BHF 1:7). Afterwards, the fan-shaped structures are etched for a short time in KOH to indicate the exact <110> directions [32]. Then, narrow slits of 25 by 1250 µm are aligned to <110> directions and patterned using the steps mentioned above. The wafer is immersed in 25 % KOH solution at 75 °C to etch the silicon until ca. 50 µm is left. An alloy layer of Pd -Ag at wt % with a thickness of 1 µm is deposited by co-sputtering [33] through a shadow mask on the bottom side of the silicon wafer, using titanium (Ti) as an adhesion layer. In this study, Pd-Ag is used as separation element because it has shown higher resistance to hydrogen embrittlement than pure Pd [34]. Next, KOH is used to etch silicon in the trenches until the SiO_2 layer is reached. Finally, this oxide layer and the Ti are removed in BHF to reveal the back surface of the Pd-Ag membranes.

A powder blasting technique [35] was used to create a flow channel on glass wafer. Finally, the silicon wafer is bonded between the two thick glass wafers by a four-electrode anodic bonding technique.

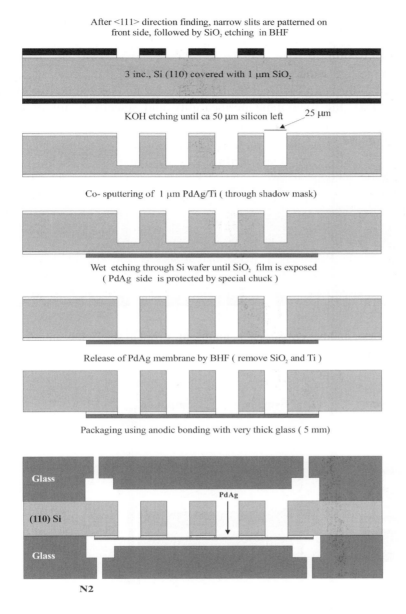

After <111> direction finding, narrow slits are patterned on front side, followed by SiO_2 etching in BHF

3 inc., Si (110) covered with 1 μm SiO_2

KOH etching until ca 50 μm silicon left   25 μm

Co- sputtering of 1 μm PdAg/Ti (through shadow mask)

Wet etching through Si wafer until SiO_2 film is exposed (PdAg side is protected by special chuck)

Release of PdAg membrane by BHF (remove SiO_2 and Ti)

Packaging using anodic bonding with very thick glass (5 mm)

Glass

PdAg

(110) Si

Glass

N2

Fig. 16 Cross-section of the manufacturing of the Pd-Ag membrane.

The process is performed in two steps, because adequate bonding requires that a positive electrical voltage be applied to the silicon and a negative voltage to the glass. This process results in a tight seal between each glass wafer and the silicon wafer.

Results and discussion

To determine hydrogen permeation and selectivity of the membranes, they were positioned in the stainless steel membrane holder and heated up to elevated temperatures. Next, the membrane permeability and selectivity for hydrogen were determined as a function of the hydrogen partial pressure (0–0.3 bar), and temperature (350–450°C).

Fig. 17 Left: topview, middle: cross-section, and right: detail of the membrane structure with a width of 25 μm.

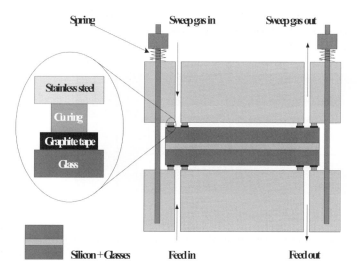

Fig. 18 Stainless steel membrane holder. Graphite rings are used to make a gas-tight connection.

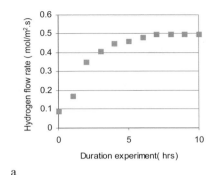
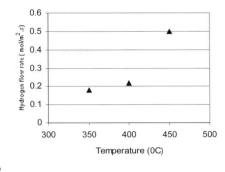

a   b

Fig. 19 a) Hydrogen permeation as a function of time and b) hydrogen flow rates through the membrane at different temperatures from 350°C to 450°C.

During these experiments the retentate and permeate sides of the membrane were continuously flushed at atmospheric pressure; the retentate side with a mixture of hydrogen and helium, the permeate side with pure nitrogen. The flux and selectivity were determined by measuring the hydrogen and helium concentration in the nitrogen stream with a gas chromatograph equipped with a Thermal Conductivity Detector (TCD).

Fig. 19b shows thatthat hydrogen flow rate increases with increasing temperature. However, the dependence is stronger than expected by theory [36]. More experiments are needed in order to get a better understanding of the hydrogen separation process. From the measured data, a minimal separation selectivity factor of 550 for hydrogen to nitrogen was calculated. The final determination of selectivity was limited by the sensitivity of our equipments.

The preliminary results indicate that high hydrogen production rates can be obtained using an industrial module that consists of a stack of glass/membrane plates.

REFERENCES

[1] A. van den.Berg, T.S.J. Lammerink, V.L. Spiering and W. Olthuis, Modular concept for miniature chemical systems, Dechema monograph series, microreactors, Vol. 32 (1995) 109-123. Weinheim, Germany: Ehrfeld, VCH Gesellschaft. ISBN 3-527-10226-4.
[2] F.C.M. van de.Pol, P.C. Breedveld and J.H.J. Fluitman, Bond graph modeling of an electro-thermo-pneumatic micropump, [MME '90], Berlijn (1990) 19-24.

[3] A. van den. Berg, E.B. van Akker, R.E. Oosterbroek, R.W. Tjerkstra and I. Barsony, Technologies and Microstructures for (Bio)Chemical Microsystems, Microreaction Technology, (1997) 91-104. Berlin: Springer verlag.

[4] R.Miyake, T.S.J. Lammerink, M.C. Elwenspoek and J.H.J. Fluitman, Micro mixer with fast diffusion, [Proc. Micro Electro Mechanical Systems (MEMS '93)]., Fort Lauderdale, USA (1993) 248-253.

[5] J.C.C. van Kuijk, Numerical modelling of flows in micro mechanical devices, Universiteit Twente, 1997.

[6] N.R.Tas, J. Wissink, A.F.M. Sander, T.S.J. Lammerink and M.C. Elwenspoek, Modeling, Design and testing of the electrostatic shuffle motor, Sensors and actuators A (Physical), No. 70 (1998) 171-178. ISSN 0924-4247.

[7] R. Legtenberg, J.W. Berenschot, M.C. Elwenspoek and J.H.J. Fluitman, A fabrication process for electrostatic microactuators with integrated gear linkages. Journal of microelectromechanical systems 6 No. 3 (1997) 234-241. ISSN 1057-7157.

[8] J.F.Burger, M. van der Wekken, J.W. Berenschot, H.J. Holland, H.J.M. ter Brake, H. Rogalla, J.G.E. Gardeniers and M.C. Elwenspoek, High pressure check valve for application in a miniature cryogenic scorption cooler.,[Proceedings, MEMS'99 Conference] (1999) 183-188. Orlando, USA. ISBN 0-7803-5194-0.

[9] R.E. Oosterbroek, T.S.J. Lammerink, J.W. Berenschot, A. van den Berg and Elwenspoek, M.C. Designing, realization and characterization of a novel capacitive pressure / flow sensor, [Proceedings of Transducers 1997] (1997) 151-154. Chicago, USA. ISBN 0-7803-3829-4.

[10] T.S.J. Lammerink, V.L. Spiering, M.C. Elwenspoek, J.H.J. Fluitman and A. van den Berg,. Modular concept for fluid handling systems: a demonstrator micro analysis system,[Proceedings of IEEE Micro electro mechanical systems (MEMS '96)] (1996) 389-394. San Diego, USA.

[11] J.H.J. Fluitman, A. van den Berg and T.S.J. Lammerink, Micromechanical components for micro TAS,[Proceedings of Micro TAS '94 Workshop, (Invited speaker)] (1994) 73-83.

[12] A. van den Berg, E.B. van Akker, R.E. Oosterbroek, R.W. Tjerkstra and I. Barsony,. Technologies and Microstructures for (Bio)Chemical Microsystems, Microreaction Technology, (1997) 91-104. Berlin: Springer verlag.

[13] A. van den Berg and T.S.J. Lammerink, Micro Total Analysis Systems: Microfluidic aspects, integration concept and applications. Topics in Current Chemistry, (1997) 21-49. Berlin: Springer verlag.

[14] J.Wissink, A. Prak, H. Leeuwis, R. Mateman and F. Eckhard, Novel Hybride mTAS using Modular Assembly Technology (MATAS), 3rd Round Table on, Micro/Nano-Technologies for Space, 15-17 May 2000, ESTEC, Noordwijk, The Netherlands.

[15] A. van den Berg, T.S.J. Lammerink and J.H.J. Fluitman, Modular concept for miniature chemical systems.,[Proc. PACIFICHEM '95]. Honolulu, Hawai (1995)

[16] Y.Fintschenko, V.L. Spiering, J.N. van der Moolen, G.J. Burger and A. van den Berg, Glass channels and capillary injectors for capillary zone electrophoresis.,[Sensor Technology in the Netherlands: State of the Art]. (1998) 77-84. University of Twente, Enschede, the Netherlands. ISBN 0-7923-5010-3.

[17] C.J.M. van Rijn, Membrane filter as well as a method of manufacturing the same, PCT patent application 95/1386026,

[18] http://www.fuse-network.com/fuse/demonstration/333/529

[19] Designed and produced by TMP subcontractor.

[20] J. Wissink, A. Prak, M. Wehrmeijer, H. Leeuwis and R. Mateman, Novel Low Cost Modular Assembly Technology for mTAS Using PCB-Technology, Proc. Vol. 2, MICROTEC 2000, Hannover.

[21] K.F. Jensen, Smaller faster chemistry, Nature, 393, 735-737 (1998).

[22] D.Qin, Y.N. Xia, J.A. Rogers, R.J. Jackman, X.M. Zhao, G.M. Whitesides, Microfabrication, microstructures and microsystems. Microsystem technology in chemistry and life science,vol. 194 (pp. 1-20). Berlin: Springer Verlag (1998).

[23] W.Ehrfeld, V. Hessel, H. Lowe, Microreactors: New technology for modern chemistry. Weinheim: Wiley-VCH (2000).

[24] A.Franz, K.F.Jensen, M.A. Schmidt, Palladium based micromembranes for hydrogen separation and hydrogenation/de-hydrogenation reactions. In W. Ehrfeld (Ed.), Microreaction technology: Industrial prospects (pp. 267-276). Berlin: Springer (2000).

[25] H.D. Tong, E. Berenschot, M. de Boer, H. Gardeniers, H. Wensink, H.V. Jansen, W. Nijdam, M. Elwenspoek, F.C. Gielens and C.J.M. van Rijn, Microfabrication of Palladium-Silver Alloy Membranes for Hydogen Selparation, J. of MEMS, 12 (2003) 622-629.

[26] Goodfellow data, Catalogue, 1991/1992.

[27] R.Dittmeyer, V.Hollein and K.Daub, J.Molecular Catalysis A: Chemical 173 (2001) 135-184.

[28] J.Shu, B.P.A.Grandjean,A.VanNeste and J.Kaliaguine, Can.J.Chem.Eng. 69, (1991) 1036.

[29] R.Hughes, Memb.Tech. 131 (2001) 9-13.

[30] S.Tosti, L.Bettinali, S. Castelli, F. Sarto, S. Scaglione and V. Violante, J. Membr. Sci., 196 (2000) 241-249.

[31] A. J.Frank, K.F.Jensen, M.A.Schmidt, Proc.IEEE Conf. MEMS (1999) 382-387.

[32] M.Elwenspoek and H.V.Jansen, Silicon Micromachining, Oxford Press, 1999.

[33] J.L.Vossen and W.Kern, Thin Film Processing II, Academic Press, INC., 1991.

[34] Z.Y.Li, J.membr.Sci. 78 (1993) 247-254.

[35] H.Wensink, M.Elwenspoek, H.V.Jansen and E.Berenschot, Proc. IEEE Conf. MEMS (2000) 769-774.

[36] T.L.Ward and T.Dao, J.Mem.Sci.153 (1999) 211-231.

Nanotechnology
and Nano Engineered Membranes

"The most appealing nanodevice is still the dairy cow, able to replicate a lot more than 1 billion self assembled casein micelles a day."

[Rolf Bos (2002)]

1. INTRODUCTION NANOTECHNOLOGY

If nanotechnology means the engineering of nanometer scale objects, then advanced semiconductor and laser interference lithography with feature sizes less than 100 nm is already nanotechnology. Also microcontact printing, an elegant alternative to classic lithography pioneered by Whitesides [146], is capable of printing 100 nanometer features. Moreover electron beam lithography has for decades been able to print features of 100 down to 10 nanometer size. It is apparent that despite their utility in direct '3D fabrication' of nano-sized structures, these technologies are apparently pioneering versions of "true" nanotechnology.

If nanotechnology means controlling individual atoms and molecules at precise location, either chemically or physically, then Scanning Probe Microscopy techniques are nano-technologies. For instance, the manipulation of molecules and atoms has become possible using Atomic Force Microscopy (AFM) and the chemical modification of molecules and surfaces could be achieved by Scanning Tunneling Microscopes (STM).

If the final product of nanotechnology is any nanoscopic device then quantum electronic devices satisfy the criteria for being called so. As the properties of those products depend on how the atoms are arranged, the Solid State Physics approach aims at building atomically precise structures at Angstrom scale. For instance, the (Palo Alto) Foresight Institute will award the Feynman prize for a digital computing device that fits into a cube no larger than 50-100 nanometers in any dimension. Whether quantum electronics can deliver such a device remains a challenging question.

Self Assembly

Self assembly of molecules or large groups of molecules is a key feature in nanotechnology. A natural example of this is the casein micelle. Most, but not all of the casein proteins exist as in the form of a colloidal particle known as the casein micelle. Its biological function is to carry large amounts of highly insoluble CaP to mammalian young in liquid form and to form a clot in the stomach for more efficient nutrition. Besides casein protein, calcium and phosphate, the micelle also contains citrate, minor ions, lipase and plasmin enzymes, and entrapped milk serum. The "casein sub-micelle" model has been prominent for the last several years, but there is no universal acceptance of this model and mounting research evidence to suggest that there is not a defined sub-micellar structure to the micelle at all.

In the submicelle model, it is thought that there are small aggregates of whole casein, containing 10 to 100 casein molecules, called submicelles. It is thought that there are two different kinds of submicelle; with and without kappa-casein. These submicelles contain a hydrophobic core and are covered by a hydrophilic coat, which consists at least partly of polar moieties of kappa-casein. A hydrophilic part of the kappa-casein extends as a 'flexible hair' (see Fig. 1).

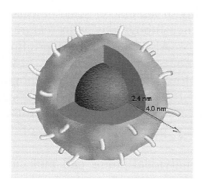

Fig. 1 Schematic outlook of a self assembled casein molecule.

The open model also suggests there are more dense and less dense regions within the middle, but there is less of a well-defined structure.

Colloidal calcium phosphate (CCP) acts as a cement between the hundreds or even thousands of submicelles that form the casein micelle. Binding may be covalent or electrostatic. Submicelles rich in kappa-casein occupy a surface position, whereas those with less are buried in the interior. The resulting hairy layer, at least 7 nm thick, acts to prohibit further aggregation of submicelles by steric repulsion.

Milk (not only cow milk, but all milk) is an essential integrated part for the function of our biosphere, and so is the production of milk. Molecules and aggregates of milk are nano sized, and thus examples of a product which belong firmly within the synergy of modern nanotechnology and biotechnology. What does it take to manufacture artificial milk? Or in other words: What does an artificial dairy cow look like? A dairy cow is an assembly of molecular devices. It is in fact nano technology designed by Nature.

How small can you make things that work? Schrödinger (of the Schrödinger equation and of Schrödinger's cat paradox) wrote in the forties a small book [1] entitled "What is Life". He poses in this book the question, why are the atoms so small? His answer is that since we are very, very complicated, we must necessarily contain many, many parts. Hence, each part must be very small seen relative to the size of us - and the smallest parts we consist of are the atoms.

Think of the glass of milk. It contains some 10^{24} molecules. If it takes one second for a molecular machine to synthesize each molecule in the milk, the time for the manufacturing process would take would be 10^{16} years, which is a million times the age of the universe.

Self replication

Another key concept is self-replication. One (or a few) molecular machines is not enough. We need lots of them. The only way to be able to produce enough is to have the molecular machines produce copies of themselves. With self-replication, a process that would take a million ages of the universe, would now only take 80 generations, which is little more than a minute if each generation takes one second to produce.

So one could think of the cow as an assembly of nano-devices. Proteins are in fact nano machines that do mechanical work in spite of rattling around due to thermal noise. How they in fact do this is still an open question. This question must be answered in order to be able to construct our own nature like machines thus bypassing Nature's self evolution.

Motility

Moreover, another essential element of the envisaged nanotechnology products is their motility. Foresight Institute requires for the award to be given the fabrication of a second device 'a robotic arm or other positional device that initially fits into a cube no larger than 100 nanometer in any dimension'. The intent of this robotic arm is a device demonstrating the controlled motions needed to manipulate and assemble individual atoms or molecules into larger structures, with atomic precision.

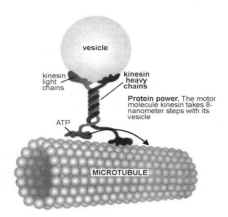

Fig. 2 Molecular motor [2] anno 1993.

Some molecular machines act as motors. One such motor is the protein myocin which makes muscles contract. Another one is the protein kinesin that move things along the microtubules network found in cells. A molecular motor is a very different device from those motors we come across in our daily lives, be it the car engine, the watch or the electrical motors that run lifts. These engines function by converting energy in one form (such as chemical fuel, wound spring or electrical energy) into mechanical energy, and the way they do this is by having forces pushing or pulling masses around. Although the most successfull path was assumed to be protein engineering, some other directions have been pursued including other sorts of biochemical engineering. The main features required for a working nanodevice would be self-replication and hence low cost. Investigations to date strongly support the feasibility of programmable self-replicating systems. But these are exactly the characteristics of natural DNA molecules. More examples of self-assembling systems exist within the biological world, for example protein self-assembly has an essential role in creating and sustaining the normal functioning of the cell. Understanding how protein's work, and knowing how proteins with new shapes and functions can be obtained, might be another step toward building nanomachines.

New technology based on the devices described above should allow us to build economically a broad range of complex nanomachines. Such devices might for example revolutionise medicine, through giving means to intervene in a sophisticated and controlled fashion at the cellular and molecular level. The nanotechnology community believes that virtually every manufacturing domain will benefit from it.

Table 1

Nanotechnology: the revolution started more than 70 years ago

1931: Electron microscope	Sub-nm imaging
1959: Richard Feynman's lecture	Thinking at nano-scale
1968: Molecular beam epitaxy	Deposit single atomic layers on a surface
1974: 'Nanotechnology'	use of word (Nori Taniguchi, Japan)
1981: Scanning Tunnel Microscope	To image single atoms and surfaces
1985: Buckminster-fullerenes	(Richard Smalley, cages of 60 C atoms Rice University, US)
1986: Atomic Force Microscope	Indenting surfaces
1989: Use of individual atoms	(D. M. Eigler, IBM, US)
1991: Carbon nanotubes	(Sumio Iljima, strength, electrical properties NEC, Japan)
1993: Virtual reality system	(W. Robinett, University of North Carolina, US)
1997: Nanomech device using DNA	(Seeman, NYU, US)
1998: Nanotech transistor	(Cees Dekker, Delft University, The Netherlands)
1999: Single molecule organic switch	(James Tour. Rice University, US)
2001: CN-based devices,	Logic gates (IBM, US)

Current opinion in nanotechnology is, that it deals with structures having at least one dimension that is sufficiently small, of the order of about one to several hundred nanometers. It is concerned with materials and systems whose structures and components exhibit novel and significantly improved physical, chemical, and biological properties – and that enable the exploitation of novel phenomena and processes due to their nanoscale size.

In the next sections some examples will be given of nano and micro engineered membrane technology and its use for nano engineered devices.

2. ZEOLITES, SELF ASSEMBLED MOLECULAR SIEVES

In 1756 Cronstedt discovered a selection of natural minerals, which produced steam when heated. Due to this strange phenomenon he called them Zeolites from the Greek for Boiling Stone (zeo - from zein - to boil, lithos - stone).

Not until the early 1900's and in particular the 1940's did zeolites become of interest to man. In 1949, Milton working at Union Carbide produced the world's first man-made zeolites, the most important being Linde A and X. Linde A has gone on to become one the most widely used zeolites.

Zeolites are highly crystalline alumino-silicate frameworks comprising $[SiO4]^{4-}$ and $[AlO4]^{5-}$ tetrahedral units. T atoms (Si,Al) are joined by an oxygen bridges. Introduction of an overall negative surface charge requires counter ions e.g. Na⁺, K⁺ and Ca²⁺. Zeolites are microporous crystalline solids with well-defined structures. Generally they contain silicon, aluminium and oxygen in their framework and cations, water and/or other molecules within their pores. Many occur naturally as minerals, and are extensively mined in many parts of the world. Others are synthetic, and are made commercially for specific uses, or produced by research scientists trying to understand more about their chemistry.

Synthetically they are made by a crystallization process with use of a template (e.g. Tetra Propyl Ammonium) that builds a self assembling skeleton together with the silicon, aluminium and oxygen atoms. Afterwards the template is removed (e.g. by heating, the calcination process) leaving small nanosized holes and tubes along the crystal planes of the zeolite structure.

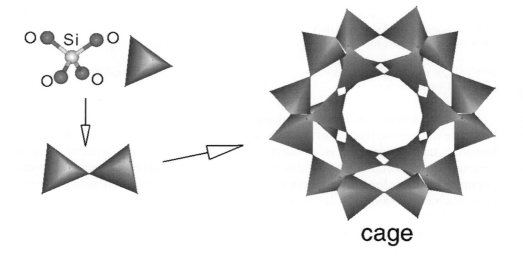

Fig. 3 Zeolite molecular sieve structure build up from unit blocks. The open structure is obtained with the use of appropriate template molecules during zeolite formation.

Because of their unique porous properties, zeolites are used in a variety of applications with a global market of several million tonnes per annum. In the western world, zeolites are mainly used for petrochemical cracking, ion-exchange (water softening and purification), and in the separation and removal of gases and solvents. Other applications are in agriculture, animal husbandry and construction. They are often also referred to as molecular sieves.

A defining feature of zeolites is that their frameworks are made up of 4-connected networks of atoms. One way of thinking about this is in terms of tetrahedra, with a silicon atom in the middle and oxygen atoms at the corners. These tetrahedra can then link together by their corners (see illustration) to from a rich variety of beautiful structures. The framework structure may contain linked cages, cavities or channels, which are of the right size to allow small molecules to enter - i.e. the limiting pore sizes are roughly between 3 and 10 Å in diameter.

Over 130 different framework structures are now known. In addition to having silicon or aluminium as the tetrahedral atom, other compositions have also been synthesised, including the growing category of microporous aluminophosphates, known as ALPO's.

Zeolites have the ability to act as catalysts for chemical reactions which take place within the internal cavities. An important class of reactions is that catalysed by hydrogen-exchanged zeolites, whose framework-bound protons give rise to very high acidity. This is exploited in many organic reactions, including crude oil cracking, isomerisation and fuel synthesis. Zeolites can also serve as oxidation or reduction catalysts, often after metals have been introduced into the framework. Examples are the use of titanium ZSM-5 in the production of caprolactam, and copper zeolites in NO_x decomposition.

Underpinning all these characteristic reactions is the unique microporous nature of zeolites, where the shape and size of a particular pore system exerts a steric influence on the reaction, controlling the access of reactants and products. Thus zeolites are often said to act as shape-selective catalysts.

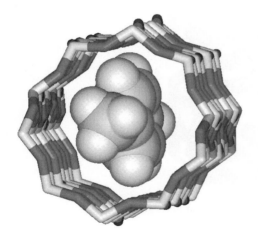

Fig. 4 Para-xylene molecule can diffuse freely in the open channels of silicalite.

Increasingly, attention has focused on fine-tuning the properties of zeolite catalysts in order to carry out very specific syntheses of high-value chemicals e.g. pharmaceuticals and cosmetics.

The shape-selective properties of zeolites are also the basis for their use in molecular adsorption. The ability preferentially to adsorb certain molecules, while excluding others, has opened up a wide range of molecular sieving applications. Sometimes it is simply a matter of the size and shape of pores controlling access into the zeolite. In other cases different types of molecule enter the zeolite, but some diffuse through the channels more quickly depending on surface affinity properties, leaving others stuck behind, as in the purification of para-xylene by silicalite.

Zeolites containing cations are extensively used as desiccants due to their high affinity for water, and also find application in gas separation, where molecules are differentiated on the basis of their electrostatic interactions with the metal ions. Conversely, hydrophobic silica zeolites preferentially absorb organic solvents. Zeolites can thus separate molecules based on differences of size, shape and polarity. The loosely bound nature of extra-framework metal ions (such as in zeolite NaA) means that they are often readily exchanged by other metal ions (in solution).

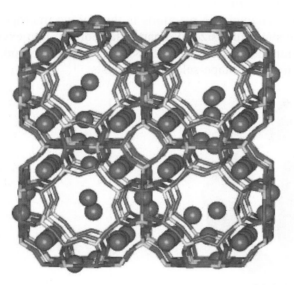

Fig. 5 Sodium Zeolite A, used as a water softener in detergent powder.

This is exploited in a major way in water softening, where alkali metals such as sodium or potassium prefer to exchange out of the zeolite, being replaced by the "hard" calcium and magnesium ions (double ion charge) from the water. Many commercial washing powders thus contain substantial amounts of zeolite. Commercial wastewater containing heavy metals, and nuclear effluents containing radioactive isotopes can also be cleaned up using such zeolites.

2.1 ORIENTED (SILICALITE-1) MOLECULAR SIEVES ON SILICONNITRIDE MEMBRANES [3] *.

In the past decades, the interest in zeolite-on-support configurations has grown worldwide [4,5]. Research on the formation of supported zeolite layers has been mainly focused thus far on the medium-pore (10-membered oxygen ring} tectosilicate MFI (silicalite-l/ZSM-5). This zeolite is known to grow relatively easily on various surfaces, whereas its growth is controlled by a broad range of concentrations, pH, and temperature without the formation of other zeolites, allowing thorough investigations on its growth behaviour. The growth of MFI [6,7] on porous supports like alumina or stainless steel generally results in a random orientation of crystals, due to the roughness of the supports applied. Until now, oriented films have only been prepared on nonporous silicon wafers [8,9]. For permeation studies, continuous layers without pinholes are essential to allow selective passage of gas molecules through the zeolite pores only. To obtain defect-free MFI layers on porous supports, relatively thick layers are required. As a result, fluxes through the membrane decrease. For a full understanding of the mass transport through zeolite layers, uniformity of the layer is essential. In modelling studies [10] on permeation behaviour, exact values of the open area of the support and the (uniform) thickness of the zeolite layer are required. A new preparation method will be reported for the synthesis of oriented MFI layers on silicon wafers. Crystal orientation on trenches and pores, etched in the wafers, will be discussed. As an extension of this work, crystal growth over silicon nitride windows inside the wafers while, by selective etching procedures, the nitride windows are removed to form an oriented and locally non-supported layer of MFI will be discussed. Such layers make investigation of stress created only by the zeolite crystals possible. It will be shown that the removal of the organic template causes stress in the obtained layer, depending on the crystal orientation.

Silicon wafers were placed in 50 cm^3 autoclaves with tetrapropylammonium hydroxide (TPAOH, 40%, CFZ Zaltbommel), demineralised water, and tetraethoxysilane (Janssen Chimica). The applied molar composition for obtaining b-oriented crystal layer was:

H_2O : TPAOH : TEOS = 99.0 : 0.14: 0.84

The obtained synthesis mixture was then aged at ambient temperature for 1 hour under vigorous shaking before being placed in the autoclaves (filling: 2/3 of its volume). After synthesis, the autoclaves were rapidly cooled with cold water and the wafers were washed with demineralised water and dried at 80°C.

---

*M.J. den Exter et al., Zeolites (1997). Partially reprinted with permission of Elsevier Science.

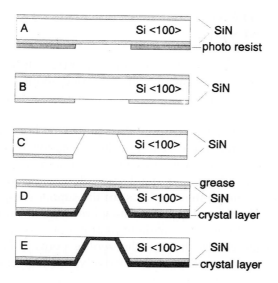

Fig. 6 Preparation of molecular sieve layer on silicon/siliconnitride composite.

Silicon as well as silicon nitride can be obtained as thin layers. In the latter case, thin layers of nitride can be prepared as windows inside the silicon wafer. These window-containing wafers were prepared using semiconductor technology. Fig. 6, A-E, shows the sequence of preparation steps to obtain the substrates.

On both sides of the <100> wafer a silicon nitride layer is grown by low-pressure chemical vapour deposition at 846°C, and a pressure (P) of 150 mtorr using dichlorosilane and ammonia (DCS/NH3 = 170/30 standard cm3/min [sccm]) is obtained. On stoichiometric silicon nitride a tensile stress is imposed due to the silicon substrate. Therefore low-stress nitride ($Si_x N_y$) was used. Such a stress-free behaviour is limited to thin layers only, thus nitride layers of 500 nm were grown. A mask was made in a photoresist film, covering one side of the substrate (Fig. 6A), using a lithographic process. The uncovered nitride parts are removed by means of plasma etching (Drytek 384 T, RF-power = 400 Watts, SF_6 = 100 sccm, p = 250 mtorr), and the photoresist was removed in an oxygen plasma (Fig. 6B). Next, the uncovered silicon is removed by etching in a 9 M KOH-solution at 83°C for 5.5-6 h (Fig. 6C). Due to differences in etching rates for the <100> and <111> direction (approximately 100 times faster in the <100> direction), windows are formed under an angle of 54.74°. The etching stops at the nitride layer due to the rate difference in etching for silicon (100 μm/h) and nitride (0.02 μm/h), resulting in a selectivity ratio of 5,000 to 1. The obtained substrates were placed in the autoclaves with the silicon side facing slightly downward. On this side crystals of silicalite-1 are grown at elevated temperature (165 °C) for 19 hours. The side facing upward, is shielded from the synthesis mixture with a layer of vacuum grease (Fig. 6D). After crystallisation, the grease is removed by boiling in refined petroleum ether and subsequent washing in HNO_3 (100%). Next, the remaining nitride is removed in boiling H_3PO_4 (85%)20 at a temperature not exceeding 157°C (etch rate SiN = 1,500 Å/h, selectivity SiN/SiO2= 32, selectivity SiN/Si = > 100) to obtain a non-supported zeolite film (Fig. 6E).

The obtained SiN layers can be perforated using a lithographic process on the nitride side of the substrate. After plasma etching (Alcatel 300 GIR) and removal of the photoresist, holes limited to a size of 4 x 4 $\mu m2$ and a total open area of 24% can be obtained.

The obtained crystal layers were investigated with light microscopy, scanning electron microscopy, and XRD analysis (Siemens Kristalloflex D5000), using CuK_α-radiation without α2-stripping. The thickness of the crystal layer was determined using a surface profiler (TENCOR 500). The removal of the organic template was monitored using FTIR spectroscopy (Broker IFS 66).

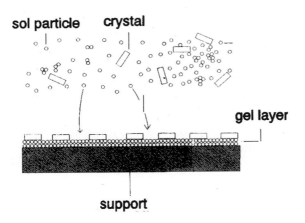

Fig. 7 Synthesis of b-Oriented silicalite-l layers on silicon/silicon nitride substrates.

The following model describes the growth of the crystal layer under the synthesis conditions applied: under aging, the silicon source is hydrolysed and small sol particles are formed. These silicon oxide sol particles try to lower surface energy by precipitation on the support surface and form a thin gel layer during synthesis. This tendency increases due to a pH drop, resulting from Hofmann degradation of TPA, causing a decrease in solubility of the silicon species.

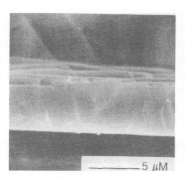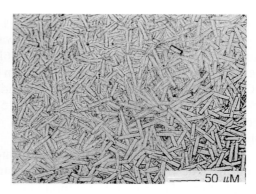

Fig. 8 Left: Cross-section of free hanging zeolite membrane, Right: overview of oriented silicalite crystals on the surface after removal of the template.

Crystals start nucleating on the interface of the gel and liquid where the gel supplies the silicon and the liquid supplies the template for zeolite growth. The crystals consume the gel until they reach the support surface. In the mother liquid sol, particles form larger units by coagulation and move to lower parts in the autoclave due to gravitational forces forming a thicker gel. In the mother liquid crystals nucleate as well. To form a uniform gel layer, no cations should be present in the synthesis mixture. Metal ions (including sodium) stimulate gel formation in the entire synthesis volume rather than selectively on the surface of the support. For that reason, a sodium-free synthesis is used while the initial pH and TPA concentration is high, so as to stimulate the pH drop while maintaining enough template for the actual growth of the zeolite.

This procedure, resulting in a b-oriented crystal layer, was applied on a l-cm^2 silicon substrate containing silicon nitride windows.

Calcination, removal of the template

Two types of calcination procedures were used on (a,b)-oriented layers, both performed in air: (I) 4 h at 400°C with a ramping of 0.5°C/min (one step); and (2) a sequential increase in temperature with steps of 50°C. After each step (ramping 0.5°C/min and 4 h at a constant temperature), the composite was cooled down to 25°C and again heated to a temperature, increased with 50°C steps.

3. PATTERNING OF NANOSTRUCTURES

With laser interference lithography it is quite easy to downscale pore sizes (in microsieves) to the nanometer regime, herewith giving birth to 'nanosieves' [11]. A known technique to pattern surfaces is to evaporate a target material through a thin membrane (shadow mask) with well-defined perforations [12]. As a poor mans technique shadow masks may be used to make patterns on a substrate if photo-lacquer patterned layers are to elaborate or if the substrate is not sufficiently flat to do photolithography. Shadow masks with micro or nano-sized perforations may also be used in other applications as reactive ion etching, ion beam etching [13], electron beam patterning, near field optics (NFO), etc.

Self assembly (nano) mask preparation

Reactive ion etching through a shadow mask has been used for direct etching of a (nano)pattern in a substrate (see Fig. 9).

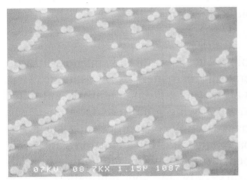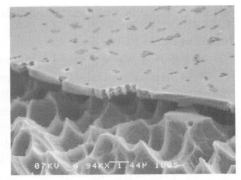

Fig. 9 Photographs of a membrane made with a self assembled mask. Left: nanosized particles on a silicon nitride membrane layer. Right: after chromium deposition over the particles and subsequent removal of the particles, the pores in the membrane have been etched (van Rijn 95 [14]).

The micro/nanopattern may, for instance, be formed by using particles with a uniform size with particle sizes ranging from 5 nm to 5 μm, for example a silica dispersion or a latex suspension. A more or less ordered distribution of particles will then be found in the pattern forming layer on the support. When using a wet or dry etching process with a higher etching rate for the particles than in comparison to the pattern forming itself, perforations in the pattern forming layer and subsequently in the membrane layer will be etched at the former location of the particles. Suspensions with small or large particles (e.g. liquid crystals with assembled molecules) may also directly be deposited on a substrate using, for instance, spin coating or evaporation techniques. After the evaporation of the solvent of the suspension a very thin metal layer, typically a

10 nm chromium layer, may be deposited, e.g. by means of vapour deposition, on the substrate and on top of the particles (e.g. silica particles with diameter 30 nm). The silica particles are then dissolved in a buffered HF solution resulting in a perforated chromium layer with perforations of approximately 20-30 nm in diameter depending on the chromium deposition conditions (correction for shadow effect of deposition). The chromium layer may be used as a membrane layer, or alternatively, as a mask layer for the (dry) etching procedure of a membrane layer underneath the chromium layer (see Fig. 9).

3.1   STENCILLING WITH NANOSIEVES COATED WITH SELF ASSSEMBLED MONOLAYERS [*][15]

A few years ago Aquamarijn Research was asked to collaborate in projects on shadow mask applications [16] using nanosieves and became also supplier of nanosieve stencils for other research groups [15, 17].

Introduction

Shadow mask evaporation technology has several advantages. It is a resistless technique that allows generation of patterns in a single step. Integration of an x,y-stage for substrates allows generation of a nearly unlimited number of patterns using a single mask. Employing a multi-material source, a broad variety of materials can be deposited within one processing step. The nanostencil technique allows fast and parallel replication of patterns in a way comparable to optical lithography. Expensive and low-throughput techniques such as LIL, EBL or FIB are necessary only for the fabrication of the nanostencil. Last but not least, the nanostencil technique is environmentally friendly since no etchants and processing liquids are needed and less waste is produced. In principle, evaporated feature sizes are limited only by the aperture sizes of the stencil and thus by the limitations in their fabrication.

Evaporation through a shadow mask (nanostencil) may be accompanied by gradual clogging of the apertures due to adhesion of evaporated material. In order to reduce this effect, nanostencils may be coated with alkyl and perfluoroalkyl self-assembled monolayers (SAMs).

The formation and properties of SAMs on planar silicon nitride substrates were studied by contact angle goniometry, XPS, and AFM. The SAMs are stable under evaporation of gold under various angles. SAM-coated nanostencils showed considerably less adhesion of gold compared to naked Si_xN_y stencils.

---

[*] Partially reprinted with permission from Nanoletters (2003).

The spiralling progress information technology (IT) has seen over the last decades was essentially driven by ever faster hardware miniaturization. In the near future, however, the semiconductor industries will reach the resolution limits of optical lithography, their current central technology. These limitations on one hand and the development of scanning probe techniques and nanotechnology in general on the other, have led to various novel patterning methods, e.g. laser interference lithography (LIL), electron beam lithography (EBL), focussed ion beam (FIB), X-ray lithography, soft lithography and nanostencil technique [18,19].

A nanostencil consists of a thin silicon nitride membrane (thickness 100-350 nm) with aperture patterns obtained by optical lithography, EBL or FIB and is used as a shadow mask for the physical vapour deposition (PVD) of metals, semiconductors, insulators or certain organic molecules onto various substrates under high vacuum (HV) or ultra high vacuum (UHV) conditions. The fabrication of nanostencils with circular or line-shaped apertures having widths

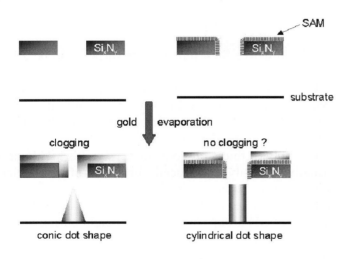

Fig. 10 Schematic representation of the nanostencil technique, the associated clogging problem and the expected effect of SAM coatings.

down to 200 nm is now state of the art and further reduction of the aperture size down to 80 and 50 nm has been demonstrated.

The schematic representation of the formation of circular metal dots given in Fig. 10 above shows a potential problem which may occur when working with a nanostencil: evaporated material is deposited not only on the substrate but also inside the apertures of the stencil, leading to their clogging. Frequent cleaning of the nanostencil using etchants or plasma treatment may solve this problem but is undesirable since it requires a time-consuming interruption of the HV process. Preventing adhesion of material to the aperture walls and thus increase of the performance and throughput of the nanostencil technique is envisaged by coating the nanostencil with an anti-adhesion layer. A choice was made to coat the nanostencils with organic self-assembled monolayers (SAMs) and to investigate their influence on the adhesion properties of the stencil. The SAMs are formed by an irreversible chemical surface reaction of silanes with surface OH groups.

SAMs on silicon nitride

Literature on the surface chemistry of silicon nitride is mainly limited to the functionalisation of silicon nitride AFM tips and their interactions with various substrates. Although chemical modification of planar Si_xN_y surfaces with ω–alkenyl-, CH_3-, NH_2- and SO_3H-terminated monolayers has been reported in two articles on tip-sample interactions [20], few details on experimental conditions and monolayer stability that could be of potential value for our objective are given. Therefore some initial experiments on the formation of organic monolayers on Si_xN_y was carried out and to understand their physical and chemical properties [21].

Silicon nitride has a thin native SiO_2 hide and its surface can be activated by treatment with piranha solution (H_2SO_4/30 % aq. H_2O_2, 3:1). Afterwards, the surface is composed mainly of silanol and silazane groups [22]. Their ratio depends on various parameters, but at low pH and in the presence of water mostly hydroxyl groups are found [23]. These surface groups react with silane or germane molecules to give the desired monolayers.

Whereas the self-assembly of thiols on gold surfaces yields highly ordered monolayers and can usually be easily reproduced, formation of monolayers on silicon oxide (as bulk material or as native oxide layer on Si or Si_xN_y) is less defined [24]. The silane (or germane) compounds used are usually trifunctional and react not only with surface OH groups but also with neighbouring molecules in the monolayer. This cross-linking increases the stability of the layers. However, since the degree of cross-linking is influenced by various factors (temperature, water content), reproducibility is poor. Especially water plays an important role in the surface reaction: traces of water appear to be necessary in order to hydrolyse the Si-Cl or Si-OR bonds of the silanes before the condensation reaction with surface OH groups can occur [25]. However, silanols generated by this hydrolysis can also react with chloro or alkoxy silanes in solution, giving polymer threads and globules which can contaminate the surface by physisorption or chemisorption (via non-polymerised functionalities). Thus, working under exclusion of moisture is indispensable in order to avoid contamination and to suppress the competing bulk polymerisation reactions, but complete removal of water from solvents, equipment, and substrates can inhibit the desired surface reaction. The optimal conditions for the reproducible formation of high quality monolayers must therefore be established by a trial and error approach.

Reproducible and good quality SAMs were obtained by immersion of freshly cleaned Si_xN_y substrates in 0.2 % v/v solutions of the silanes or germanes (for compounds see figure below) in dry CH_2Cl_2 at room temperature over a period of 16 h, followed by extensive rinsing with organic solvents in order to remove physisorbed material.

		θ_{adv} [°]	θ_{rec} [°]
C12-SiCl₃	Cl—Si-C₁₂H₂₅ (Cl, Cl)	115 118[a]	91 84[a]
C12-Si(OEt)₃	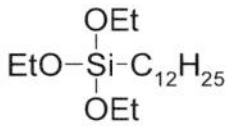 EtO—Si-C₁₂H₂₅ (OEt, OEt)	113	93
CF-Si(OEt)₃	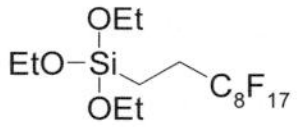 EtO—Si C₈F₁₇ (OEt, OEt)	120	95
CF-SiMe₂Cl	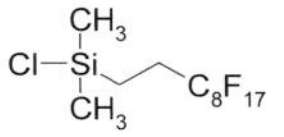 Cl—Si C₈F₁₇ (CH₃, CH₃)	113[a]	88[a]
HOOC-GeCl₃	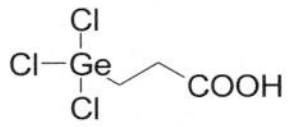 Cl—Ge COOH (Cl, Cl)	<20	<20

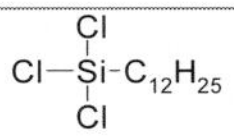

Fig. 11 Dynamic advancing and receding water contact angles for silane and germane SAMs on Si_xN_y. [a] **deposition from the gas phase.**

Compounds C12-SiCl₃ and CF-SiMe₂Cl (because of its lower boiling point compared to CF-Si(OEt)₃) were employed in an attempt to use vapour phase formation of a SAM. The resulting monolayers were studied by dynamic contact angle goniometry, tapping mode AFM, and XPS.

The water contact angle data indicate hydrophobic surfaces with a hysteresis of 20-25° for the alkyl and perfluoroalkyl SAMs deposited from solution. This is in accordance with values reported for silane SAMs on silicon oxide. Also the higher hydrophobicity of perfluorinated SAMs compared to hydrocarbon SAMs has been observed before [26]. For CF-SiMe₂Cl a slightly less hydrophobic and less ordered SAM compared to CF-Si(OEt)₃ was observed. This is ascribed to a lower surface density of perfluoroalkyl chains due to the space required by the methyl groups. Nevertheless, the difference in SAM quality is small (also for C12-SiCl₃) and gas phase deposition can thus be regarded as a valuable, alternative preparation method equivalent to solution deposition. The reduced demand of solvents and the exclusion of solution polymerisation may make it preferable for some applications from a technical point of view. The low contact angle values measured for the SAM of the trichlorogermane HOOC-GeCl$_3$ indicate successful formation of a closely packed highly hydrophilic SAM.

AFM showed smooth surfaces with very little thread-like or globular aggregates (height ~5 nm), most likely resulting from some solution polymerisation. The roughness was the same as for bare silicon nitride surfaces (3 nm). Despite the XPS penetration depth of only 5 nm, intense signals for Si and N were observed which can be ascribed to the Si$_x$N$_y$ substrate. This confirms the absence of multilayers. The atomic concentrations of C and/or F are in the expected range, although somewhat obscured by excess carbon, probably from atmospheric contamination [27].

Apolar compounds C12-SiCl$_3$, C12-Si(OEt)$_3$, and CF-Si(OEt$_3$) for coating the nanostencils were chosen. It was found on contact angle measurements) that these monolayers can be stored at ambient conditions for months without changes in their properties. They are stable against repeated washings with organic solvents like CH$_2$Cl$_2$, ethanol, hexane, and toluene. No change in contact angles was detected after 24 h immersion in water and after storing the samples in an oven at 100 °C for 2 h. They are also fully stable in KI/I$_2$ gold etching solution for hours. In basic cyanide gold etch solution the contact angle starts to decrease slowly after immersion for >10 min. Surprisingly, the SAMs are also relatively stable in piranha solution (contact angles decreased <10° over 1 h for CF-Si(OEt)$_3$). In order to check the stability of the SAMs under the conditions of metal evaporation out an experiment with evaporation followed by etching was carried out. Hereto, SAMs of C12-SiCl$_3$ and CF-Si(OEt)$_3$ were prepared on silicon nitride substrates. The advancing and receding water contact angles were found to be as listed in figure above. One half of the substrate was masked with a silicon square. Then 40 nm gold was evaporated onto the unprotected region under an angle of 90°. The evaporation was carried out using a thermal evaporator. The change in thickness of the gold layer formed was monitored by means of an internal quartz crystal microbalance (QMC). After this process, the contact angles of the uncoated region of the monolayer had not changed significantly, whereas the gold-coated area showed strongly reduced hydrophobicity. The gold was removed by selective etching (K$_3$[Fe(CN)$_6$]/ K$_4$[Fe(CN)$_6$]/KOH/Na$_2$S$_2$O$_3$, 15 min). Hereafter, the advancing contact angle was about ~12° lower on both areas. This change can probably be ascribed to the etch process.

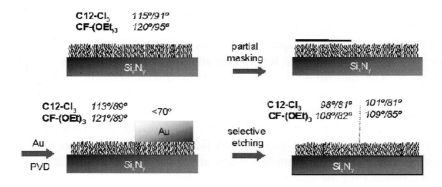

Fig. 12 Experiment on bombardment of SAMs with high-energy gold atoms; ascending/receding dynamic water contact angles for SAMs of C12-SiCl3 and CF-Si(OEt)3 after individual steps of the procedure (for details see text).

The absence of a difference between gold-covered and uncovered areas led to the conclusion that bombardment with gold under HV conditions has no destructive impact on an alkylsilane or perfluoroalkylsilane monolayer.

In order to better model evaporation inside the apertures, planar substrates with and without CF-Si(OEt)$_3$ SAM were placed at angles of 45°, 30°, and 0° with the direction of the incoming gold. The last case represents lateral bombardment of the SAM, with the high-energy metal atoms attacking the molecules in the SAM not at their end, but perpendicular to their chain, as likely is the case in the nanostencil apertures. After evaporation of 40 nm gold, the substrates placed at angles of 45° and 30° were covered with a layer of gold visible to the naked eye, for the SAM-coated substrates as well as for the untreated controls. For the substrates placed at 0°, however, no gold deposition was apparent. Water contact angle measurements at the SAM-coated sample showed no decrease in hydrophobicity (θ_{adv}=120°), but no distinct value for the receding contact angle could be measured, indicating that the order of the SAM had somewhat decreased. The latter observation can be ascribed to adhesion of small amounts of gold to the SAM-coated surface (3 % according to XPS measurements). These results indicate that the SAMs are stable also under lateral bombardment.

Experiments with Nanostencils

The nanostencils consist of silicon nitride membranes with hexagonal arrays of holes having nominal diameters of 500 nm, and spacings of 1 µm. The thickness of the membranes is 800 nm. The holes have a tapered cross-section. Whereas on one side their diameters correspond to the nominal value, they amount to 605±35 nm on the other side.

The SAM coating of the nanostencils was carried out in exactly the same manner as described above for planar substrates.

In order to be able to clearly visualize differences in material deposition inside the stencil apertures large amounts of gold (thickness 1µm) were evaporated. The nanostencils were placed on silicon substrates (1x1cm). The face with the smaller aperture diameters was positioned towards the gold source, the opposite face towards the substrate. In order prevent the micrometer dots from growing into the stencil apertures, a photoresist spacer ring of 1.5 µm height was fabricated on the substrate. A flexible metal clamp was used to hold the samples in place on a sample holder, an aluminum plate with a diameter of 3'' and windows for the samples.

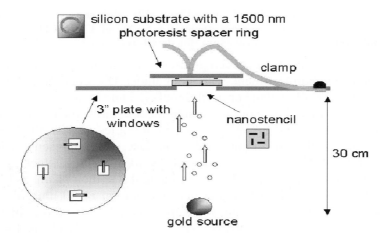

Fig. 13 Experimental setup for nanostencil evaporation experiments.

After the evaporations, the nanostencils were investigated by SEM. FIB etching was used to cut slits into the gold-covered membranes in order to image the cross-sections of the apertures by FIB itself (not shown) [28] and by SEM under high tilt angles.

Figure shows SEM images of the nanostencil after evaporation of 1µm gold through 500-nm aperture. The holes of the untreated nanostencil seem to contain more gold than those of the CF-Si(OEt)$_3$-coated stencils. The pictures show that the gold forms long, leaf-shaped deposits in the holes of the former, indicating diffusion controlled growth, whereas thinner lines of gold seem deposited inside the CF-Si(OEt)$_3$-coated apertures. The same was observed for an evaporation experiment with 1 µm gold through 500- nm apertures coated with C12-Si(OEt)$_3$ [29].

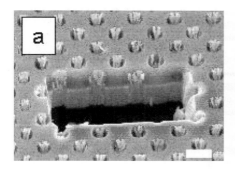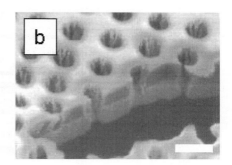

Fig. 14 SEM images of stencils after evaporation of 1 μm Au; a) uncoated (22 Kx), b) CF-Si(OEt)3-coated (15.3 Kx); scale bars: 1 μm.

This proves that SAM formation has taken place inside the apertures as well as on the planar areas. Obviously, the highly a-polar monolayers strongly reduce the attractive interactions of the surface of the aperture walls to gold. Whereas gold is deposited readily inside the uncoated holes, with clusters deposited early in the process apparently acting as nucleation centres for the rapid growth of crystalline gold, only insignificant amounts of metal adhere to the a-polar organic coatings. An observation was made that gold peels off easier from the SAM-coated stencils than from the uncoated ones.

The nanostencils could be cleaned easily by immersion in I_2/KI gold etch solution (0.2 g I_2, 3.15 g KI in 50 mL H_2O) for ~10-20 min at room temperature. As mentioned earlier, the SAMs are stable under these conditions. No difference in etch rate between coated and uncoated nanostencils was observed.

Nano Stencil Fabrication.

The used nanostencils were made by silicon micro machining by Aquamarijn Research, shown in Fig. 15.

First on a one side polished <100> silicon wafer a thin layer (thickness 500nm) of low stress silicon nitride was deposited by Low Pressure Chemical Vapour Deposition (SiH_2Cl_2:NH_3 = 70:18 at 850°C, 250mbar). On the silicon nitride (step 2), at the polished side, a pattern with 500nm pores in photo resist was made using mask lithography (wafer stepper ASML PAS 5000/50). After development, this pattern was subsequently etched using Reactive Ion Etching (CHF_3:O_2=25:5, Power=75 Watt, Pressure = 10mBar, 6 minutes) making pores in the silicon nitride (step 3).

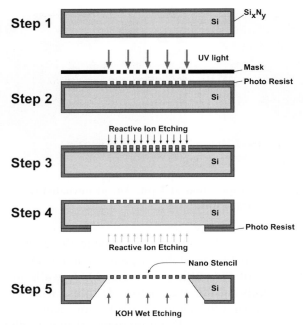

Fig. 15 Manufacturing of a nanostencil.

In step 4, the process for lithography and reactive ion etching is being repeated on the back side of the wafer, defining large rectangular openings in the silicon nitride. Next the silicon between the nanostencil layer and the back side is being removed with an anisotropic etch along the <111>-planes with a 25% KOH solution at 70°C until the membrane layer is reached, step 5. The nanostencils are diced (2mm/sec to avoid damage of the nanostencil) into square pieces of 5*5 mm and cleaned in a 70°C, 25% KOH and 40°C, 5% H2SO4 solution to remove all silicon particles that are being released during dicing.. All glass-ware was immersed in fresh piranha of ~80 °C for at least 1 h, rinsed three times with milliQ water, blown dry with nitrogen and stored in a glove box under N_2 (not longer than 2 h). The planar silicon nitride substrates (150 nm Si_xN_y on Si, 1 cm x 1 cm) or nanostencils were treated with fresh piranha for 30 min, rinsed three times with milliQ water, blown dry with nitrogen and also stored in a glove box under N_2 prior to use (not longer than 1 h). Solutions were prepared immediately before use. 0.05 mL silane or germane were added to 25 mL CH_2Cl_2 (freshly distilled from $CaCl_2$) in a glass dish inside the glove box. The substrates were immersed in the solution and the dish covered with a lid. After 16 h the substrates were transferred into a beaker-glass containing 20 mL distilled CH_2Cl_2 and rinsed under shaking. The solvent was discarded and the rinsing repeated three times. After addition of another 20 mL CH_2Cl_2, the substrates were taken out of the glove box and rinsed again with two 20-mLportions CH_2Cl_2, abs. EtOH and CH_2Cl_2 each. Then the substrates were blown dry with N_2. For gas phase deposition, the piranha-cleaned substrate was placed top-down above a dish containing 1 mL $CF\text{-}Me_2Cl$ in a desiccators. The desiccators was evacuated to ~10 mbar, and after standing for 16 h at room temperature the substrate was removed and treated in the same manner as described for solution deposition.

Conclusions

Silane SAMs on silicon nitride can be prepared and have properties comparable to those known from SAMs on silicon dioxide. The SAMs are stable against various physical and chemical influences and against vertical and lateral bombardment with gold vapour in HV. SAM coatings were expected to reduce adhesion of gold inside the apertures of silicon nitride nanostencil membranes. This effect is confirmed by SEM on nanostencils after gold evaporation: the images clearly show that less gold is deposited inside the pores of SAM-coated stencils compared to uncoated ones. These results show that SAM coatings can be used to improve performance and lifetime of nanostencils, thus increasing the throughput of this promising new nanopatterning technique.

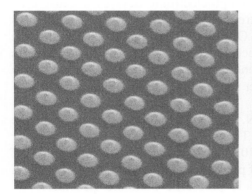

Fig. 16 SEM micrographs of gold spots (300 nm) and pillars (bottom 300 nm) evaporated through a nanostencil (courtesy Micro and nano engineering conference 2001).

4. NANOSIEVES FOR PHOTONIC STRUCTURES

Eli Yablonovitch proposed in 1989 [33] to study a new class of photonic structures, i.e. photonic crystals.

Photonic crystals are artificially 3-dimensional (or 2- or 1-dimensional) structures fabricated in an optical material (crystal or amorphous) with unit cells whose dimensions are comparable to the optical wavelength. If the artificial structure has an appropriate symmetry, it can exhibit a photonic band gap forming what is called, a photonic bandgap (PBG) material or crystal. This bandgap in the photon energies is analogous to electron bandgaps in semiconductors. This is accomplished by nanofabrication of a structure which has 3-dimensional periodicity in the dielectric constant (compare this to the 3-D periodicity in the potential for electrons in semiconductors). These structures make good reflectors since photons with energies within the bandgap are not allowed and so will not propagate inside the crystal. A 3-dimensional photonicbandgap has a bandgap for photons travelling in any direction.

Photonic crystals are interesting for optoelectronic devices because they make it possible for photons to be controlled and confined in small structures (on the scale of one wavelength cubed). One of the most interesting application of photonic crystal structures is to project novel light-emitting devices such as light emitting diodes (LED's) and semiconductor lasers. These devices should be able to achieve greater efficiency, smaller size and greater modulation speeds.

In 2000 van Rijn was asked by N. van Hulst and K. Kuipers from the Applied Optics Group of University of Twente to transfer knowledge on laser interference lithography engineering [30,31] to enable the manufacturing of photonic structures with a high degree of periodicity. In fact free hanging silicon nitride nanosieves themselves are already prototype photonic structures. W. Nijdam from Aquamarijn together with M.Sc. student L. Vogelaar (joined Aquamarijn in 2001-2002) performed an excellent job in the transfer of this knowledge and a nice article about the nanosieve/photonic structure technology was written.

4.1 SILICON NITRIDE PHOTONIC CRYSTALS [*][32]

Recently, there has been growing interest in photonic band-gap materials, which have a spatially periodic refractive index with a lattice constant on the order of the wavelength of light, because of potential scientific and technological applications based on their unique light propagation properties

---

[*] Partially reprinted with permission form Advanced Materials, Wiley-VCH Verlag (2001).

[33]. However, experimental surveys of the optical properties of photonic band-gap materials have been limited due to the difficulty in preparing well-constructed samples except in the long wavelength region [34,35].

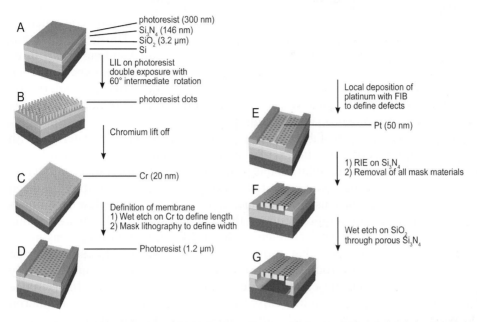

Fig. 17 Schematic representation of procedure, which combines large area patterning through Laser Interference Lithography and local introduction of defect structures through FIB-assisted deposition. The dimensions in these pictures are chosen for illustrative clarity and are not to scale. Note that a slab of 100 μm x 4 mm contains over 4.5 million holes. A photo resist layer on top of a Si₃N₄ (146 nm), SiO₂ (3.2 μm) and Si layer stack (A) is converted into a large area triangular dot pattern with LIL (B). By chromium lift off a thin, highly selective etch mask containing holes is formed (C). Next, the etch mask is divided into smaller sections by mask lithography (D). Local deposition of platinum is used to define defect structures of any shape in the photonic crystal slabs. Here the introduction of a line defect (E) is demonstrated. By RIE the structure is etched into the Si₃N₄ layer, after which Pt, Cr and photo resist are removed (F). By a wet etch through the holes the underlying SiO₂ layer is removed and the freestanding Si₃N₄ photonic crystal slabs are formed (G) (Courtesy Adv. Mat. [36]).

Photonic crystals are novel materials with unique optical properties [37]. The crystals have a periodic modulation of the refractive index. As a result, the dispersion of light will be described by a band structure analogous to those of electron waves atomic crystals. Under the right conditions a photonic crystal can exhibit a photonic bandgap: light in a certain range of optical frequencies is forbidden in the crystal [38]. The existence of a photonic bandgap enables an unprecedented control of spontaneous emission and propagation. By locally disturbing the periodicity, a defect-associated photon state is created which can be used to guide light.

A photonic crystal slab is a thin film with a two-dimensionally periodic refractive index modulation in the plane [39,40]. In a photonic crystal slab the light is confined to the crystal plane by a classical slab waveguide construction. For in-plane wave vectors a bandgap can be created. Thus, the slab has applications in light guiding without the need for a full bandgap in all three dimensions. Optimal performance of the photonic crystal slab is expected when the slab is mirror symmetric in the vertical direction, i.e. when the material on both sides of the slab has equal refractive index at least in the region of the near-field tail of the in-plane propagating light [41].

Photonic crystals are fabricated by periodically arranging materials with highly dissimilar refractive index. To obtain a bandgap in the visible the periodicity of the index modulation has to be in the submicron range, < 350 nm. For the fabrication of photonic crystal slabs electron beam lithography has been by far the most frequently used technique, since it is one of the few techniques that fulfils the accuracy requirements on these length scales [42]. However, due to its 'direct sequential write' character e-beam lithography is time consuming for large area structures plus that uniformity errors in periodicity can occur due to errors (or quantization roundings) in the beam alignment. It would be highly advantageous to define the entire periodic pattern at once. Laser Interference Lithography (LIL) has proven its ability to generate uniform periodic submicron patterns over very large areas (in the order of square centimetres) [43].

 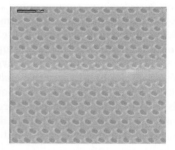

Fig. 18 Left: SEM-image of triangular photo resist dot pattern with a period of 300 nm after Laser Interference Lithography. The uniform pattern extends over an area of the order of a centimetre squared. Right FIB-image of an approximately 50 nm thick platinum line covering holes in a 20 nm thin chromium mask. The platinum line defines a defect structure in the photonic crystal slab and is deposited locally with a Focused Ion Beam (Courtesy Adv. Materials).

A combination of large area patterning with LIL, and local patterning with a direct writing method, e.g. a Focused Ion Beam [44], is ideal for future fabrication of photonic crystal slabs with arbitrarily shaped light-guiding defects.

Here a combination of LIL with Focused Ion Beam (FIB) assisted deposition is presented. Highly uniform freestanding photonic crystal slabs have been fabricated, extending over areas as large as a few hundred micron squared, with the possibility to introduce any kind of defect structure. In this way is realized, to our knowledge, the largest photonic crystal slab (100 μm x 4 mm) for visible radiation. Long line defects, extending over more than 1 mm, were also introduced.

Fig. 19 Left; SEM micrograph of the end face of a photonic crystal slab between the support bars. This 146 nm thin slab is freestanding across 100 μm and has a length of 4 mm. Right: SEM micrograph, close-up of the defect line in the photonic crystal slab. The highly uniform triangular hole pattern has a period of 297 nm. The axial widths of the elliptical holes are 212 and 164 nm. The shape of the holes appears unharmed by the presence of the defect line. The defect is expected to form a multimode waveguide for visible light (estimated central wavelength 670 nm). (left: Courtesy Applied Optics Group, University of Twente [36], right: Courtesy Adv. Materials).

In the photonic crystal slabs Si_3N_4 (n=2.16 at λ=670 nm) and air acted as high and low index material, respectively. A triangular lattice of air holes in the Si_3N_4 layer was fabricated in order to obtain a bandgap for TE modes (polarization in the plane of the crystal). By making the slab freestanding, i.e. embedded in air on both sides, one obtains the desired vertical mirror symmetry of the crystal slab.

The advantage of LIL is that the entire periodic pattern is created at once. A high accuracy and flexibility of FIB-deposition to introduce line defects in the extended crystals has been used. In principle, any defect is possible. The combination of LIL with FIB-deposition opens avenues for more optical applications of photonic crystals. Due to its simplicity and scope, FIB-assisted LIL is a promising alternative for e-beam lithography in the fabrication of photonic materials.

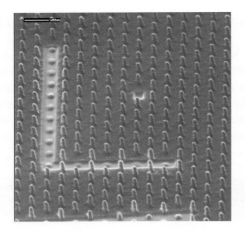

Fig. 20 SEM micrograph of a FIB assisted removal of a number of pillar structures obtained with Laser Interference Lithography (Courtesy Applied Optics Group, University of Twente).

4.2 ALUMINA PHOTONIC CRYSTALS [45]

Porous alumina, made by anodising aluminium, consisting of an ideally ordered air-hole array with a lattice constant of 200 to 250 nm exhibits a two-dimensional (2D) photonic band gap in the visible wavelength region.

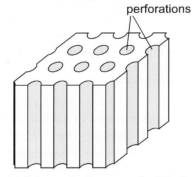

Fig. 21 Schematic diagram of a 2D photonic crystal using anodic porous alumina.

The air-hole preparation process was based on Al pretexturing by imprinting with a mould, followed by the anodisation of Al in acid solution.

The depth limit of the ordered air holes is due to the disturbance of the arrangement of the air holes along with the growth of the oxide layer during anodisation.

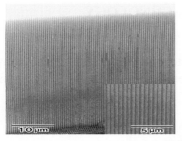

Fig. 22, SEM micrograph of the highly ordered anodic porous alumina. Anodisation was conducted in 0.1 M phosphoric acid solution at 0°C at 200 V for 360 min. The lattice constant and filling factor were 500 nm and 0.3 μm respectively. The inset shows an enlarged view of the air-hole array at the bottom region of the anodic porous alumina (Courtesy Jpns. J. Appl. Phys. [45]).

Anodization in a phosphoric acid solution under a constant voltage condition of 195 V yielded an air-hole configuration with a lattice constant of 500 nm.

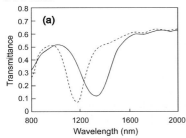 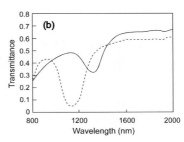

Fig. 23 Transmission spectra of anodic porous alumina for H (a) and E(b) polarization light for the Γ-J (- - -) and Γ-X (---) directions. The lattice constant and filling factor were 500 nm and 0.1 respectively (Courtesy Jpns. J. Appl. Phys. [45]).

Fig. 23 shows the typical transmission spectra of the anodic porous alumina with a 500 nm lattice constant. The filling factor of the air holes was 0.1. In Fig. 23 a), which presents the transmission spectra for H polarization light, distinct dips in the transmission were observed for both the -J, and -X directions. The observed position and width were almost in good agreement with the theoretical prediction based on the plane-wave method for the triangular lattice of air holes in the alumina matrix [46].

For a wider range of scientific and technological applications an air-hole array with a higher aspect ratio's will be required for 2D photonic crystals.

5. NANO MEMBRANE SEPARATION TECHNOLOGY

Introduction

Nano and microsieves can be used as stencils for patterning, as photonic crystals, but they can also be used as a proper solid support for thin functional membranes (see Fig. 24).

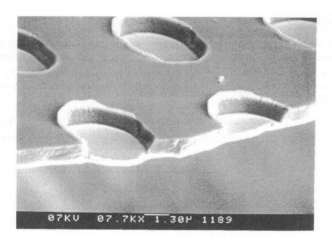

Fig. 24 Pd membranes with a thickness of 200 nm and diameter 4 μm hanging in the pores of a silicon nitride microsieve with a thickness of 1 μm [van Rijn, 47].

Advantage of this nano architecture is that a decoupling is made between the thickness of the functional membrane layer and the strength of the support membrane layer. This nano architecture will be of use in separation processes like gasseparation and ultra filtration, in which the flux through the functional membrane layer should be high and not limited by the permeation resistance of the support. It is even possible to obtain gas separation/molecular sieve membrane layers (e.g. silicon oxide, silicalite) with a thickness of a few nanometers that would enable gas molecules to pass the membrane ballistically [48] without colliding with the walls inside the membrane layer. Very high fluxes may be expected in this case.

An estimate of the maximum pressure load q of such a thin functional membrane with thickness h, width l, Youngs Modulus E and yield stress σ may be obtained with an earlier derived expression (see chapter 5, eq.15):

$$\sigma_{total} = \sigma_{tensile} + \underset{\substack{u \gg l \\ x=0}}{\sigma_{bend}} = 0.29(1+1.47/0.37)\sqrt[3]{\frac{q^2 l^2 E}{h^2}}$$

Using current values for a siliconoxide (see table 1, chapter 5) membrane layer with a thickness of only 1 nm and width 1000 nm a maximum pressure load is obtained of 16 bar, a value comparable with (e.g. a siliconnitride layer with a thickness of 1 μm and width 200 um) the maximum pressure load of the membrane support layer itself.

Thin defect free micro-porous silica membranes show high fluxes for small molecules, such as natural gas purification, molecular air filtration, selective CO_2-removal and industrial H_2-purification [49]. The resistance of a silica membrane to transport of small (fast) molecules should be as small as possible, which requires a very thin layer. Such thin defect-free layers have been achieved by dip-coating silica onto a porous supporting structure, usually consisting of a g-alumina layer on top of an a-alumina layer [50,51].

In order to decrease further the overall resistance to gas transport of ceramic membranes functional membrane layers can be e.g. dip or spin coated onto silicon nitride microsieves (see Fig. 25).

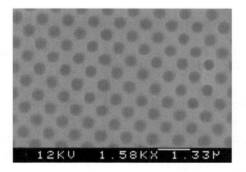

Fig. 25 SEM micrograph of a nanosieve (500 nm) spincoated with a mesoporous silica layer with a mean thickness of 500 nm.

Micro engineered membranes have also been subject of study at other institutes for e.g. air filtration and particle counting applications [52]. Following earlier work [47] palladium membranes for hydrogen separation have also been recently made using a microsieve structure as a support for a very thin palladium membrane (<200nm) for e.g. small-scale fuel cell applications [53]. Also membranes with straight nanometer sized pores (20 nm) have recently been made for cell immuno-isolation and biomolecular separation applications [54,55,56] following formerly presented nanopore construction technologies [47] with use of conventional lithographic techniques (see Fig. 26).

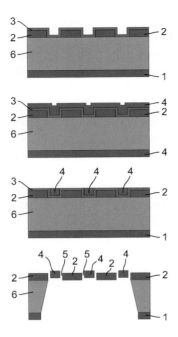

Fig. 26 Schematic drawing showing in cross section subsequent stages of a method of manufacturing a membrane filter with nano sized pores (van Rijn 95 [47]).

In a membrane layer 2 of silicon nitride with thickness 1 μm parallel grooves with a depth of 1 μm and a length of 10 μm are being etched with reactive ion etching techniques. A sacrificial layer 3 of silicon dioxide is grown uniformly with a uniform thickness of 20 nm by means of chemical vapour deposition of tetra ethoxysilane at elevated temperature. Square to the grooves a line pattern (not shown in the figures) is etched in the sacrificial layer 3 with a depth of 20 nm, a width of 1 μm, a length of 1000 μm and a spacing of 10 μm with standard lithographic techniques using a buffered HF solution as a selective etchant. Over the sacrificial layer 3 and in the grooves a groove filling layer 4 with a thickness of 2 μm is deposited, e.g. a polysilicon layer. The groove filling layer 4 is then planarised by polishing or etching such that the grooves remain at least partially filled with groove filling material 4. The groove filling material 4 is then at least partially directly connected to the membrane layer 2 near the line pattern. Subsequently, the sacrificial layer 3 is removed from using the buffered HF solution such that channel like pores 5 with a width of 20 nm, a depth of 1 mm and a lateral length of 10 μm are realized. Openings are etched in the support 6 using KOH as an etchant and the patterned silicon nitride layer 1 as an etch mask. With the process it is possible to make very small pores without alignment of the mask for the parallel grooves and the mask for the line pattern.

An intermediate layer between the membrane layer 2 and the support 6 may be applied as a selective etch stop, for example silicon nitride or alumina.

The membrane may also be used as a separation filter (potential low diffusion times), for instance for `in vivo` screening of blood cells, or in a sensor [3,57], or actuator system. This will give a clear advantage for microsensors and actuators that are processed with use of thin film technology.

Desai [56] has studied the use of nanoporous membrane filters with pore sizes of 25 and 50 nm for albumin retention and diffusion characteristics.

Table 2

Time (min)	Whatman (albumin concentration)		Millipore (albumin concentration)		Micromachined (albumin concentration)	
	Absolute	(g/dl)	Absolute	(g/dl)	Absolute	(g/dl)
0	0.381±	3.88±0.02	0.423±0.003	4.31±0.03	0.395	3.980±0.003
420	0.002	3.58±0.02	0.398±0.003	4.05±0.03	0.394	3.970±0.002
Change in albumin concentration	0.352± 0.004	0.30		0.26		0.01

Interaction of albumin with membranes is also measured by looking at the difference between the initial albumin concentration and the final concentration; the differences can be attributed to both protein diffusion and adsorption [56].

Some typical diffusivity parameters found with their membranes in comparison with conventional membranes are listed in below.

Table 3

Membrane parameters and calculated diffusivities [56].

	Whatman	Millipore	Micromachined
Membrane Thickness (μm)	1 (asymmetric pores)	105	5
Porosity (%)	50	70	0.86
Effective area (mm²)	4	5.6	0.041
Effective difusivity	4.00E-07	5.292E-05	1.025E-08
Absolute diffusivity (D_{eff}/A_{eff})	1.00E-07	9.45E-06	2.50E-07

6. NANO ENGINEERED BIOMEMBRANES[*]

6.1   INTRODUCTION

Research goal in biomembrane science is to investigate how natural biomembranes function and to engineer new types of biomembrane materials on the nanometer scale [58,59,60]. Natural biomembranes, in particular lipid bilayer cell membranes, have an important function in biological molecular exchange (metabolism) and signal transduction processes (e.g. immune reactions, hormone detection) of the living cell. Generally, these processes are taking place across the cellular membrane wall by transmembrane channels and pores for molecular exchange [61,62,63] and signal transduction [64,65].

Most gram-negative bacteria possess an inner and an outer cellular membrane wall. The inner one contains a 'dense' lipid bilayer, which is electrically closed, so that no ions or other hydrophilic substrates can cross the barrier without the help of highly specific membrane proteins. This inner cell wall is separated from the outer one by an aqueous phase filled with water-soluble polymers, the peptidoglycan. In contrast, the outer cell wall is fairly permeable to smaller solutes below a molecular weight of about 400 Da. Such substances can freely permeate under a concentration gradient through general diffusion channels, called porins, in the outer cell wall. The most prominent of the general diffusion porins is OmpF (Outer membrane protein F) [66,67], which is very stable and does not denaturate in e.g. 4 M GuaHCL, 70° C or in 2% SDS (=ionic surfactant). In case of a lack of nutrition, the pure diffusion process is too slow and the bacteria need to enhance the efficiency of the translocation functions. For those cases, nature has created a series of rather specific and highly sophisticated membrane channels. An extensively studied example is the malto-oligosaccharide specific channel LamB, or Maltoporin [68] of *Escherichia coli*. Nano engineered porins in artificial lipid or block copolymer biomembranes may allow observation of the crossing of a single substrate molecule through the porin channel. A well studied example [69] is the α-hemolysin porin in a diphy-tanoylphosphatidylcholine (lipid) bilayer.

High-resolution measurements of ion currents through single porin channels probing among others e.g. neutral solute transport [70], have already been reported. Also the screening of facilitated transport of a wide variety of molecules [71], for instance to study antibiotic uptake through porins and to test for toxins etc., is an subject of ongoing research.

---

[*] This section has been cowritten with W. Nijdam, Aquamarijn Research.

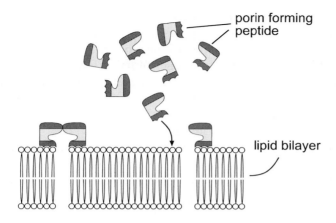

porin forming
peptide

lipid bilayer

Fig. 27 Schematic illustration of the assembly of a pore channel with (α-hemolysin) peptide oligomers in a (diphy-tanoylphosphatidylcholine) lipid bilayer.

Surface layers

In micro-organisms of almost every taxonomic group of walled eubacteria and archaeobacteria [72,73] the outer cell wall has a nanoporous crystalline surface layer (S-layer). This surface layer is composed of a single protein or glycoprotein species with molecular weights ranging from 40,000 to 200,000 Da and can exhibit either oblique, square or hexagonal lattice symmetry [74,75,76].

 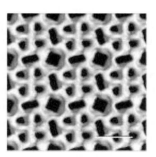

Fig. 28 SEM photograph of a liposome with a nanoporous cell-envelope (Surface-layer). Right a reconstruction of an S-layer structure (bar = 10 nm). S-layers are formed by self-assembly and are made of protein and glycoproteins. (MW 40.000 - 200.000 Da) with an oblique, square or hexagonal 2D crystal structure with typical lattice constants between 5 and 30 nm and a thickness of 5-10 nm. (Courtesy Nanosearch membrane GmbH).

An important feature of these nanoporous structures are their repetitive physicochemical properties, i.e. pores passing through S-layers show identical size and morphology and are in the range of ultrafiltration membranes. Functional groups on the surface and in the pores may be well aligned and are accessible for specific molecules in a very precise fashion. S-layer proteins also reveal the ability to self-assemble into two-dimensional crystalline arrays in suspension, on solid supports, at the air/water interface and on lipid films [77].

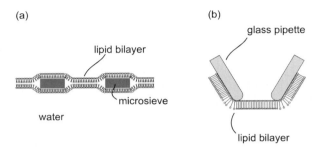

Fig. 29(a) Schematic illustration of a 'black' lipid membrane generated by the method of Montal and Mueller on a micro/nanosieve. Originally this method involved the generation of a lipid bilayer over a septum with an orifice of 40-800 micron [78,79]. For this purpose either a small drop of lipid dissolved in alkane is placed on the opening of the septum [80] or the membrane is formed from two lipid monolayers at an air/water interface by the position of the hydrocarbon chains through an aperture made in a hydrophobic partition which separates the two monolayers. (b) Schematic illustration of a lipid bilayer, generated on the tip of a patch clamp pipette.

Modification of S-layers [81] would allow several possibilities including the manipulation of pore permeation properties, the introduction of switches to open and close the pores, and the covalent attachment to surfaces or other macromolecules through defined sites on the S-layer protein. S-layer proteins can also be produced in large amounts by continuous cultivation of S-layer-carrying organisms. S-layer proteins can be considered biopolymeric membranes with properties perfectly tailored by nature for many biotechnological applications [82], such as matrices for the immobilization of a variety of materials including biologically active macromolecules [83,84] and for functionalisation of synthetic materials. In contrast with the broad pore size distribution of phase inversion membranes the regular pore properties of S-layers enable the use of these artificial biomembranes for nanosieving and separation applications with excellent molecular mass cut-off specifications [85].

One of the most common organisms used for producing S-layer-carrying cell wall fragments and S-layer protein for biotechnological applications is Bacillus stearothermophilus PV72, for which a synthetic growth medium has been developed [86,87]. S-layers have also been used for the production of S-layer based Ultrafiltration Membranes (SUM's) [88]. SUM's are produced by S-layer self-assembly or by depositing S-layer-carrying cell wall fragments on normal microfiltration membranes. Adsorption studies and contact angle measurements have confirmed that these SUM's are net negatively charged and basically hydrophobic [89].

S-layers may also be used as matrices for covalent binding of biologically active macromolecules, such as enzymes (invertase, glucose oxidase, glucuronidase, L-glucosidase, naringinase, peroxidase), ligands (protein A, streptavidin) or mono- and polyclonal antibodies. S-layer-carrying cell wall fragments with immobilized protein A could be applied as escort particles in affinity cross-flow filtration for isolation and purification of human IgG from serum or of monoclonal antibodies from hybridoma cell culture supernatants [90,91]. SUM's with immobilized monoclonal antibodies were also used as reaction zones for dipstick-style immunoassays [92]. For this purpose, human IgG was either directly coupled to the carbodiimide-activated carboxyl groups of the S-layer protein, or it was adsorbed onto a SUM with covalently bound protein A . Alternatively, human IgG was biotinylated and bound to a SUM onto which streptavidin was immobilized in a monomolecular layer [93].

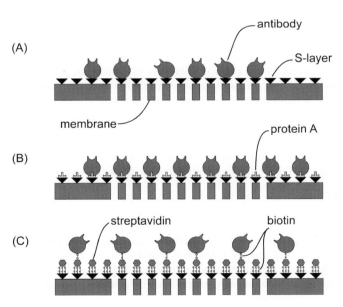

Fig. 30 Schematic drawing illustrating: (A) the immobilization of IgG to carbodiimide-activated carboxyl groups of the S-layer protein of SUMs, (B) protein A covalently bound to the S-layer lattice and (C) after biotinylation to a streptavidin-modified SUM.

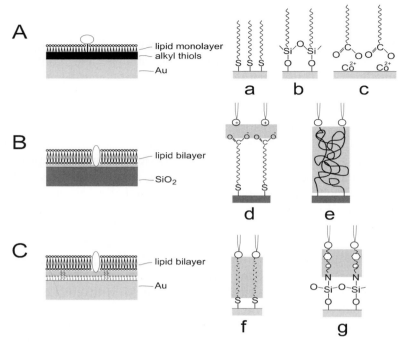

Fig. 31 Architectures of supported membranes [62]. Three basic strategies are sketched here, together with some possibilities for their realization.

(A)   A phospholipid monolayer on a hydrophobic base layer, which in most cases consist of rigidly packed alkyl chains. This base layer may be produced by the self-assembly of alkylthiols on gold (a), of alkylsilanes on oxide surfaces (b), by the LB transfer of a monolayer of fatty acid salts to a hydrophilic surface (c), or by the coupling of a rigid hydrophobic polymer to the surface (not shown). The phospholipid monolayer on top is either formed by entropy-driven self-assembly in solution, or transferred horizontally from the air-water interface. This monolayer configuration is suited for the observation of processes on the membrane surface or for anchoring peripheral membrane proteins.

(B)   A phospholipid bilayer on a hydrophilic surface. This surface may be a clean glass, quartz, or silicon oxide surface, on which bilayers may be formed by vesicle spreading or by two-step Langmuir Blodgett transfer. Gold surfaces must be functionalised with hydroxy-, carboxy- or amino-terminated thiols to confer hydrophilicity; for bilayer spreading, surface charge can be exploited by using charged lipids (d). The formation of bilayers supported on a hydrophilic polymer (e). The incorporation of transmembrane proteins in such a `coating' bilayer has been demonstrated.

(C)   An anchored phospholipid bilayer on a hydrophilic surface. To mimic the natural situation more closely, phospholipid derivatives must be used as hydrophobic anchors for the membrane, and the interstices on the surface must either expose a hydrophilic material or be correspondingly functionalised.

Anchor lipids may be thiolipids on gold (f), lipids coupled by a crosslinker to oxide surfaces (g), His-lipids, succinimidyl-lipids etc. In this configuration, the membrane must be formed by self-assembly from vesicle suspensions or detergent solutions; it is well suited for the accommodation of transmembrane proteins (Courtesy S. Heyse, Biochimica et Biophysica Acta).

The experimental use of crystalline bacterial surface layer proteins (S-layers) as combined carrier/adjuvants for vaccination and immunotherapy has since 1987 progressed in three areas of application: immunotherapy of cancers, antibacterial vaccines and antiallergic immunotherapy.

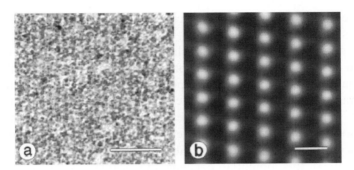

Fig. 32 (a) Transmission electron micrographs of a nanometric point pattern of CdS particles obtained by biomineralisation on an S-layer with oblique lattice symmetry. Protein appears white, CdS particles dark. Bar, 60 nm. (b) Corresponding computer image reconstruction to (A). Bar = 10 nm (Courtesy of FEMS Microbiology Reviews).

Block copolymers membranes

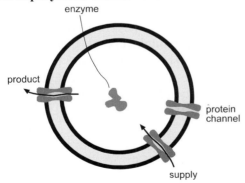

Fig. 33 Schematic view of a bioreactor/nanovesicle formed by a poly 2-methyloxazoline-poly dimethylsiloxane-poly 2-methyloxazoline, PMOXA-PDMS-PMOXA triblock copolymer with en-capsulated enzyme. Channel proteins in the shells of the bioreactors control the exchange of substrates and products (Courtesy W. Meier, Rev. Mol. Biology [94]).

Self-assembly of reactive amphiphilic block copolymers may also be used to prepare nanostructured hydrogel membranes with exceptional permeability properties [94]. Although the block copolymer membranes (PMOXA-PDMS-PMOXA triblock) are two to three-fold thicker than conventional lipid bilayers, they can be regarded as mimetic of biological membranes and can further be used as a matrix or scaffold for membrane-spanning proteins [95]. Surprisingly it has been found that the proteins remain functional, despite the thickness of the membranes and even after polymerisation of the reactive block copolymers [96]. The unique combination of block copolymers with membrane protein based channels allows the preparation of mechanically stable, defect-free membranes and nanovesicles [97] that have highly selective permeability and/or specific recognition sites.

The block copolymer/protein hybrid shells of the nanovesicles, see Fig. 33 can be regarded as a semi-permeable membrane separating their internal volume from the external solution. This property opens a convenient approach to trigger the gating transition of OmpF (Outer membrane protein F). Large polyelectrolyte ions, such as polystyrenesulfonate, will not permeate, and therefore the sodium counter ions will be distributed inside and outside the nanovesicles according to Donnan equilibrium conditions, giving rise to a Donnan potential. If this potential exceeds the critical value necessary for closure of OmpF, the substrates can no longer enter the interior of the nanoreactor vesicle, i.e., the reactors are then deactivated. The closure is a reversible process, and decreasing the potential below 100 mV may reactivate the nanoreactor vesicles. This could be carried out by diluting the system with buffer or by increasing the Na^+ concentration in the system.

Peptide nanochannels or nanotubes have also been formed by cyclo([L-Trp-D-Leu]$_3$-L-Gln-D-Leu) in phospholipid multilayers obtained by air drying on a substrate [98]. Nanotubes formed from cyclo([L-Trp-D-Leu]$_4$) have also been allowed to assemble into preformed monolayers of dodecanethiol or octadecyl sulfide deposited on gold films [99]. Bonding of the bilayer can be adjusted by coating of the membrane using a matching precoat, for instance high-quality alkyl monolayers. The stability of the precoat is of very high importance for the final stability of the total stack [100,101].

6.2 BIOSENSORS

Micro and nano engineering technologies offer revolutionary possibilities for biosensors and sensor arrays for drug and chemical screening and environmental monitoring. When a cell membrane detects (senses) a target molecule, it can turn electrical currents on or off by opening or closing molecular channels. When the channels are open, charged ions can pass in or out of the cell. These ions would not pass through the otherwise insulating membrane. When these channels open, the ion flow creates a potential difference across the membrane, which in turn creates a current. Such biosensors have a huge range of potential uses, especially in medicine, for detecting for instance drugs, hormones, viruses, pesticides and to identify gene sequences for diagnosing genetic disorders. The ability to detect the modulation of cell membrane channel activity by the binding of therapeutic agents is considered paramount for rational and efficient drug discovery and design. Since combinatorial libraries of potential therapeutic compounds are rapidly growing, fast and highly sensitive methods for functional drug screening are required. An attractive possibility is the use of self-assembled tethered membranes containing specific channel receptors as the sensing element in an otherwise solid-state biosensing device. Massive arrays of individually addressable microsensors with integrated fluid handling are conceivable. Even very simple sensor designs offer valuable advances in low-cost sensing for clinical medicine, the food and health care industry. The much-discussed 'artificial nose' containing a dense array of receptor sites affording unambiguous identification of molecular species could analyse the breath of patients for known chemical signatures of diseases such as liver cirrhosis and lung cancer.

Scaffolds

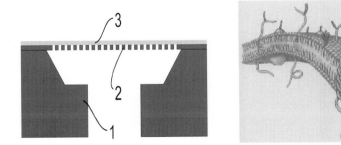

Fig. 34 Left: Support structure 1 with a nano engineered scaffold 2 for a functional membrane layer 3. Right: detail of e.g. 3.

Micro engineered membranes coated with titanium with a pore size of 500 nm have also been used [102] as a scaffold for a functional mono/bi-lipid layer obtained by the coalescence and spreading of corresponding lipid vesicles on the membrane surface. Micro and nano engineered membranes can further be used as a scaffold for the construction of functional 2D nano architectures to probe for example selective molecular transport and controlled release properties of biomedical molecules.

Cornell et al. [103] have constructed tethered supported lipid bilayers into devices that are able to sense both large and small analytes. The impedance of lipid bilayers depends upon the density of open gramicidin channels. In one manifestation, an analyte (such as a protein) with two antigenic sites binds to antibodies at the membrane surface. In a second manifestation, channels are opened when an analyte, which can in this case be a small molecule. Another promising application is stochastic sensing with nano engineered pores [104]. Stochastic sensing is based on the detection of individual binding events between analyte molecules and a single pore. The read-out is the single-channel electrical current. The frequency of the binding events is determined by the concentration of the analyte. The nature of the binding events (e.g. the magnitude and duration of the associated signal) is determined by the properties of the analyte. The ability to identify an analyte by its characteristic signature is a distinctive feature of stochastic sensing. Engineered versions of α-hemolysin were for example first used to detect and quantify divalent metal ions.

7. NANOTUBES[*]

7.1 INTRODUCTION

Nanotubes can be defined as hollow cylindrical structures and constructs with an inner diameter smaller than 100 nm. The current worldwide research is mainly focused on:

Peptide nano channels and tubes that can be found in cell membranes acting as size or charge selective barriers between the inner an outer lipid bilayer cell membrane and can also be artificially constructed and be used as molecule specific bio sensors (see section on nanoengineered biomembranes.

Carbon and carbonlike dense and open nanotubes that can be grown selectively on appropriate prepared substrates and can be used for molecule specific adsorption/separation processes or for electrical nanowiring applications.

Templated nanotubes or tubules where e.g. an anodised alumina membrane is used as a template for the deposition of appropriate materials (carbon, gold etc.) inside the long cylindrical pores (diameter 50-200 nm) of the alumina membrane. Procedure: immerse an alumina membrane in a solution of organometallic nickel, evaporate solvent, leaving Ni film in pores, form nanotubes in membrane pores via CVD of carbon from ethylene decomposition [105]. Increasing deposition time yields solid carbon nanofibres rather than hollow tubes. Recently silica nanotubes [106,107] have also been obtained. Other tubular structures formed by heavier element compounds have been predicted, such as GaSe [108], and synthesized, such as WS_2 and MoS_2 [109].

7.2 CARBON NANOTUBES

Two types of carbon nanotubes exist: those originally observed by Iijima [110,111] were multi-wall nanotubes (MWNTs), formed by concentric shells of apparently seamless cylinders of graphene, having a separation between them similar to that in graphite. More recently, single-wall nanotubes (SWNTs) have also been synthesized. As their name indicates, these consist of a single seamless cylinder of graphene [112]. Soon after the discovery of carbon nanotubes it was proposed that other compounds forming graphite-like structures, such as BN [113], BC_3 [114], BC_2N, and CN [115], could also form nanotubular structures and indeed BN, BC_3 and BC_2N have already been synthesized [116]. Single-wall

---

[*] This section has been cowritten with W. Nijdam, Aquamarijn Research.

carbon nanotubes SWNT were first synthesized in an arc discharge in a presence of a transition metal catalyst [117,118]. Since then significant efforts have been directed at optimising conditions for the arc production of SWNT [119,120]. For example, a modification of the arc discharge process allowing production of high-quality nanotube material within a restricted region of the apparatus has been reported [121]. However, the overall nanotube yield for this process remains relatively low. Other methods for SWNT synthesis have also been developed. An efficient SWNT production method is the laser vaporization of graphite/transition metal catalyst targets in a heated oven, which can produce nanotube yields of >70%. SWNT syntheses by metal-catalysed disproportionation of carbon monoxide and catalytic decomposition of acetylene have also been reported [122].

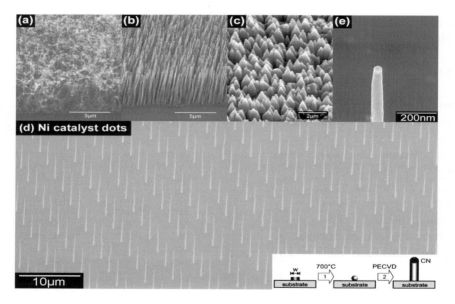

Fig. 35 Showing the possibilities of the dc-PECVD process (Courtesy K.B.K. Teo, Nanotechnology [123]): Figure (a), A substrate containing thin film Ni catalyst is exposed to C_2H_2 and NH_3 gases at 700°C (i.e. standard thermal CVD, with the exception that the substrate is biased at –600V (i.e. to initiate dc glow discharge). Figure (b), the degree of nanotube/nanofibre vertical alignment depends on the magnitude of the voltage applied to the substrate during growth and –600V. The thickness of the thin film Ni catalyst was found to determine the average diameter and inversely the length of the nanofibres. The yield and density of the nanofibres were controlled by the use of different diffusion barrier materials under the Ni catalyst. Figure (c), the shape of the structures could be varied from very straight nanotube- like to conical tip-like nanofibres by increasing the ratio of C_2H_2 in the gas flow. Figure (d), example of patterned growth of CN. Figure (e), close-up of vertically grown nanotube.

Ren et al. [124] discovered that vertically aligned multiwall carbon nanotubes could also be grown using Plasma Enhanced Chemical Vapour Deposition (PECVD). Other plasma techniques nowadays are hot filament CVD [124], dc (glow discharge) PECVD [125], microwave PECVD [126] and inductively coupled plasma PECVD [127].

Structural studies with TEM/SEM revealed that these nanofibres contained the catalyst at their tips which suggest a tip-growth mechanism. Over a large range of diameters (30-400nm), the structures were hollow (i.e. large inner diameters, hence "tube"- like) with graphene planes running parallel to the length-axis, and bamboo type defects along its length.

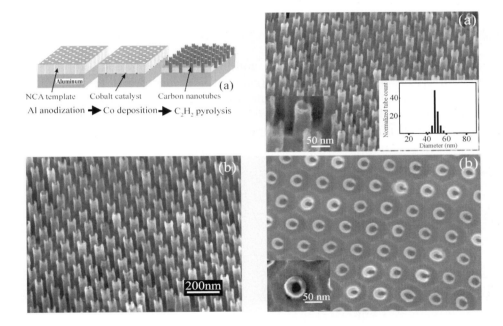

Fig. 36 Highly-ordered carbon nanotube arrays. Figure top left, Process scheme of fabrication. Figure bottom left, SEM image of the resulting hexagonally ordered array of carbon nanotubes fabricated using the method in (a). SEM image top right, showing oblique view of periodic carbon nanotube array. The inset at the lower left is an enlarged view of the tubes. The inset at the lower right is a histogram of the nanotube diameter showing a narrow size distribution around 47 nm. Top-view SEM image bottom right, carbon nanotubes showing hexagonal close-packed geometry. The hexagonal cells have sides approximately 57 nm long and the intercell spacing is 98 nm. The slight splitting of the tube ends and the apparent increase in tube wall thickness is an artefact of the non-specialised ion-milling apparatus that was used in the experiments. The inset shows a close-up view of a typical open-ended carbon nanotube in its hexagonal cell (Courtesy Applied Physics Letters [128]).

The field of prospective applications of carbon nanotubes continues to expand in nanofluidics and nanofiltration applications. Carbon nanotubes might be used as nanopipets [129], as sieves for DNA sequencing applications, for enantiomer separation etc. A fundamental issue is the ability of a fluid to wet the interior of a CNT as this would facilitate solution chemistry inside CNT [130].

The surface of pure graphite is hydrophobic, the contact angle of water on graphite is 80-90° [131]. Several independent studies reveal different results [132,133,134] on the wettability of pure carbon nanotubes The addition of impurities during manufacturing of the nanotubes or surface modification afterwards might be a solution to overcome serious wetting problems.

7.3 TEMPLATED NANOTUBES AND NANOFILTRATION

Classical filtration membranes and in particular anodised alumina membranes [135] and nanosieves [136] may also be transformed to nanotubulur structures by deposition of appropriate materials in the filtration pores or channels. They will enable new ways of separating and detecting analytes for applications in chiral separations and single-molecule sensing.

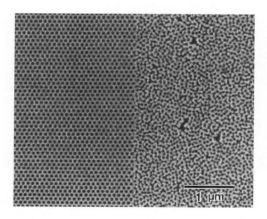

Fig. 37 SEM micrograph of ordered nanochannels in anodic porous alumina. Ideally ordered hole development was observed only in a pretextured area (left), while random development of holes took place in the untreated area (right). The hole interval corresponded to that of the pretextured concaves of Al. From this result, it was concluded that concave features formed by indentation could act as an initiation point and guide the growth of channels (Reproduced by permission of The Electrochemical Society, Inc [137]).

The anodising of an Al substrate was conducted under a constant voltage condition in an oxalic acid solution. The hole interval of anodic porous alumina, in other words, the cell size, was determined by the applied voltage used for anodisation. It was reported that the cell size has a good linear relationship with the applied voltage, where the proportionality constant of cell size for a specific applied voltage is approximately 2.5 nm/V. In the case of the sample shown in Fig. 37, anodising was conducted under a constant voltage of 40 V after pretexturing of a 100 nm period.

Nanotubule membranes are easy to make and each pore in the membrane essentially is a nanobeaker in which chemistry can occur. Metals can be deposited inside the pores, either electrochemically or through so-called electroless plating by using a reducing agent to plate the metal from the solution. Nanotubules of inorganic materials such as silica or titania can be prepared through sol-gel chemistry, and carbon nanotubules can be made through chemical vapour deposition of ethylene inside the pores.

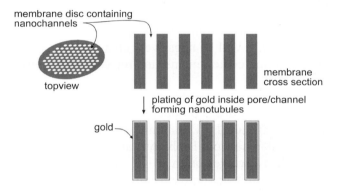

Fig. 38 Schematic overview of the formation of nanotubules.

The nanotubules come out embedded in a membrane, aligned, and monodisperse - that is, of uniform diameter and of a uniform length equal to the thickness of the membrane. Depending on deposition reaction times, thin- or thick-wall tubules are formed. The tubules can be capped on both ends to create a cluster of individual confined spaces. Or the membrane can be removed to release individual nanotubules. The generality of this template-based synthetic methodology broadens the scope of nanomaterials that can be prepared. One of the earliest separation applications to be explored was selective-ion transport. A simple experiment with a gold nanotubule membrane in a U-tube cell demonstrates this phenomenon [138]. On one side of the membrane is a feed solution containing KCl, which is colourless, and the cationic dye methylene blue. On the other side is a receiver solution of KCl. After a while, the colourless receiver solution turns blue, indicating transport of the cationic dye across the membrane. But when the feed contains permanganate anion, which is red, the receiver remains colourless. The anion, although smaller than the cationic dye, can not cross the membrane. Only cations pass because the nanotubule walls have excess negative charge due to adsorbed chloride ions. The excess charge can also be controlled by applying a potential to the membrane provided that the membrane material is electrically conductive. If a positive potential is applied, the nanotubule wall will have excess positive charge and the nanotubule membrane will reject cations and transport only anions. At negative applied potentials, the membrane will transport cations and reject anions. At the specific potential where the wall is electrically neutral, the membrane will be non-selective. Thus, the nanotubules together form a switchable ion-exchange membrane [139]. Having shown these membranes to have charge- and size-based selectivity's, attempts were made to induce chemical selectivity by modifying the chemistry of the nanotubule walls in which alkyl and other groups are attached to the nanotubule walls through gold-sulphur bonds [140,141]. Cysteine is a thiol as well as an amino acid and therefore has both amino and carboxyl groups. Consequently at low pH, it will be positively charged; at high pH, it will be negatively charged. Lee and Martin have shown [142] that the cysteine-coated membrane rejects cations at low pH and anions at high pH.

Chiral separation is important in the pharmaceutical industry, as it has become clear that the enantiomers of drugs produced as racemic mixtures could have very different properties. Brinda B. Lakshmin [143] used a "sandwich" assembly to separate D-phenylalanine from L-phenylalanine. Yoshio Kobayashi et al. [144] achieved a highly sensitive detection at 10^{-11} M with unadorned gold nanotubule membranes through which an ionic current passes.

8. NANO PRINTING AND ETCHING WITH PHASE SEPARATED MEMBRANES [145]

Aloys Senefelder used in 1796 a porous stone (in greek, lithos) as a tool for printing by patterning the stone with ink attracting (hydrophobic) and ink repelling (hydrophilic) regions. Lithography for semiconductor mass fabrication and other micro system and nano technology applications has nowadays regained interest on inexpensive micro printing methods as an alternative or complement on current high tech optical wafer stepper technology. There is a need for a convenient, inexpensive, and reproducible method of plating or etching a surface according to a predetermined pattern. The method would ideally find use both on planar and nonplanar surfaces, resulting in patterns having features in the micron and submicron domain. Additionally, the method would provide for convenient reproduction of existing patterns. Additionally, a need exists for the fabrication of surfaces that can pattern local regions (e.g. SAMs) amenable to attachment of biological species, such as antibodies, antigens, proteins, cells, etc., on the (sub)micrometer scale.

The study of self-assembled monolayers (SAMs) is an area of significant scientific research. Such monolayers are typically formed of molecules each having a functional group that selectively attaches to a particular surface, the remainder of each molecule interacting with neighbouring molecules in the monolayer to form a relatively ordered array. Such SAMs have been formed on a variety of products including metals, silicon dioxide, gallium arsenide using relief printing with a moulded stamp made from polydimethylsiloxane (PDMS) [146]. The upper relief part of the stamp provided with a suitable SAM coating is then being contacted with a product with a high affinity for the SAM species and a conformal SAM pattern is formed on the product, e.g. alkanethiol pattern on a goldcoated product. PDMS is a rather elastic and relatively strong material very well suited for reproducible contacting purposes on non planar surfaces, however it lacks a microporous microstructure for enabling functional fluid(ink) transport to the product domains or to enable other functional properties and other deposition techniques as will be described.

Microporous stamps 21 with (alternating) regions with a dense skin layer and adjacent regions with a (porous) layer without the skin layer can easily be made with a phase separation process by locally removing the skin layer by e.g. oxygen plasma etching with the aid of a perforated mask shielding the remaining dense skin layer regions. The stamps are made with a phase separation process (see chapter 6.4) using a mould having patterned regions with sharp protrusions penetrating the microporous layers and patterned regions without such sharp protrusions where a dense skin layer is formed. In case the skin layer is not dense but nanoporous the skin layer can of coarse first be hermetically sealed without sealing the microporous part of the stamp with e.g. a hydrophobic coating (e.g. aliphatic and cyclic olefin-based polymers, or fluoropolymers or silicon based polymers). The stamps may also be sub-patterned through use of photosensitive precursors in the casting solution of the product.

In Fig. 39 three basically different printing techniques are represented.

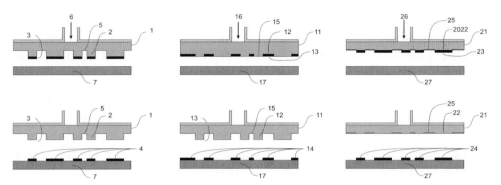

Fig. 39 Left: the art of relief printing. The upper relief part 2 of the stamp 1 provided with a suitable ink coating 3 is being contacted with a substrate 7 with a high affinity for the ink species and a conformal pattern 4 is formed on the substrate 7. The lower relief part 5 may be made ink repelling with a suitable coating (e.g. PVA, PVP) in order to avoid smearing of the pattern 4 of ink originating from sections 5. The upper relief part of the stamp 1 is provided with a macro or nanoporous structure to contain ink or to transport ink from an injection point 6 for reproduction or continous printing of the pattern 4 on the substrate 7.

Middle: the art of gravure printing The engraved part 12 of the stamp 11 provided with a suitable ink coating 13 is then being contacted with a substrate 17 with a high affinity for the ink species and a conformal pattern 14 is formed on the substrate17. The non engraved part 15 may be made ink repelling with a suitable coating (e.g. PVA, PVP) in order to avoid smearing of the pattern 14 of ink from sections 15. The engraved part of the stamp 11 is provided with a macro or nanoporous structure (cf. SEM picture below, engraved part is microporous, non engraved part has a dense skin layer) to contain ink or to transport ink from an injection point 16 for reproduction or continuous printing of the pattern 14 on the substrate 17.

Right: the art of planographic printing (i.e. art lithography). The ink delivering part 22 and the non-ink delivering part 25 of the stamp 21 are not determined by a difference in height but are made by the provision of suitable ink-repelling and ink-attracting coatings. The stamp 22 with a suitable ink coating 23 on part 22 is then being contacted with a substrate 27 with a high affinity for the ink species and a conformal pattern 2024 is formed on the substrate 27. The part 25 may be made ink repelling with a suitable coating (e.g. PVA, PVP) in order to avoid smearing of the pattern 24 of ink to sections 25. The part 22 is provided with a macro or nanoporous structure to contain ink or to transport ink from an injection point 26 for reproduction or continuous printing of the pattern 24 on the substrate 27. Also in another embodiment part 25 may be microporous and be filled with an ink repelling medium (e.g. water, Senefelder 1796).

A microstructured stamp (Fig. 39, Middle) with alternating nanoporous hydrophilic and dense hydrophobic surface regions may for instance locally be filled with an aqueous chromium etch solvent and brought into contact with a substrate having a chromium layer with a thickness of for example 20 nm. Whereas the dense regions (13) locally protects the chromium layer, in the nanoporous regions (12) an interaction between the chromium layer and the etch solvent may result in a locally dissolved and patterned chromium layer. Instead of chromium, other materials or combinations of materials can be used, e.g. aluminium, metal-oxides and nitrides, metals, semiconductors, polymeric lacquer layers, etc. The chromium layer may be replaced by a one phase lacquer layer and the solvent could be replaced by a second phase vulcanising agent for the one phase lacquer layer.

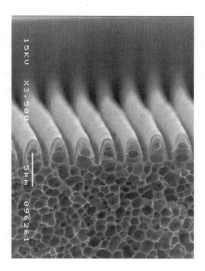

Fig. 40 SEM micrograph shows a cross-section of a polyimide microporous micro printing tool with a smooth skin layer as obtained with a phase separation process.

Instead of solvents, also reactive gases can be used to etch patterns, e.g. SF_6 to etch and pattern silicon products. Furthermore the microporous stamp can be used to dab or adsorb locally a liquid or viscous layer that has been casted on a product. Dabbing may be improved by locally compressing the microporous regions during the contact of the stamp with the product.

In order to facilitate mask alignment of subsequent mask stamping steps and to reduce thermal expansion differences between the stamps and the product, stamp regions parts may be provided on a transparent (e.g. Pyrex, borosilicate glass) support material with the same thermal expansion coefficient as the substrate. The microporous parts leading to the injection point preferentially should have a high inner porosity to reduce flow resistance. It may be useful to print an ink pattern with the stamp first on an intermediate dense or microporous transfer foil, which can be used to transfer the ink pattern subsequently to the substrate. Transfer foils to be used should exhibit a well-defined wetting contact angle with respect to the selected ink medium.

The surface condition of the substrate should also receive some attention. The substrate may be provided with a suitable adsorptive non-splattering or nanoporous (sacrificial) coating for more adsorption of the ink. This nanoporous coating may be obtained for example with a phase separation method. Also this nanoporous coating can be nanoporous alumina made by anodic oxidation of an deposited aluminium layer on a conducting (silicon) wafer. Silicon itself and many other materials with different layer thickness' can also be anodised or transformed as well with e.g. anisotropic etching techniques for similar purpose.

Microstructured stamps may of coarse also be used for the formation of microstructures or micro transfer molding on planar and non planar surfaces of polymeric, ceramic or metallic substrates.

REFERENCES

[1] E. Schrödinger , What is life, Cambridge press, ISBN 0-521-06223-3 (1944).
[2] Science 27 August 1993.
[3] M.J. den Exter, H. van Bekkum, C.J.M. van Rijn, F. Kapteijn, J.A. Moulijn, H. Schellevis and C.I.N. Beenakker, Stability of oriented silicalite-1 films in view of zeolite membrane preparation, Zeolites 19 (1997) 13-20.
[4] Y.Yan, M.E. Davis and G.R. Gavalas, Ind. Eng. Chem. Res. 1995, 34, 1652.
[5] H. van Bekkum, E.R. Geus and H.W. Kouwenhoven, Stud. Surf. Sci. Catal. 1994, 85, 509.
[6] E.R.Geus, M.J. den Exter and H. J. van Bekkum, Chem. Soc. Faraday Trans. 1992, 88, 3101.
[7] T.Matsuda, A. Sato, H. Hara, M. Kouno and K. Hashimoto, Appl. Catal. 1993, AIII, 143.
[8] J.C.Jansen, W. Nugroho and H. van Bekkum, in Proceedings ofthe 9th International Zeolite Conference (Eds. R. Von Ballmoos, J.B. Higgins and M.M.J. Treacy) Butterworth-Heinemann, Stoneham, MA (1992) 247.
[9] J.H.Koegier, H.W. Zandbergen, J.L.I.J. Harteveld, Nieuwenhuizen, J.C. Jansen and H. van Bekkum, Stud. Suf. Sci. Catal. 1994, 84A, 307.

[10] J.M. van de Graaf, F.Kapteijn and J.A. Moulijn, in Structured Catalvsts and Reactors, (Eds. A. Cybulski and J.A. Moulijn) Marcel Dekker, New York, 1996.

[11] C. J. M. van Rijn, G. J. Veldhuis, Nanosieves obtained with laserintereference lithography, Nanotechnology 9 (1998) 343-345.

[12] G.J. Burger, E.J.T. Smulders, J.W. Berenschot, T.S.J. Lammerink, J.H.J. Fluitman and S. Imai, High resolution shadow mask patterning in deep holes and its application to an electrical wafer feed-through.,[Proceedings Transducers '95]. 573-576. Stockholm.

[13] B. de Heij, Thesis, University of Twente, 1997.

[14] C.J.M. van Rijn, PCT patent application 95/13860, 'Membrane filter as well as a method of manufacturing the same'.

[15] M. Kölbel, R.W. Tjerkstra, J. Brugger, C.J.M. van Rijn, W. Nijdam, J. Huskens and D.N. Reinhoudt, Nanoletters, 0, A (2003), in press.

[16] J.W. Brugger, J.W. Berenschot, S. Kuiper, W. Nijdam, B. Otter and M. Elwenspoek, Microelectronic Engineering 53 (2000) 403-405.

[17] R. R. Schlittler, J. W. Seo, J. K. Gimzewski, C. Durkan and M. S. M. Saifullah, M. E. Welland *Science,* 292, (2001) 1136-1139.

[18] R. Lüthi, R.R. Schlittler, J. Brugger, P. Vettiger and M.E. Welland, J. K. Gimzewski Appl. Phys. Lett., 75 (1999) 1314-1316.

[19] M.M. Deshmukh, D.C. Ralph and M. Thomas, J. Silcox Appl. Phys. Lett., 75 (1999) 1631-1633.

[20] T. Ito, M. Namba, P. Bühlmann, Y. Umezawa Langmuir, 13 (1997) 4323-4332; V. V. Tsukruk, V. N. Bliznyuk Langmuir, 14 (1998) 446-455.

[21] M. Kölbel, J. Huskens, J. Brugger and D.N. Reinhoudt, Proceedings Micro and Nano Engineering Conference 2001, Grenoble, Sept. 16-19, (2001) 350-351.

[22] G. Busca, V. Lorenzelli, M. I. Braton and R. Marchand J. Mol. Struct. 1986, 143, 525

[23] L. Bergström, E. Bostedt Coll. Surf. 1990, 49, 183-197; M. Okabe, S. Okazi Nippon Kagaku Kaishi 1989, 10, 1802-1806; Y.Li, Y. Gao, Y. Liang, F. Zheng, K. Xiao, Z. Hu Wuli Huaxue Xuebao, 11 (1995) 886-889.

[24] A. Ulman Chem. Rev., 96 (1996) 1533-1554.

[25] M. E. McGovern, K. M. R. Kallury, M. Thompson Langmuir, 10, (1994) 3607.

[26] M.-W. Tsao, J.F. Rabolt, H. Schönherr, D.G. Castner Langmuir, 16, (2000) 1734-1743.

[27] The atomic concentration of C in Si_xN_y substrates when measured without pre-treatment was up to 40 %. XPS, carried out after 30 min cleaning in piranha still gives ~8% C.

[28] FIB cross-sections did not give reliable information since it was difficult to distinguish between gold deposited during the PVD process and gold redeposited during the FIB etching. Our conclusions are based exclusively on SEM images of apertures to which no FIB had been applied.

[29] C12-(OEt)$_3$ and C12-Cl$_3$ give SAMs with nearly identical properties. The results obtained with C12-Cl$_3$ on stencils are, however, less reproducible. Assumed this is due to the higher reactivity and moisture sensitivity of the compound compared with C12-(OEt)$_3$ and recommend always to use alkoxy silanes.

[30] C.J.M. van Rijn, G.J. Veldhuis, Nanosieves with laser interference lithography for microfiltration applications, Nanotechnology 9 (1998) 343-345.

[31] C.J.M. van Rijn, W. Nijdam, S. Kuiper, G.J. Veldhuis, H. van Wolferen and M. Elwenspoek, Microsieves made with laser interference lithography for micro-filtration applications, J. Micromech. Microeng. 9 (1999) 170-172.

[32] L.Vogelaar, W. Nijdam, H.A.G.M. van Wolferen, R.M. de Ridder, F.B. Segerink, E. Flück, L. Kuipers and N.F. van Hulst, Large area photonic crystal clabs for visible light with waveguiding defect structures: Fabrication with FIB-assisted Laser Interference Lithography, Advanced Materials 13 (2001) 1551-1554.

[33] E. Yablonovitch and T. J. Gmitter: Phys. Rev. Lett. 63, 1950 (1989)

[34] K. Inoue, M. Wada, K. Sakoda, A. Yamanaka, M. Hayashi and J.W. Haus, Jpn. J. Appl. Phys. 33, 1463 (1994)

[35] A. Rosenberg, R.J. Tonucci, H.-B. Lin and A.J. Campillo, Opt. Lett., 21, 830 (1996)

[36] L. Vogelaar, W. Nijdam, H.A.G.M. van Wolferen, R.M. de Ridder, F.B. Segerink, E. Flück, L. Kuipers and N.F. van Hulst, Adv. Materials 13, 1551 (2001)

[37] J.D. Joannopoulos, R.D. Meade, J.N. Winn, Photonic Crystals, Molding the Flow of Light, Princeton University Press, 1995

[38] E. Yablonovitch, Phys. Rev. B 58, 2059 (1987)

[39] M. Lončar, T. Doll, J. Vučković and A. Scherer, Journ. Lightw. Technol. 18, 1402 (2000)

[40] T.F. Krauss, R.M. De La Rue and S. Brand, Nature 383, 699 (1996)

[41] S.G. Johnson, S. Fan, P.R. Villeneuve and J.D. Joannopoulos, Phys. Rev. B., 60, 5751 (2000)

[42] See for example, M.D.B. Charlton, M.E. Zoorob, G.J. Parker, M.C. Netti, J.J. Baumberg, S. Cox, H. Kemhadjian, Mater. Sci. Eng. B-Solid State Mater. Adv. Technol. 74, 17 (2000)

[43] V. Berger, O. Gauthier-Lafaye and E.Costard, Electron. Lett., 33, 425 (1997)

[44] C. Peeters, E. Flück, A.M. Otter, M.L.M. Balistreri, J.P. Korterik, L. Kuipers and N.F. van Hulst, Appl. Phys. Lett., 77, 142 (2000)

[45] H. Masuda, M. Ohya, K. Nisahio, H. Asoh, M. Nakao, M. Nohtomi, A. Yokoo and T. Tamamura, Jpn. J. Appl. Phys. Vol. 39, 1039 (2000)

[46] H. Masuda, M. Ohya, H. Asoh, M. Nakao, M. Nohtomi and T. Tamamura, Jpn. J. Appl. Phys. 38, 1403 (1999)

[47] C.J.M. van Rijn, Membrane filter as well as a method of manufacturing the same, PCT patent application 95/13860

[48] Current STW project, Inorganic Material Science, University Twente.

[49] J. Zaman, and A. Chakma, Inorganic Membrane Reactors, J. Membrane Sci., 92, 1-28 (1994)

[50] R.M. deVos and H. Verweij, High-Selectivity, High-Flux Silica Membranes for Gas Separation, Science, 279, 1710-1711, (1998)

[51] R.M. deVos and H. Verweij, Improved performance of silica Membranes for Gas Separation, J. Membrane Sci. 143 37-51 (1998)

[52] X. Yang, J.M. Yang, Y. Tai and C. Mo, Sensors and Actuators, 73, 184 (1999)

[53] A.J.Frank, K.F.Jensen and M.A.Schmidt, Proc. IEEE Conf. MEMS, 382 (1999)

[54] T.A. Desai, D.J. Hansford and M. Ferrari, Biomolecular Engineering, 17, 23 (2000)

[55] T.A. Desai, Medical Eng. & Physics, 1 (2000)

[56] T.A. Desai, D.J. Hansford, L.Leoni, M. Essenpreis and M. Ferrari, Biosensors & Bioelectronics, 15, 453 (2000)

[57] M. Essenpreis, T.A. Desai, M. Ferrari and J. Hansford, PCT Patent Application, 'Implantable analyte sensor'.

[58]S.M. Bezrukov, I. Vodyanoy and V.A. Parsegian, Nature, 370, 279 (1994)

[59]C. Nardin, J. Widmer, M. Winterhalter and W. Meier, J. Phys. E, 4, 403 (2001)

[60] M. Winterhalter, C. Hilty, S.M. Bezrukov, C. Nardin, W. Meier and D. Fournier, Talanta, 55, 965 (2001)

[61] H. Bayley, Current Opinion in Biotechnology, 10, 94 (1999)

[62] S. Heyse, T. Stora, E. Schmid, J. H. Lakey and H. Vogel, Biochimica et Biophysica Acta, 85507, 319 (1998)

[63] P. van Gelder, F. Dumas and M. Winterhalter, Biophys. Chem., 85, 153 (2000)

[64] C.-H. Heldin, M. Purton, Signal Transduction, Chapman and Hall, London 1996.

[65] J.T. Hancock, Cell Signalling, Addison Wesley Longman, Essex 1997.

[66] T. Schirmer, T.A. Keller, Y.F. Wang and J.P. Rosenbusch, Science, 267, 512 (1995)

[67] N. Saint, K.L. Lou, C. Widmer, M. Luckey, T. Schirmer and J.P. Rosenbusch, J. Biol. Chem. 271, 20676 (1996)

[68] S.M. Bezrukov and M. Winterhalter, Phys. Rev. Lett., 85, 202 (2000)

[69] B. Schuster, D. Pum, O. Braha, H. Bayley and U. B. Sleytr, Biochimica et Biophysica Acta, 1370, 280 (1998)

[70] S.M. Bezrukov, I. Vodyanoy and V.A. Parsegian, Nature 370, 279 (1994)

[71] S.M. Bezrukov, J. Membr. Biol., 174, 1 (2000)

[72] U.B. Sleytr, P. Messner, D. Pum, M. Sara, Crystalline Bacterial Cell Surface Proteins, Academic Press Austin TX, 1 (1996)

[73] P. Messner and U.B. Sleytr, Advances in Microbial Physiology, 33, 213 (1992)

[74] T.J. Beveridge, Curr. Opin. Struct. Biol. 4, 202 (1994).

[75] U.B. Sleytr and P. Messner, Electron Microscopy of Subcellular Dynamics, CRC Press Boca Raton, 13 (1989)

[76] U. B. Sleytr, H. Bayley and M. Sára, FEMS Microbiology Rev. 20, 151 (1997)

[77] D. Pum, U.B. Sleytr, Crystalline Bacterial Cell Surface Proteins, Academic Press Austin TX, 175 (1996)

[78] M. Montal and P. Muëller, Proc. Natl. Acad. Sci. USA, 69, 3561 (1972)

[79] H. Schindler, Methods Enzymol. 171, 225 (1989).

[80] W. Hanke and W.-R. Schlue, Biological Techniques Series, Academic Press, London, 9 (1993)

[81] M. Sára and U.B. Sleytr, J. Bacteriol. 169, 4092 (1987)

[82] M. Sára and U.B. Sleytr, J. Membrane Sci. 33, 27 (1987)

[83] M. Sára, S. Kuëpcuë and U.B. Sleytr, Crystalline Bacterial Cell Surface Proteins, R.G. Landes Academic Press Austin TX 133 (1996)

[84] M. Sára and U.B. Sleytr, Micron 27, 141 (1996)

[85] S. Nakao, J. Membrane Sci. 96, 131 (1994)

[86] K.C. Schuster, H.F. Mayer, R. Kieweg, W.A. Hampel and M. Sára, Biotechnol. Bioeng., 48, 66 (1995)

[87] H. Kuhn, U. Friederich and A. Fiechter, Appl. Microbiol. Biotechnol., 6, 341 (1979)

[88] U.B. Sleytr and M. Sára, US Pat. No. 4,886,604 (1989)

[89] S. Weigert and M. Sára, J. Membrane Sci., 106, 147 (1995)

[90] C.Weiner, M. Sára and U.B. Sleytr, Biotechnol. Bioeng., 43, 321 (1994)

[91] C.Weiner, M. Sára, G. Dasgupta and U.B. Sleytr, Biotechnol. Bioeng. 44, 55 (1994)

[92] Breitwieser et al., BioTechniques 21, 918 (1996)

[93] U.B. Sleytr et al., FEMS Microbiology Reviews 20, 151 (1997)

[94] C. Nardin and W. Meier, Rev. in Mol. Biotechn. 90, 17 (2002)

[95] C. Nardin, M. Winterhalter and W. Meier, Langmuir 16, 7708 (2000)

[96] C.Nardin, W. Meier, Chimia 55, 142 (2001)

[97] C. Nardin, T. Hirt, J. Leukel and W. Meier, Langmuir 16, 1035 (2000)

[98] H.S. Kim, J.D. Hartgerink and M.R. Ghadiri, J Am Chem Soc 120, 4417 (1998)

[99] K. Motesharei and M.R. Ghadiri, J Am Chem Soc 119, 1306 (1997)

[100] A.B. Sieval, A.L. Demirel, J.W.M. Nissink, M.R. Linford, J.H. van der Maas, W.H. de Jeu, H. Zuilhof, and E.J.R. Sudhölter, Langmuir 14, 1759 (1998)

[101] A.B. Sieval, R. Linke, H. Zuilhof, and E.J.R. Sudhölter, Adv. Mat. 12, 1457 (2000).

[102] R. Kraayenhof, C.J.M. van Rijn, see e.g. R. Kraayenhof, G.J. Sterk et al., Biomembr., 1284, 191 (1996)

[103] B.A. Cornell, V.L.B. Braach-Maksvytis, L.G. King, P.D.J. Osman, B. Raguse, L. Wieczorek and R.J. Pace, Nature 387, 580 (1997)

[104] O. Braha, B. Walker, S. Cheley, J.J. Kasianowicz, L. Song, J.E. Gouaux and H. Bayley, Chem. Biol., 4, 497 (1997)

[105] G. Che et al., Chem. Materials, 10, 260 (1998)

[106] W. H. Chu et al., SPIE Proceedings, 2593, 9

[107] J. L. Gole, R. P. Gao, Z. L. Wang and J. D. Stout, Adv. Mat., 12, 1938 (2000)

[108] M. Côté, M.L. Cohen and D.J. Chadi, Phys. Rev. B, 58, 4277 (1998)

[109] R. Tenne, L. Margulis, M. Genut and G. Hodes, Nature 360,444 (1992)

[110] S. Iijima, Nature 354, 56 (1991)

[111] S. Iijima and T. Ichihashi, Nature 363, 603 (1993)

[112] See e.g. P.M. Ajayan and T.W. Ebbesen: Rep. Prog. Phys., 60, 1025 (1997); M.S. Dresselhaus, G. Dresselhaus and P.C. Eklund, Carbon Nanotubes, Preparation and Properties CRC Press, Boca Raton (1997).

[113] A. Rubio, J.L. Corkill and M.L. Cohen: Phys. Rev. B, 49, 5081 (1994); X. Blase et al., Europhys. Lett. 28, 335 (1994); X. Blase et al., Phys. Rev. B 51, 6868 (1995)

[114] Y. Miyamoto, A. Rubio, S.G. Louie and M.L. Cohen, Phys. Rev. B, 50, 360 (1994)

[115] Y. Miyamoto, M.L. Cohen and S.G. Louie, Solid State Commun. 102, 605 (1997)

[116] N.G. Chopra, R.J. Luyken, K. Cherrey, V.H. Crespi, M.L. Cohen, S.G. Louie and A. Zettl, Science, 269, 966 (1995)

[117] S. Iijima and T. Ichihashi, Nature, 363, 603 (1993)

[118] D.S. Bethune, C.H. Kiang, M.S. de Vries, G. Gorman, R. Savoy, J. Vazques and R. Beyers, Nature 363, 605 (1993)

[119] P.M. Ajayan, J.M. Lambert, P. Bernier, L. Barbedette, C.Colliex and J.M. Planiex, Chem. Phys. Lett. 215, 50 (1993)

[120] J.M. Lambert, P.M. Ajayan, P. Bernier and J.M. Planiex, Chem.Phys. Lett. 226, 364 (1994)

[121]C. Journet, W.K. Maser, P. Bernier, A. Loiseau, M. Lamy de la Chapelle, S. Lefrant, P. Deniard, R. Lee and J.E. Fischer, Nature 388, 756 (1997)

[122]H. Dai, A.G. Rinzler, P. Nikolaev, A. Thess, D.T. Colbert and R.E. Smalley, Chem. Phys. Lett. 260, 471 (1996)

[123]K.B.K. Teo, M. Chhowalla, S.B. Lee, D.G. Hasko, H. Ahmed, G.A.J. Amaratunga and W. I. Milne, University of Cambridge, United Kingdom

[124]Z. F. Ren, Z. P. Huang, J. W. Xu, J. H. Wang, P. Bush, M. P. Siegal and P. N. Provencio, Science 282, 1105 (1998)

[125]V. I. Merkulov, D. H. Lowndes, Y. Y. Wei, G. Eres and E. Voelkl, Appl. Phys. Lett. 76, 3555 (2000)

[126]C. Bower, W. Zhu, S. Jin and O. Zhou, Appl. Phys. Lett. 77, 830 (2000)

[127]L. Delzeit, I. McAninch, B. A. Cruden, D. Hash, B. Chen, J. Han and M. Meyyappan, J. Appl. Phys 91, 6027 (2002)

[128]J. Li, C. Papadopoulos, J. M. Xu and M. Moskovits, Appl Phys. Lett, 75, 367 (1999)

[129]M.-H. Hong, K.H. Kim, J. Bae and W. Jhe, Appl. Phys. Lett. 77, 2604 (2000)

[130]D. Ugarte, T. Stöckli, J.M. Bonard, A. Châtelain and W.A. de Heer, Appl. Phys. A 67, 101 (1998)

[131]A.W. Adamson, A.P. Gast, Physical Chemistry of Surfaces, 6th ed., John Wiley & Sons (1997)

[132]E. Dujardin, T.W. Ebbesen, A. Krishnan, M.M.J. Treacy, Adv. Mater. 10, 1472 (1998)

[133]T. Werder, J. H. Walther, R. L. Jaffe, T. Halicioglu, F. Noca and P. Koumoutsakos, Nano Letters 1, 687 (2001)

[134]A.Gil, J. Colchero, M. Luna, J. Gómez-Herrero and A. Baró, Langmuir 16, 5086 (2000).

[135]Anopore membranes Whatman ltd.

[136]Aquamarijn Micro Filtration BV, www.microsieve.com.

[137]H. Asoh, K. Nishio, M. Nakao, T. Tamamura and H. Masuda, J. of The Electrochemical Soc., 148, 152 (2001)

[138]C. R. Martin, M. Nishizawa, K. Jirage, M. Kang and Sang Bok Lee, Adv. Materials 13, 1351 (2001)

[139]C.R. Martin et al., Science 268, 700 (1995)

[140]S. Yu, Sang Bok Lee, M. Kang and C.R. Martin, Nano Letters 1, 495 (2001)

[141]Sang Bok Lee and C.R. Martin, Chemistry of Materials 13, 3236 (2001)

[142]Sang Bok Lee and C.R. Martin, Analytical Chemistry 73, 768 (2001)

[143]B.B. Lakshmi et al., Nature 388, 758 (1997)

[144]Y. Kobayashi, C.R. Martin et al., Anal. Chem. 71, 3665 (1999)

[145]C.J.M. van Rijn, L. Vogelaar, W. Nijdam, J.N.Barsema, M. Wessling, WO0243937 Publication date: 6 June 2002, Method of making a product with a micro or nano size structure and product.

[146]Y.Xia and G.M. Whitesides, Soft Lithography, Angew. Chem. 37 (1998) 550-575.

Chapter 10

Atomisation

> *"Flow through conducts is governed by the principle of least dissipation."*
>
> [Lord Rayleigh, Royal Society, Phil. Trans. (1895)]

1. INTRODUCTION THEORY

When a liquid is fed through an orifice under a sufficiently applied pressure a liquid jet with a relatively high kinetic energy emerges from the orifice. The surface of the jet is then subjected to disruptive forces caused by the friction of the jet in air and the surface tension γ of the liquid, resulting in disintegration of the jet into droplets. This process is referred to as *primary* atomisation [1]. If the so formed droplets have still enough kinetic energy and are subjected to further perturbations (air friction), they further disintegrate into even smaller droplets with a relatively broad droplet size distribution, a process known as *secondary* atomisation.

Rayleigh break-up
Lord Rayleigh [2] employed the method of small disturbances to predict the conditions necessary to cause the break-up of a liquid jet with diameter d into droplets with a diameter D. The surface energy disturbance E_s is given by:

$$E_s = \frac{\pi\gamma}{2d}\left(\left(\frac{d\pi}{\lambda}\right)^2 + n^2 - 1\right)b_n^{\,2} \tag{1}$$

Where λ is wavelength of disturbance, b_n is a constant in the Fourier series expansion, and n is a positive integer (including zero). For $n = 0$ and $d\pi/\lambda < 1$ it follows that $E_s < 0$, which means that for those cases the disturbances will grow.

The conclusion to be drawn from Rayleigh's analysis of the break-up of a non-viscous liquid jet under laminar flow conditions is that all disturbances on a jet with wavelengths greater than its circumference will grow.

By assuming that b_n is proportional to $\exp(qt)$, where q is the exponential growth rate of disturbance, Rayleigh showed that the exponential growth rate of the fastest-growing disturbance is given by:

$$q_{max} = 0.97 \left(\frac{\gamma}{\rho d^3} \right)^{0.5} \tag{2}$$

with corresponding $\lambda = 4.51d$.

After break-up, the jet part with length $4.51d$ becomes a spherical drop, so that

$$4.51d * \frac{\pi}{4} d^2 = (\pi/6)D^3 \text{ and hence } D = 1.89d \tag{3}$$

Thus for the Rayleigh mechanism of break-up the average droplet-size is nearly twice the diameter of the undisturbed jet.

Weber [3] extended Rayleigh's analysis to include viscous contributions:

$$\lambda = \sqrt{2}\pi d \left(1 + \frac{3\eta}{\sqrt{\rho\gamma d}} \right)^{0.5} \tag{4}$$

Weber introduced the Weber number [4], which is a dimensionless ratio between inertial and surface tension forces:

$$We = \frac{\rho v^2 d}{\gamma} \tag{5}$$

The most commonly quoted criteria for classifying jet disintegration are proposed by Ohnesorge [5], who classified experimental data according to the relative importance of gravitational, inertial, surface tension, and viscous forces. The break-up mechanisms of a jet could be expressed in three stages, each stage characterised by the Reynolds number and a dimensionless number Z :

$$Z = \frac{\eta}{\sqrt{\rho\gamma d}} \tag{6}$$

often referred as the stability or the Ohnesorge number(Oh). Note the first appearance (1931) of this number in the original work of Weber [3] cf. eq. (4).

The Ohnesorge number is a dimensionless number obtained by dividing the square root of Weber number by Reynolds number, which eliminates velocity from both; and has been regarded as an indicator of jet or sheet (in)stability.

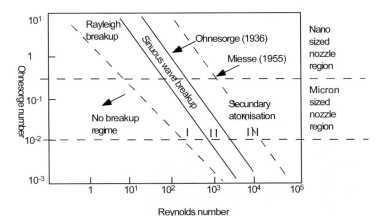

Fig. 1   **Three atomisation regions can be discerned:**
I   **Rayleigh break-up at low Ohnesorge and low Reynolds number,**
II   **At intermediate Reynolds number, the break-up of the jet is by jet oscillations with respect to the jet axis. A wide range of drop sizes may be produced,**
III   **At high Reynolds numbers, secondary atomisation occurs within a short distance from the orifice.**

The depicted lines in Figure 1 describe a scaling relation between the relevant fluidic parameters. The upper right line can also be fitted, Miesse [6], to

$Oh=100Re^{-0.92}$ and can be rewritten: $We^{0.5}=150.Oh^{-0.087}$ (7)

The latter relation shows that the transition to the (third) atomisation regime is almost fully determined by a critical Weber number and nearly independent of the Ohnesorge number! In figure 1 only atomisation results have been depicted of atomisation processes in ambient air at 1 bar. For atomisation processes in which viscous contributions of the ambient air at other pressures contribute to the jet break-up other critical Weber numbers are being found [7].

Clearly a minimum jet velocity or a minimum kinetic energy E_{kin} is required to form a jet with length l and an enhanced surface energy E_γ :

$$E_{kin} > E_\gamma \quad or \quad \frac{mv^2}{2} = \frac{\rho \pi d^2 l v^2}{8} > \gamma \pi d l \qquad (8)$$

This relation can also remarkably be rewritten to a minimum critical Weber number, We > 8 (see figure 1, Breakup/No breakup regime is indicated by dotted line ($Re=8^{0.5}/Oh$).

So appropriate Weber numbers mainly determines the onset as well as the end of each atomisation regime. This means that the viscosity of the jet fluid (implicitly in the Ohnesorge number) has nearly no impact on the location of the atomisation regimes. The viscosity has only an influence on the size of the droplet (see eq.4) as originally proposed by Weber.

Effect of jet velocity

Through the modulation of injection chamber pressure or liquid flow rate, researchers [8], have been able to study the effect of fixed wavelength perturbations on the behaviour of low speed jets. At moderate inflow velocities, the presence of the surrounding gas can be neglected, and the jet break-up length increases as the inflow velocity is increased.

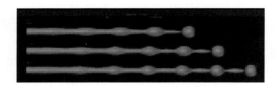

Fig. 2 Increasing the jet velocity will lead to longer break-up lengths

For fixed orifice radius and liquid density, increasing the Weber number corresponds to either decreasing the surface tension or increasing the inflow velocity. As the jet velocity is increased, the wave formed on the surface of the jet from the unsteady inflow should tend to grow more slowly, leading to longer break-up lengths.

Effect of perturbation magnitude

Liquid jet break-up simulations can be performed for various perturbation magnitudes, wave numbers, and Weber numbers. All calculations begin with a long column of fluid, usually about 15 to 20 jet radii, outside the orifice in order to insure that perturbations from the assumed spherical end cap do not influence the development of waves imposed by perturbations. Increasing the size of the perturbation initiated by the nozzle outlet normally decreases the breakup length of the jet. Increased friction with the air may result in droplets with a *squashed* shape.

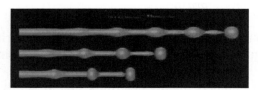

Fig. 3 Effect of perturbation magnitude and friction with air.

Effect of wave number on droplet size

From eq. (3) and (4) it is seen that an increased viscosity or a reduced surface tension will lead to an increased wave number and thus enlarged droplets.

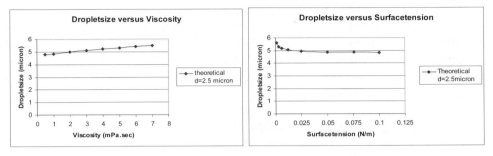

Fig. 4 Droplet size dependences on viscosity and surface tension for a jet of water leaving a nozzle with diameter 2 .5 micron.

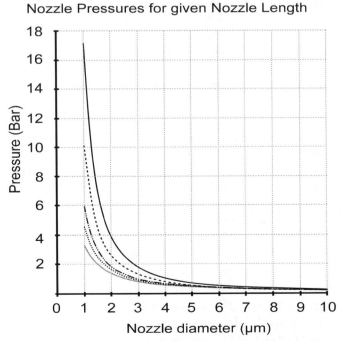

Fig. 5 Required minimum trans nozzle pressure to induce Rayleigh breakup as a function of nozzle diameter plotted for nozzle lengths: 100 nm (lower curve), 1, 2, 5 and 10 micron (upper curve). The minimum nozzle pressure has been calculated by eq. (1) in section 2.

2. ATOMISATION WITH MICROMACHINED NOZZLES

For conventional nozzles with diameters in the micron area with relatively large lengths Rayleigh theory never becomes applicable, because the necessary pressures are too large, and that requires complex devices and/or in nozzle break-down. The pressure build up P_{vis} by fluid viscosity through a micro nozzle with length L and diameter D is characterised by Dagan [9], who proposed a transition formula between theory of Poiseuille and Stokes [10]. At larger fluid velocities (several meters per second) also the kinetic pressure build up P_{kin} must be added, so total pressure P_{tot} is given by P_{vis} + P_{kin}, depending on L, d, fluid velocity v, density and dynamic viscosity of the fluid.

$$P_{tot}=P_{vis}+P_{kin} = 6\eta\pi[1+16L/3\pi d]d^{-1} \ v \ + \ 0.5 \ \rho \ v^2 \tag{1}$$

For nozzles with a d of 2.5 µm and L of 5 µm at a trans-nozzle pressure of 2 bar the P_{kin} / P_{vis} ratio becomes 1, indicating that this is the transition point between viscous and kinetic behaviour, see Fig. 5. For d = 2.5 µm and L = 1 µm the P_{kin} / P_{vis} ratio is already 5, indicating that a small nozzle length L compared to diameter D leads to kinetic pressure behaviour, independent of viscosity. In other words, it is possible now to produce micro nozzles that use most (pressure) energy for accelerating the liquid instead of overcoming fluid friction. Perfect control of nozzle dimensions is crucial here; small variations in geometry lead to different oscillations in the jet and hence different break-up behaviour.

Medical application
Portable, reliable and inexpensive atomisation devices preferably exclude the use of external vaporisation chambers and electronic hardware like piezos. Rayleigh break-up, or jet break-up due to capillary instability, is a very elegant way to produce monodisperse droplets using pressurised liquid as actuation mechanism. However so far relatively little effort has been put in dedicated development of suitable micro nozzles with small orifice diameters (<10 µm) and small orifice lengths (< 1 µm). This can be explained by the lack of production technologies and fast micro nozzle fouling.

In the last decades it became gradually accepted that exact control over drug particle size, size distribution and dose, means control over total drug deposition in the lung, and hence drug intake. Deep pulmonary drug delivery technologies today have many requirements: monodisperse 1-5 µm droplets, flexible dosing, easy to use, portable, low cost, low energy input and liquid parameter independence [11]. None of the technologies present on the market today for atomising liquids or generating monodisperse nebulae can meet all of these requirements. User-friendly atomisation devices should in principle be as simple as a Metered Dose Powder Inhaler [12,13].

Aerosol Deposition

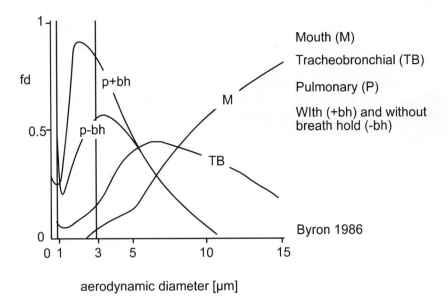

Fig. 6 Efficiency of pulmonary inhale as a function of the droplet diameter [14].

When using micron sized nozzles, fouling becomes a serious problem, since all micro (biological) particles, such as bacteria, potentially could block micro nozzles completely or ruin jet characteristics. To avoid nozzle fouling an absolute particle filter is co-manufactured with the nozzle: A micro sieve with a pore sizes of 450 nm, meaning 100 % of the particles with a diameter larger than 450 nm are blocked by the sieve. In case of sieve blockage the smart nozzle (nozzle with filter included) is replaced by a new one, only possible due to the low cost price.

Micro system technology (MST) for micro nozzle and micro sieve production is used to fine-tune all important nozzle parameters, such as number, length, diameter, spacing, material and coating in micrometer range with sub micrometer precision, an advantage that out performs all present laser drilling and electroplating techniques. High numbers and low product cost are characteristic of this technology, 6 inch wafer scale production is presently running in most silicon micro machining foundries.

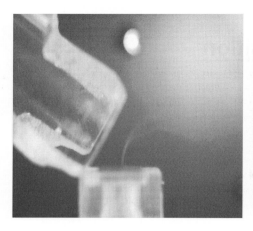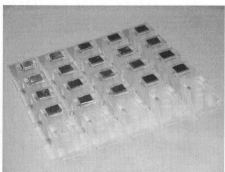

Fig. 7 Left: nozzle head with spraying nozzle die, Right: Array of nozzle and microsieve dies.

Experiments

Experiments with smart micro machined nozzles show surprisingly good Rayleigh jet break-up for the generation of monodisperse droplets of aqueous solutions. Typical micro nozzle diameters of 1-10 µm are chosen larger than the orifice lengths 500 nm -1 µm leading to a very low fluid resistance [15]. Due to the short nozzle length, the interaction time of the fluid flow with the nozzle is small and dissipation forces are minimised. The second important feature is that micro nozzle fouling can be preventedby co-manufacturing and assembly of micro sieves (absolute filters) with pore size smaller than 450 nm (sterile pre-filtration) with the same manufacturing technology.

Perfect monodisperse droplets (2-20 µm) are already formed at trans nozzle pressures below 5 bar. Droplet coalescence has been prevented by acceleration of air near the point of jet break-up to a velocity of about 10 m/s, see Fig. 7.

First droplet experiments [16]

Up to now, nozzles with orifices of 2.5 and 10 µm have been characterized to confirm theory, the latter visible in the high-speed photograph Fig. 8. Perfect Rayleigh break-up occurs in a wide trans nozzle pressure range varying from 0.5 to 4 bar. Looking at the droplet jets in detail it can be observed that all droplets have nearly the same size and perfect round shape. Some droplets are still in the process of breaking up, with the risk of joining in a later stage.

100 micron

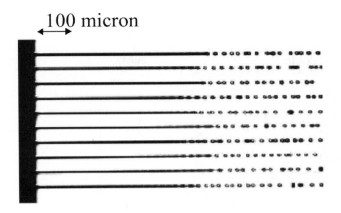

Fig. 8 Picture of Rayleigh break-up (shutter time 3 microsec) with a micromachined nozzle.

The droplet jets do not coalesce directly, due to drag of preceding droplets. Data on size distributions (lognormal) obtained optically indicate perfect droplet monodispersity with geometric standard deviation below 1.2 after break-up.

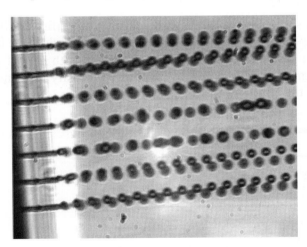

Fig. 9 High speed photo of a monodisperse aerosol generator at 0.5 bar. 14 droplet jets 2 in-line, moving to the right. Orifice diameter is 10 μm. Due to slow shutter (3 μs) the jet diameter d_j appears thicker than in reality (The jets oscillate faster than 3 μs.). No air co-flow was used. Acknowledgement J. Marijnissen TUdelft, The Netherlands.

Jet break-up lengths vary between 75 and 400 μm and depend very much on applied pressure, which confirms theory. A more advanced set-up with higher magnification is being developed for droplet experiments with droplets below 10 μm to monitor faster break-up faster with a higher optical magnification.

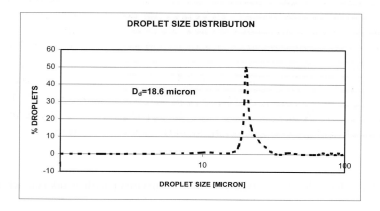

Fig. 10 Droplet size distribution of jetting 10 μm orifices indicating the droplet monodispersity, with (lognormal) σ = 1.09 < 1.2.

Conclusion: A paradigm shift?

Smart micro nozzles show perfect Rayleigh break-up with orifices <10 μm at trans nozzle pressures below 5 bar, with jet characteristics shifting from viscous to kinetic. This is in contrast to conventional nozzles with $L \gg 1$ μm, where jet velocity patterns are determined by dynamic viscosity, leading to parabolic flow patterns in the jets.

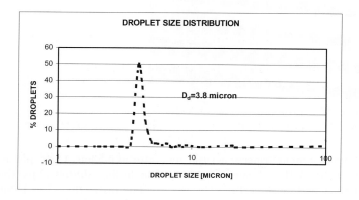

Fig. 11 Droplet size distribution of jetting with 1.8 μm orifices indicating the droplet monodispersity, with (lognormal) σ = 1.12 < 1.2.

Minimizing viscous friction by making nozzle membranes extremely thin, typically < 1 μm, instead of (overcoming viscous friction by) increasing nozzle pressure, brings about great advantages for the practical use of micro nozzles in pulmonary drug delivery devices: (i) low pressures below 5 bar can now be employed to atomise monodisperse droplets, thereby eliminating the risk of macromolecular drug   degeneration and making portable low pressure devices possible, (ii) the generation of droplets becomes a kinetic process, and therefore the droplet size and size distribution becomes more independent of drug viscosity and temperature. Also the drug dose generated is considerably higher than with other systems. A nozzle array with 100 orifices of $d = 3$ μm generates around $1.3 \; 10^9$ droplets per second, which is almost 1 ml per minute.

Future

The smart micro nozzle drug delivery platform gives great control over aerosol parameters, like tuneable droplet size distributions and dose output, enabling drug device engineers to optimise their pulmonary drug deposition [11]. Medspray [17] is a small Twente start-up company focusing on implementation and commercialisation of smart micro nozzles in portable nebuliser and inhaler concepts. Future work will focus on implementation of smart micro nozzles in portable nebuliser and inhaler concepts and on production fine-tuning.

REFERENCES

[1] I.Colbeck, Physical and Chemical properties of aerosols, Thomson Science, (1998) ISBN 0 7514 0402 0
[2] Lord Rayleigh, On the instability of jets, Proc. London Math. Soc. 10, (1878) 4-13.
[3] C.Weber, Disintegration of liquid jets, Z.Angew.Math.Mech.,Vol. 11 No. 2 (1931) 136-159.
[4] A.H.Lefebvre, Atomisation and Sprays, (1989) ISBN 0-89116-603-3.
[5] W.V. Ohnesorge, Drop formation at nozzles and the decomposition of liquid jets, Zeitschrift fur angewandte mathematik und mechanik 16 (1936) 355-358.
[6] C.C.Miesse, Correlation of Experimental Data on the Disintegration of Liquid Jets, Ind.Eng.Chem.,Vol. 47, No. 9, 1955, 1690-1701.
[7] A.M. Sterling, C.A. Sleicher, The instability of capillary jets, J. Fluid. Mech., 68, 477-495, (1975).
[8] Yim and Pyongwon, The role of surface tension oxidation in the break-up of laminar liquid metal jets, Thesis MIT (1996).

[9]   Z. Dagan, S. Weinbaum and R. Pfeffer, Chem. Eng. Sci. 38 (1983) 583-596.

[10] Z. Dagan, et al., Chem. Eng. Sci. 38 (1983) 583-596.

[11] K. Corkery, Inhalable drugs for systemic therapy, Respir. Care 45(7), (2000) 831-835.

[12] A.H. de Boer, G. Molema, H.W. Frijlink, Pulmonary drug delivery: delivery to and through the lung, In: G. Molema, D.K.F.Meijer, eds. Drug targeting. Organ-specific strategies. Methods and principles in medicinal chemistry. Weinheim (Germany): Wiley-VCH, 2001: 53-87 (chapter 3).

[13] P. Zanen, Aerosol formulation and clinical efficacy of bronchodilators, Thesis University Utrecht, 1998 ISBN 90 393 2027 6.

[14] P.R. Byron, Prediction of drug residence times in regions of the human respiratory tract following aerosol inhalation, J Pharm Sci. 75 (1986) 433-438.

[15] C.J.M. van Rijn, W. Nijdam and M.C. Elwenspoek, Microfiltration membrane sieve with silicon micromachining for industrial and bio-medical applications, IEEE proc. MEMS .., (1995) 83-87.

[16] J.M.Wissink, W.Nijdam, I.Heskamp and C.J.M.van Rijn, Smart microengineered nozzles for monodispers aerosol generation with Rayleigh break-up, Respiratory Drug Delivery VIII, Tuscon, Arizona 12-16 may 2002.

[17] http://www.medspray.com

Chapter 11

Membrane Emulsification

> *"If it has been possible to bring into suspension in a liquid a large number of particles all of the same nature, we say that an emulsion has been produced. This emulsion is stable if the particles in suspension do not stick together when the hazards of the Brownian movement bring them into contact, and if they re-enter the liquid when these hazards bring them against the walls or to the surface."*
>
> [Jean Baptiste Perrin (1926)]

1. INTRODUCTION[*]

An emulsion is a dispersion of droplets of one immiscible liquid within another liquid. An emulsifier may be added to stabilise the dispersion. An emulsifier is a molecule consisting of a hydrophilic and a hydrophobic (lipophilic) part and will concentrate at the interface between the immiscible liquids, where they form interfacial films. The hydrophobic part of the emulsifier may consist of a fatty acid, whereas the hydrophilic part of the emulsifier may consist of glycerol, possibly esterified with e.g. acetic, citric or tartaric acid.

Emulsions are indispensable in the modern food, pharmaceutical and cosmetic industry and are categorized as: oil-in-water (o/w) emulsions, such as liquid cream or milk, in which oil is the dispersed phase and water the continuous phase, and water-in-oil (w/o) emulsions, such as margarine, in which water is the dispersed phase and oil the continuous phase.

---

[*] This section has been cowritten with Remko Boom, Univ. of Wageningen & Jo Janssen, Unilever Research, Vlaardingen.

Many commercial emulsions today are obtained by stirring or by use of high pressure homogenisers. A new technique to produce emulsions is cross flow membrane emulsification (XME): a process in which the to-be-dispersed phase, for instance oil, is pressed through a membrane and droplets formed at the membrane surface are carried away with the continuous phase, for instance water, flowing across the membrane, resulting in an oil in water (o/w) emulsion [1], see Fig. 1 left.

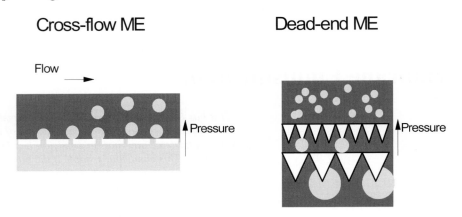

Fig. 1 Left, cross flow membrane emulsification (XME). Right, dead end membrane emulsification (DME) using a coarse pre-emulsion.

The simple permeation of a coarse pre-emulsion through a porous membrane (DME, see Fig. 1 right) has also been studied [2,3], which resulted in the formation of a good-quality emulsion with a relatively narrow droplet size distribution. The required pressures are about one decade lower (10-30 bar) than for the hight pressure homogeniser. This technique may be particularly employed as an energy saving alternative of current homogenisers in the dairy industry. However, during preliminary experimental studies carried out with this technique, severe problems were experienced with depth-fouling of the membrane, especially when more practical systems, for instance systems stabilised with proteins, were used. This seems to indicate that cross-flow membrane emulsification is currently another interesting candidate for production of food-related products [4].

Two important advantages of cross-flow membrane emulsification (XME) are depicted in Fig. 2. In this figure, the energy consumption per m^3 product of XME is compared with that of some existing emulsification methods: the high-pressure homogeniser, the diaphragm homogeniser and the microfluidizer. These methods work through the application of intense shear fields, elongational flow or even cavitations. By contrast, cross-flow ME is based on direct droplet formation at the pores. Clearly, the energy consumption of cross-flow ME for a given volume fraction φ of dispersed phase and given final droplet size d is much less than for the existing homogenisers.

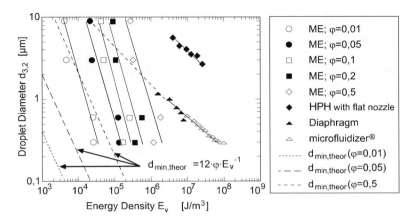

Fig. 2 Different emulsification processes and energy dissipation.

From Fig. 2 it can be seen that the slope of the droplet size versus energy consumption correlation is much steeper for ME, which suggests that ME is a much more promising method for achieving very small droplets than the conventional emulsification processes. Moreover, the high shear and the temperature rise due to viscous dissipation in the conventional homogeniser methods may have an adverse effect on the quality and functionality of delicate ingredients.

Indeed there are strong indications that ME-produced emulsions differ positively from homogeniser-produced emulsions in terms of the functionality and stability of proteins, which is an example of improved ingredient functionality/quality via a mild processing route. Finally, the flow fields in conventional equipment are often far from uniform, which means that different parts of the emulsion experience different levels of shear and elongation. This inevitably results in a broad droplet size distribution. Under ideal conditions, ME was shown to produce quasi-monodisperse emulsions, albeit at very low production rates.

The exploration of the lower limits in droplet sizes that can be produced with membrane emulsification (much smaller than 1 μm), necessary for the production of the primary emulsion, is an interesting subject. This is in itself an industrially very useful deliverable, for instance for encapsulation of micronutrients, or for products with specific properties, it is also the first step towards the production of mechanically stable duplex emulsions with small overall sizes; traditionally these emulsions are made with a large outer diameter, but this has serious implications for their applicability.

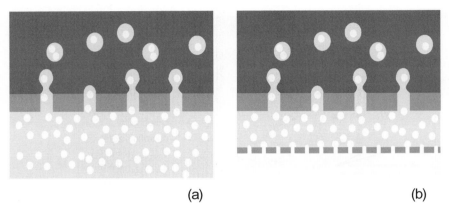

(a)   (b)

Fig. 3 Schematic principle of producing duplex emulsions; (a) from a pre-emulsion made beforehand; (b) in one step using a dual nozzle plate.

On of the routes towards the creation of these duplex emulsions is the use of two emulsification membranes, with differing pore sizes and shapes, and surface properties, see Fig. 3. By dual permeation, a duplex emulsion can be created.

Current membranes

A common feature of all current commercial membranes for cross-flow membrane emulsification (e.g. sintered powders and stretched polymer sheets) is the fact that they are not uniform on a microscopic scale, i.e. they have a wide distribution of pores/channel sizes, shapes and spacing, see Fig. 4.

Membrane emulsification has been studied with sintered ceramic membranes [5,6], porous glass membranes [7], microchannels [8,9,10] and microsieves [32]. Originally ceramic and porous glass membranes were developed for filtration applications, and have not been optimised for emulsification processes. The high pore density of such membranes may promote the initiation of large droplets, as well as coalescence of growing droplets at the membrane surface, causing a broadening of the droplet size distribution, especially at high trans membrane pressures. With micro engineering techniques optimised designs become possible. Also microscopic observations [11,32] during the droplet forming process will enhance the understanding of membrane emulsification.

 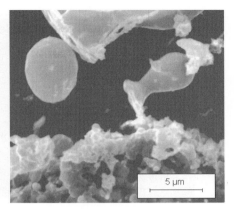

800nm Membran nach Gebrauch gereinigt, Querbruch   30µm

5 µm

Fig. 4 ~Left, SEM micrograph of an alumina membrane with a pore size of 0.8 μm. Right, micrograph of droplet detachment, courtesy V. Schröder [6].

Assuming a relatively rigid, spherical droplet, the droplet size will be mainly determined by the cross-flow velocity of the continuous phase, the pore diameter, the proe density, the transmembrane pressure, and the (dynamic) interfacial tension [35]. It is recognised that the droplet during the formation process will be deformed; the effect of this deformation on the actual droplet size is difficult to estimate with analytical equations, however proper computer simulations would be helpful to gain more insight in this matter.

Peng and Williams [12] showed that at high dispersed phase pressures $(0.35–21\times10^5$ Pa) the droplet diameter increased with increasing transmembrane pressure. This was explained by the increased flux at a higher pressure, which increases the final droplet volume during the second stage of droplet formation, the detachment phase. These variations inevitably cause a broadening of the drop size distributions obtained, and explain the fact that there is often a considerable discrepancy between experimental results and the predictions of theoretical models in XME. The latter are all based on the strongly simplifying assumption that a membrane in XME consists of equispaced circular pores in a flat surface. Also here computer simulations would be a very helpful instrument.

Recent work carried out by Abrahamse et al. [1] on modelling of the emulsification has resulted in the first set of guidelines derived from fundamental process principles that can be used to design the membrane and process. Accurate membrane design however, is not possible with conventional membrane production procedures. Preliminary work has shown that the membranes prepared through microengineering can indeed be made according to these guidelines, see Fig. 5.

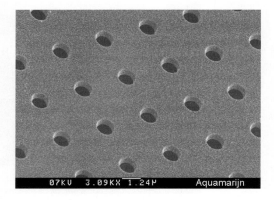

Fig. 5 Microengineered membrane with uniform pore size for cross flow emulsification.

Functional foods

Today's food industry is putting much effort in the development and production of new food products of high quality and often with specifically optimised nutritional value. As more knowledge of nutritional needs becomes available, biofunctional ingredients are added to the products, which may ultimately lead to a better health of the population through a better fulfilment of the individual's nutritional needs.

The food industry therefore requires new processing techniques that meet all ecological and safety standards to be expected in the future, but also offers the possibility to include a diversity of fragile biofunctional ingredients in the products, which are often degraded during the harsh production conditions in existing processes. This becomes a more challenging task when the developments in genomics, nutrition and related fields give the possibility to tailor food products to one's specific genetic make-up. This would require processing techniques that can produce taylor-made food products specific for each consumer.

Cross-flow membrane emulsification may be able to deliver this all, for emulsion based food products. Further, the technology has a much wider applicability towards other types of multi-phase food products and towards other industrial sectors important for Europe, such as the pharmaceutical and fine chemical industries (e.g. the coating industry).

Membrane emulsification may be very suitable not only for large-scale production of foods, but also for production on small scale, e.g. for final 'point-of-purchase' product assembly, where the consumer determines the exact composition of the product from his or her personal requirements or preferences. Therefore, it closely fits the developments in other scientific fields, as discussed above.

2. CFD AND MEMBRANE EMULSIFICATION

Development of reliable Computational Fluid Dynamics simulation tools for membrane emulsification processes is a complicated and time-consuming task. An overview of CFD software is available on the Internet [13]. PHOENICS [14] was the first general-purpose tool for CFD, which appeared in 1981. The numerical foundation is the finite volume method in the notation of Patankar [15]. FIDAP software package [16] is a general-purpose fluid flow package and applies finite elements for the spatial discretization. CFX software package [17] is another modern, general-purpose package addressing complex flow problems in the process and chemical industries, including turbulence, multi-phase flow, bubble-driven flow, combustion, flames, and so on. Another general-purpose CFD package is CFD2000 [18], which uses finite volumes on curvilinear grids and handles incompressible as well as compressible flows, with turbulence, heat transfer, multiple phases, and chemical reactions. The here listed CFD packages are all based on the so called finite volume or Volume-Of-Fluid (VOF) method, which describes the fluid phases as a continuum and uses the Navier-Stokes equation (see chapter 4.2) as the governing time propagator for fluid flow development. Brackbill [19] implemented surface tension in this continuum approximation and Renardy [20] further developed this concept in the context of emulsification processes.
A previous study on emulsification processes with microsieves [1] has been performed by the VOF method. However, the disadvantage of the VOF method is that is not based on kinetic behaviour at the atomic level, and hence it is intrinsically difficult to incorporate diffusion processes and surfactant dynamics, which is known to be important for emulsification in microdevices.

In order to model emulsification processes starting from the atomic scale the Lattice Boltzmann theory (see also chapter 4.2) can be used to model droplet breakup with micro engineered membranes. A complicated task is to implement mathematic tools in the CFD model to describe properly the dynamics at the interface of the dispersed and continuous phase. During the last ten years the Lattice Boltzmann Equation (LBE) method has been developed as an alternative numerical approach in computational fluid dynamics. Originated from the discrete kinetic theory, the LBE method has emerged with the promise to become a superior modelling platform, both computationally and conceptually, compared to the existing arsenal of the continuum-based CFD methods. The LBE method has been applied for simulation of various kinds of fluid flows under different conditions. The number of papers on the LBE method and its applications continues to grow rapidly, especially in the direction of complex and multiphase media. From its birth over 10 years ago [21], the lattice

Boltzmann Equation (LBE) method has been pursued at a pace that is strongly accelerating in the past few years. The method has found application in different areas of computational fluid dynamics, including simulation of flows in porous media; non-ideal, binary and ternary complex fluids; microfluidics; particulate and suspension flows [22,23]. Proponents of the LBE method consider the method to possess potentials to become a versatile CFD platform that is superior over the existing, continuum-based CFD methods. At the same time, since the method, and its variants and extensions, are still being formulated and improved, the diverse and growing body of the LBE literature suffers from controversy and lack of distillation.

Another numerical method based on a more detailed atomic description of the diffuse interface is the phase field method, which is based on the Cahn-Hilliard theory of phase separation in fluids [24,25]. In certain phase-separating polymer mixtures the diffuse interface is a physical reality, and is described by the Cahn-Hilliard theory [26]. For multiphase flow the diffuse interface can be used as a numerical tool. Whether the fluid is the dispersed or continuous phase is indicated with an order or density difference parameter $\Phi = (\rho_a - \rho_b)/\rho$. At the interface the order parameter changes value in a region of several grid spacings. Gradients in the order parameter give lead to a capillary pressure, which is incorporated as a body force in the Navier-Stokes equation. The order parameter evolves more or less as a passive scalar, which is transported by the fluid flow. Consequently, the interface does not have to be tracked explicitly. Hence, there is no need for adaptive grid refinements at interfaces as is required in Finite Element and Boundary Element methods.

Cahn-Hilliard theory[*]
The Cahn-Hilliard theory is based on a free-energy functional of the order parameter. From the functional two thermodynamic quantities are derived:
    i) the chemical potential driving the diffusion of the order parameter, and
    ii) the pressure tensor to be inserted in the Navier-Stokes equation, describing the velocity field in both the dispersed and continuous phase. The capillary pressure due to the (curved) interface is incorporated in the pressure tensor, and is related to the gradient in the order parameter. The use of the Cahn-Hilliard theory in the context of Lattice Boltzmann schemes has been pioneered by Julia Yeomans and coworker [27], and has been shown to describe droplet breakup.
The Cahn-Hilliard theory is an example of a time-dependent Ginzburg-Landau theory, which is frequently used to describe microemulsions with

---

[*] Contribution from R.G.M. van der Sman, Food Process Technology, University of Wageningen.

surfactants [28]. It can also be used at a larger scale, using the diffuse interface again as a numerical tool, to describe coalescence of emulsion droplets (Ostwald ripening). Hence, this seems a promising route to model surfactant dynamics. Also Diamant et al. [29] view the free energy approach a valuable tool to model surfactant dynamics.

For a homogeneous immiscible binary fluid (oil water mixture) with phase densities ρ_a and ρ_b the bulk free energy density ψ_0 is given by Flory-Huggins:

$$\psi_0 = kT[\rho_a \, ln\rho_a + \rho_b \, ln\rho_b + X_{ab}\rho_a\rho_b] \tag{1}$$

Instead of densities ρ_a and ρ_b, the free energy is described by total density $\rho_o = \rho_a + \rho_b$ and the density difference $\Phi = (\rho_a - \rho_b)/\rho$. A frequently used approximation of the Flory-Huggins free energy expression is the following (under the assumption of a constant density ρ_o):

$$\Psi_0 = -\frac{A}{2}\Phi^2 + \frac{B}{4}\Phi^4 \tag{2}$$

For an inhomogeneous fluid (having interfaces between the immiscible phases) one should include a gradient term in the free energy according to van der Waals, cf. [30]:

$$\Psi = \Psi_0(\rho,\Phi) + \frac{\kappa}{2}(\nabla\Phi)^2 \tag{3}$$

κ is a parameter, which is linked to the surface tension.

The chemical potential is the driving force for diffusion of the order parameter Φ:

$$\mu = \frac{\delta\Psi}{\delta\Phi} = -A\Phi + B\Phi^3 - \kappa\nabla^2\Phi \tag{4}$$

Next to the chemical potential the capillary stress tensor has to be defined, which is inserted in the Navier-Stokes equation. The scalar part of the stress tensor (the hydrostatic pressure) p_o follows from the thermodynamic definition [31]:

$$p_0 = \Phi\mu - \Psi \tag{5}$$

Substitution of the expression of the free energy density Ψ gives:

$$p_0 = \rho c_s^2 - \frac{A}{2}\Phi^2 + \frac{3B}{4}\Phi^4 - \kappa\Phi\nabla^2\Phi - \frac{\kappa}{2}(\nabla\Phi)^2 \tag{6}$$

c_s^2 is directed related to the viscosity (cf. Lattice Boltzmann method, chapter 4).

For the complete stress tensor, $P_{\alpha\beta}$ one has to ensure that it obeys the condition of mechanical equilibrium, namely that it is divergence-free: $\partial_\beta P_{\alpha\beta}=0$. A suitable choice is:

$$P_{\alpha\beta} = p_0\delta_{\alpha\beta} + \kappa(\partial_\alpha\Phi)(\partial_\beta\Phi) \tag{7}$$

Note that here no summation over double indices is implied. Hence, $P_{xx} \neq P_{yy}$. The evolution of the velocity field $\rho_0 v$ and the order parameter Φ are described respectively by the Navier-Stokes equation (with the capillary stress tensor) and the convection-diffusion equation (with the diffusive flux proportional to the gradient in the chemical potential).

CFD modelling of cross flow emulsification through a single pore [*][1]

The shape of an oil droplet forming at a single cylindrical pore will be calculated with the CFD finite volume tool CFX [17]. A simulation will be given taking into account the effect of droplet shape and on the oil flux through the pore.

Current models only assess the process of membrane emulsification with a single pore. This has given very useful results. It was found that use of modern simulation techniques such as finite volume methods in fact give a reliable simulation of the actual process, describing the influences of process parameters of membrane morphology. In case of multiple pore configurations growing droplets can push each other away, leading to premature detachment of some droplets and creation of very large droplets at other places [32].

To design an optimal membrane for emulsification, it is important to know the shape of the droplets during their formation. The shape will determine, for instance, whether droplets coalesce at a given membrane surface. However, this shape cannot be calculated using overall global force equations. The shape changes locally due to variations in viscous and inertial forces. For example, the greater the distance from the membrane, the higher is the velocity of the continuous phase. As a result, local forces on the droplet vary in place and time, resulting in a changing droplet shape. Of course, this shape also depends on the values of other parameters such as dynamic interfacial tension and transmembrane pressure.

---

[*] Partially reprinted with permission from AIChe Journal, Abrahamse et al. (2001).

Model Set-up

The CFD software package used in this research was CFX 4.2 AEA Technology, UK, which is a finite-volume code.

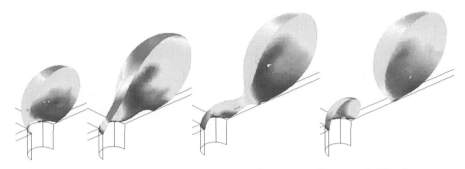

Fig. 6   3D CFD simulation of droplet break-up in a cross flow emulsification process. Only half of the droplet was modelled, since the droplet and the fluid flow is symmetrical with respect to centre plane of the pore.

Oil flowed through a cylindrical pore and water through a rectangular flow channel. The cylindrical pore had a diameter and a length of 5 μm. Water flowed through the inlet of the flow channel in the x-direction with an average velocity of 0.5 m/s. A laminar parabolic velocity profile at the inlet was generated. At the outlet of the flow channel the pressure was set at 1.0×10^5 Pa.   At the inlet of the pore the pressure was set at 1.3×10^5 Pa. The other geometry boundaries were walls, except for the symmetry planes y =0 and z=50 μm. The initial conditions were programmed in Fortran. At t=0 s the flow channel was filled with luminary flowing water and the pore was completely filled with oil. At t = 0 s the interface between oil and water was flat.

By calculating the oil flux, the droplet volume increase is calculated. From the droplet volume a corresponding equivalent sphere diameter was calculated, which could be used to iterate the Laplace pressure. This equivalent Laplace pressure is an approximation, because with the CFD calculations it was shown that the droplet is not fully spherical. However, the equivalent Laplace pressure agrees quite well within 20 % with the pressure drop over the interface in the part of the droplet that is almost spherical. The model was validated against a series of test cases, including free-surface flows and wall effects. The assumption of the validity of the calculations of the implemented interfacial tension force [19,33] seems reasonable.

3. ANALYTICAL CROSS-FLOW EMULSIFICATION MODEL [34]

Based on analysis of microscopic and experimental observations of the cross-flow emulsification process a more detailed understanding of the basic principles of droplet break-up will be presented and verified with known experimental studies and new experiments with well defined micro engineered membranes.

If an emulsion droplet grows a number of relevant forces are acting on the droplet that tries to hold or to remove the droplet from the pore. The main force F_γ holding the droplet is the surface tension $\gamma(t)$ multiplied by the circumference of the neck [6].

$$F_\gamma = \pi \cdot D_n \cdot \gamma(t) \tag{1}$$

with D_n: Diameter of the neck of the droplet.

The surface tension $\gamma(t)$ of the droplet surface may vary in time, due for instance to adsorption of added emulsifier from the continuous phase. The neck diameter of the droplet is proportional to the pore diameter and depending on the forces acting may be larger or smaller than the pore diameter D_p. During the growing of the droplet the diameter of the neck of the droplet will decrease.

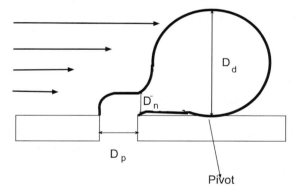

Fig. 7 Droplet with neck configuration just before neck break-up and droplet detachment from the membrane. The total force/torque   balance is relatively complicated, see also [35].

The neck will also slowly bend from a vertical towards a horizontal position with respect to the membrane surface. In case F_γ is counterbalanced by a larger force, then the neck will break and the droplet will detach from the pore.

The main force that tries to remove the droplet from the pore is the drag force F_d (see chapter 7.2, cross-flow large particle model):

$$F_{drag} = 5.1\pi\eta v_c D_d = 15.3 \frac{\pi\eta_c <v> D_d^2}{h} \tag{2}$$

With η_c Viscosity of the continuous phase
        v_c Velocity of the continuous phase at half the height of the droplet
        $<v>$ Mean velocity of the continuous phase in the channel
        h Height of the channel
        D_d Mean droplet diameter

At high cross flow velocities the wall shear stress will grow and a Lift Force F_L [36] will be induced to lift the droplet from the pore.

$$F_L = 0.761 \cdot \frac{\tau_w^{1.5} \cdot D_d^3 \cdot \rho_c^{0.5}}{\eta_c} \tag{3}$$

With τ_w Wall shear stress $\tau_w = \eta \, \partial v / \partial x$

           (laminar flow channel: $\tau_w = \dfrac{2v_c\eta}{D_d} = \dfrac{6<v>\eta_c}{h}$)

           (laminar flow tube: $\tau_w = \dfrac{2v_c\eta}{D_d} = \dfrac{8<v>\eta_c}{D_t}$)

        σ_c Mass density of continuous phase
        D_t Diameter tube

There is also a static pressure difference (Laplace pressure) between the inner droplet P_D (double curvature) and the inner neck P_N (single curvature) part:

$$P_D = \frac{4 \cdot \gamma(t)}{D_d} \qquad \text{and} \qquad P_N = \frac{2 \cdot \gamma(t)}{D_n} \tag{4}$$

The Laplace pressure of the droplet will decrease when the droplet is growing and will contribute to the rate of filling of the droplet with the dispersed phase. The rate of filling of the droplet is determined by the transmembrane pressure P_{TMP}, the pressure drop over the pore ΔP_p, the pressure drop over the neck ΔP_n, and the static pressure P_D in the droplet:

$$P_{TMP} = \Delta P_p + \Delta P_n + P_D \tag{5}$$

The flow rate Φ_d of the dispersed phase in the droplet with a mean velocity v_d will create a pressure drop over the pore (diameter D_p and length L_p) and the neck (diameter D_n and length L_n) given by:

$$\Phi_d = \frac{\pi D_p^{\ 4} \Delta P_p}{128 L_p \eta_d} = \frac{\pi D_n^{\ 4} \Delta P_n}{128 L_n \eta_d} \qquad (6)$$

with η_d the viscosity of the dispersed phase

In case of a long pore length (ceramic membrane approximation) and droplet detachment is near then the pressure drop over the membrane is substantially higher than the pressure drop over the neck, and the former equation reduces to:

$$\Phi_d = v_d \frac{\pi D_p^{\ 2}}{4} = \frac{\pi D_p^{\ 4} (P_{TMP} - P_d)}{128 L_p \eta_d} \qquad (7)$$

The inflow of the dispersed phase through the pore with mean velocity $v_d = v_p$ will result in a viscous pressure contribution that may keep the neck longer open and may retard the moment of droplet detachment. It has been observed that at larger transmembrane pressures, i.e. a higher velocity v_d of the dispersed phase, the emulsion droplets are larger than at a lower transmembrane pressure (see Fig. 8). In order to estimate the contribution of this effect a comparison has to be made between this viscous pressure contribution and the fluid drag perturbation pressure exerted on the neck.

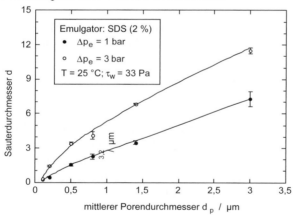

Fig. 8 Increase of dropletsize ratio at larger transmembrane pressures [6].

The minimum perturbation pressure needed to make a perturbation in the neck with a curvature with a diameter of about the length of the neck (about three times Dn) is about $2\gamma/3Dn$. So:

$$\Delta P_n = \frac{32 L_n \eta_d}{D_n^{\ 2}} v_n \qquad \text{and} \qquad P_{Perturbation,neck} = \frac{2\gamma}{3 D_n} \qquad (8)$$

A prediction can be made that substantially larger emulsion droplets will be formed in case $\Delta P_n > \Delta P_{Perturbation,,neck}$ or:

$$v_n > \frac{D_n \gamma}{12 L_n \eta_d} \tag{9}$$

If one presumes that γ may vary between 0.07 (pure water) and 0.003 N/m (SDS 2 %), that η_d may vary between 0.001 (water) and 0.7 (nut oil) Pa.s and length L_n is about 3 times the diameter D_n then one may expect such an effect at dispersed phase velocities v_n in the neck larger than 1×10^{-4} to 2 m/s.

It is difficult to verify if the increased droplet size at larger transmembrane pressure from experiments with ceramic membranes is solely due to the predicted effect. Due to unknown values of droplet forming times it is not possible to estimate mean values for the velocity of the disperse phase in the neck in this case.

In case of a short pore length (microsieve approximation) and droplet detachment is about to happen then the pressure drop over the neck is substantially higher than the pressure drop over the pore and the former equations reduces to:

$$\Phi_d = v_d \frac{\pi D_n^2}{4} = \frac{\pi D_n^4 (P_{TMP} - P_d)}{128 L_n \eta_d} \tag{10}$$

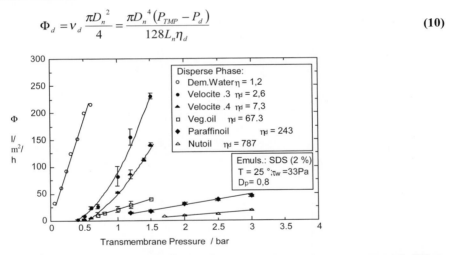

Fig. 9 Flux as a function of the transmembrane pressure for a number of different disperse phase fluxes [6].

In case the inflow velocity of the dispersed phase in the neck v_n becomes very large, a Bernoulli under-pressure drop $\Delta P_{Bernouilli,neck} = \frac{1}{2} \rho_d v_n^2$ will be created in the neck that will counterbalance the viscous pressure contribution ΔP_n in the neck and instead of larger emulsion droplets one can now expect the detachment of smaller emulsion droplets at velocities v_n:

$$v_n > \frac{64 L_n \eta_d}{D_n^2 \rho_d} \tag{11}$$

If one presumes that η_d may vary between 0.001 and 0.7 Pa.s and ρ_d is about 1000 kg/m³, the length L_n is about 3 times the diameter D_n and if D_n is about 1 micron than one may expect a $\Delta P_{Bernoulli}$ effect at dispersed phase velocities v_n in the neck larger than 200 to 135.000 m/s. Such high velocities are not realistic and this effect will be very difficult to observe at pore sizes of about 1 micron. However at very large pore diameters (>20-50 micron) this effect may be observed at much slower dispersed phase velocities.

Neck formation

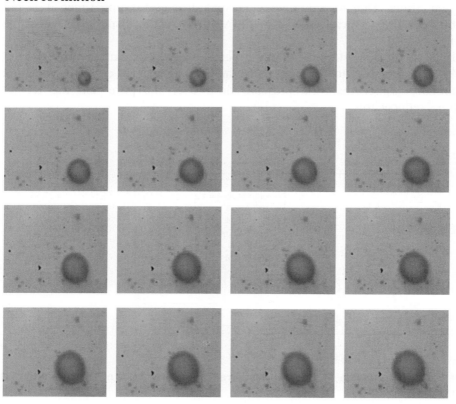

Fig. 10 Subsequent growing stages of a droplet sunflower oil in a Tween 20 (0.5%) aqueous solution. The shear rate is 5 Pa, the pore diameter is 2 micron, the pore Laplace pressure is about 0.1 bar, the applied transmembrane pressure is 0.5 bar. Time frame is 40 msec/shot; at shot 8 a neck is visible at a corresponding droplet size of 20 micron. Just before detachment at shot 16 the droplet size has increased to 25 micron diameter and a neck is visible with a diameter of approximately 3 micron.

The droplet is continuously filled through the pore. Initially the droplet is small and directly attached to the pore with the surface tension force. When the droplet grows the drag force tries to remove the droplet from the pore exit, the droplet first becomes a little a-spherical and a short neck is formed. The total interfacial surface of the droplet including the neck strives continuously towards a minimum energy and hence a minimum surface area. This would theoretically be a perfect spherical droplet with a given diameter and a (short) neck with a diameter as small as possible. However the diameter of the neck cannot be zero, because the neck serves to transport fluid and also in thermodynamic equilibrium (Laplace condition eq. (4)) its minimum diameter is also determined by the pressure profile in the neck.

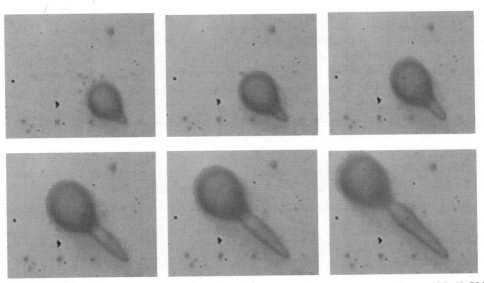

Fig. 11 Subsequent growing stages of a droplet sunflower oil in a Tween 20 (0.5%) aqueous solution. The shear rate is 5 Pa, the pore diameter is 2 micron, the transmembrane pressure is 3 bar. Time frame is 40 msec/shot, at all shots clearly a neck is visible with a maximum diameter of about 10 micron. Just before detachment at shot 6 the droplet size has increased to 30 micron diameter and the neck diameter near the droplet has strongly decreased.

The shape and length of the neck is continuously changing during filling of the droplet and is determined by the disperse phase velocity, the pressure difference as well as the value of the absolute pressures at the beginning and the end of the neck. Especially at the end of the droplet filling process the absolute pressure in the droplet has become so small that more fluid at the end of the neck may be dragged into the droplet than there is feed of fluid through the pore, the diameter of the neck near the droplet will then decrease and detachment of the droplet may occur.

Model assumptions

In order to make some qualitative and quantitative analytical predictions some model assumptions have to be made:

1) low disperse phase velocity (transmembrane pressure is only slightly higher than the pore Laplace pressure).
2) rotation (moment) forces on droplet are being neglected
3) no adhesion between the droplet and the membrane surface (perfect non-wetting condition)
4) Rayleigh break-up criterion

Initially the droplet is small and directly attached to the pore with the surface tension force $F_\gamma = \pi \cdot D_n \cdot \gamma(t)$, eq. (1). When the droplet grows the drag force, eq. (2) will counterbalance at a certain point this surface tension force. At this point a neck with a diameter equal to the pore diameter can be formed continuously with a speed at about the velocity of the continuous phase at half the height of the droplet, v_c. However as soon as the length of this neck is about three times the neck diameter a Rayleigh type break-up process will cause detachment of the droplet from the neck (see former chapter 10.1). The conclusion to be drawn from Rayleigh's analysis of the break-up of a non-viscous liquid jet under laminar flow conditions is that all disturbances on a cylindrical fluid element with a length greater than its circumference will grow and will eventually lead to break-up from a neighbouring fluid element.

The time ΔT needed to form a neck with such a length can be estimated to be

$$\Delta T = \frac{6\eta_c}{\tau_w} \qquad (12)$$

This neck forming time is typical in the submilliseconds regime for o/w emulsions, which is considerable slower than the timescale $1/q$ of the Rayleigh oscillations/perturbations (Mhz domain), but considerable fast in comparison to cross flow emulsification droplet forming times in the order of many milliseconds.

Just before droplet detachment, the drag force F_d should equal approximately the surface tension force F_γ exerted along the (horizontal) neck. In this approximation torque effects are neglected. Substituting expressions (1) and (2) one obtains

$$15.3 \frac{\pi \eta_c <v> D_d^2}{h} = \pi \cdot D_n \cdot \gamma(t) \qquad (13)$$

If one substitutes the former definition of the wall shear stress for the laminar flow regime for a channel with a height h or for a tube with diameter D_t, the approximation $D_n = D_p$, and the definition of the Laplace pressure at the pore $P_p = 4\gamma/D_p$, one obtains the simple expressions:

$$\frac{P_p}{\tau_w} = 10.2\left(\frac{D_d}{D_p}\right)^2 \qquad \textbf{for a channel and for a tube} \qquad \frac{P_p}{\tau_w} = 13.6\left(\frac{D_d}{D_p}\right)^2 \qquad (14)$$

So the presented membrane emulsification model predicts that the ratio of the Laplace pore pressure P_p / τ_w scales with the second power of the ratio between the droplet diameter D_d and the pore diameter D_p. It will be clear that this model is only valid if the droplet diameter D_d is substantially larger than the pore diameter D_p and thus in the regime that the Pascal pore pressure P_p is substantially larger than the wall shear stress τ_w.

In view of the crude approximations that have been made the constants 10, 13.6 in the second power size ratio-pressure ratio formula must be seen as an indication of its real value. Scaling [37] of the relevant droplet break-up forces will give in any way a simple relation between the relevant parameters determining the break-up process.

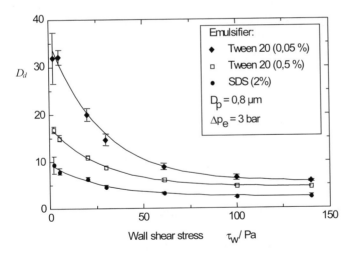

Fig. 12 Droplet size versus wall shear stress for an alumina membrane [6].

Experimental verification

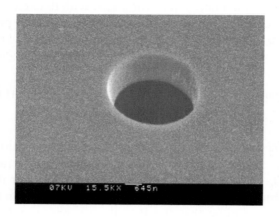

Fig. 13 SEM micrograph of a micro engineered micro-orifice, courtesy Aquamarijn.

Cross flow emulsification experiments have been performed with use of a rectangular cross flow channel with a height of 350 micron and a width of 2.5 cm at room temperature conditions. Sunflower oil has been used for the disperse phase and a 0.5 % Tween 20 aqueous solution for the continuous phase. The micro engineered silicon nitride orifice has a diameter of 4.0 ± 0.2 micron. Before each experiment the orifice was subjected to a RF oxygen plasma to obtain maximum hydrophilicity of the membrane surface. The Laplace pressure of the orifice was experimentally determined at 5000 ± 100 Pascal.

Table 1

Droplet size as a function of the shear rate for a micro-orifice membrane.

Micro-orifice	Shear rate τ_w	Trans-Membrane Pressure	Droplet size
	Pascal	Pascal	Micron
	0.4	6100	120
	0.9	6000	95
	2	5900	71
	4	5900	58
	8	5900	46
	12	5900	40
	16	5900	28

Experiments with the micro-orifice were well reproducible provided that the oxygen plasma treatment was performed, otherwise larger droplet sizes were found and smaller Laplace pressures indicating an increase of the effective pore size, possibly to spreading of the oil outside the pore on the membrane surface. In order to circumvent this complication we performed experiments with a microengineered micro-tube slightly protruding out of a shallow microwell.

Fig. 14 SEM micrograph of a micro engineered micro-tube, courtesy Aquamarijn.

Cross flow emulsification experiments have been performed under similar conditions as the former experiment. The micro engineered silicon nitride microtube has an (outer) diameter of 4.0 ± 0.2 micron. The Laplace pressure of the orifice was experimentally determined at 4800 ± 100 Pascal. Experiments with the micro-tube were well reproducible also for longer periods without the use of an additional RF oxygen plasma treatment.

Table 2

Droplet size as a function of the shear rate for a micro-tube membrane.

Micro-tube	Shear rate τ_w Pascal	Trans-Membrane Pressure Pascal	Droplet size Micron
	0,4	6000	120
	0,9	6000	102
	2	6000	59
	4	6000	48
	8	6000	29
	16	5900	19

Schröder et al. [6,38] have also obtained many experimental data with many variations in the wall shear stress τ_w, the D_d / D_p ratio of the droplet diameter and the pore diameter, the viscosity of the continuous as well as the dispersed phase, and mostly for sintered ceramic membrane tubes, see Fig. 12.

The size of the emulsion droplet may also be dependent on the rate of adsorption of the emulsifier at the droplet interface. From Fig. 15 it is seen that the scaling relation Eq.(14) reasonably fits well with the experimental data. Results obtained with the micro-tube structure seems to give better concordance with Eq.(14) than the micro-orifice structure. The latter results might be influenced by a little spreading of the oil outside the pore on the membrane surface, leading to larger droplets.

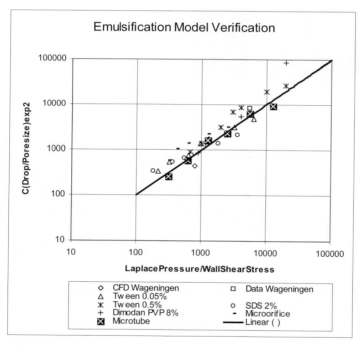

Fig. 15 Fitting of Eq. 14 (Linear line) with some experimental data, ◊: CFD Abrahamse et al. [1], : Data Abrahamse et al. [32], Δ,*,o: Schröder et al.[38], +: Joscelyne and Tragardh [4], -: Micro-orifice (Table 1), and ⊠ : Micro-tube (Table 2). At small wall shear stress values larger droplet sizes may be expected due to droplet coalescence.

Results obtained with the micro-tube (⊠) structure seems to give better concordance with Eq.(14) than the micro-orifice (-) structure, see Fig. 15. The latter results might be influenced by the already mentioned little spreading of the oil outside the pore on the membrane surface, leading to larger droplets.

Table 3 lists the dynamic interfacial tensions that have been used to fit the data and are all in reasonable agreement with the static interfacial tensions. For a very low concentration of surfactant (Tween 0.05%) it may be expected that the dynamic interfacial tension in the neck during droplet growth is higher due to the diminished rate of adsorption of the surfactant.

Table 3

Static interfacial tension forces and fitted dynamic interfacial tension values in Fig 11.

Oil/Detergent	Static Inter facial Tension mN/m	Fitted Dynamic Inter facial Tension mN/m
Tween 0.5 %	6.4	5.0
Tween 0.05 %	>10	20.0
SDS 2 %	3.6	3.6
Dimodan PVP 8 %	3.0	3.0

With this analytical model it therefore seems possible to obtain good estimated values of the dynamic interfacial tension of the neck during droplet growth. For the presented data no clear deviation between the static and fitted dynamic interfacial tension may be noticed. In a subsequent study new experimental material will be presented that may reveal such deviation and also the effect of the transmembrane pressure on droplet size will be a topic of further investigation.

REFERENCES

[1] A.J. Abrahamse, A. van der Padt, R.M. Boom, Process fundamentals of membrane emulsification: Emulsification with Computer Fluid Dynamics, AIChe Journal, Vol. 47, No. 6 (2001) 1285-1291.

[2] Suzuki et al., JP06039259.

[3] EP0452140 B1, Method for producing low-fat spread and oil-in-water-in-oil type spread, Morinaga Milk Industry.

[4] S. M.Joscelyne and G. Tragardh, Food Emulsions Using Membrane Emulsification: Conditions for Producing Small Droplets, J. Food Eng. 39 (1999) 59.

[5] V. Schroder, O. Behrend and H. Schubert, Effect of Dynamic Interfacial Tension on the Emulsification Process Using Micro-porous Ceramic Membranes, J. Colloid Interface Sci. 202 (1998) 334.

[6] V. Schroder, Herstellen von Ol-in-Wasser-emulsionen mit Mikroporosen Membranen, Ph.D-Thesis, Technische Hochschule, Karlsruhe, 1999 .

[7] S.Omi, Preparation of Monodisperse Microspheres Using the Shirasu Porous Glass Emulsification Technique, Colloids Surf. A: Physicochem. Eng. Aspects 109 (1995) 97.

[8] S. Sugiura, M. Nakajima, S. Iwamoto and M. Seki, Interfacial Tension Driven Monodispersed Droplet Formation From Microfabricated Channel Array, Langmuir, 17 (2001) 5562.

[9] S. Sugiura, M. Nakajima and M. Seki, Prediction of Droplet Diameter for Microchannel Emulsification, Langmuir, 18 (2002) 3854.

[10] S. Sugiura, M. Nakajima and M. Seki, Preparation of Monodispersed Emulsion With Large Droplets Using Microchannel Emulsification, J. Am. Oil Chem. Soc. 79 (2002) 515.

[11] T. Kawakatsu, Y. Kikuchi and M. Nakajima, Visualization of Microfiltration Phenomena Using Microscope Video System and Silicon Microchannels, J. Chem. Eng. Jpn 29 (1996) 399.

[12] R.A. Williams, S.J. Peng, D.A. Wheeler, N.C. Morley, D. Taylor, M. Whalley and D.W. Houldsworth, Controlled production of emulsions using a cross-flow membrane. Part II: industrial scale manufacture, Trans. Inst. Chem. Eng. 76 (1998) 902.

[13] http://www.cfd-online.com/Resources/soft.html.

[14] http://213.210.25.174/phoenics/d_polis/d_info/phover.htm.

[15] S.V.Patankar .Numerical Heat Transfer and Fluid Flow. Series in computational methods in mechanics and thermal sciences. McGraw-Hill,1980.

[16] http://www.fluent.com/software/fidap/.

[17] http://www.software.aeat.com/cfx/.

[18] http://www.adaptive-research.com/.

[19] J.U. Brackbill, D.B. Kothe, and C. Zemach. A continuum method for modelling surface tension. J. Comput. Phys. 100(2): 335-354 (1992).

[20] J. Li and Y. Renardy. Shear-induced rupturing of a viscous drop in a Bingham liquid. J. Non-Newtonian Fluid Mech. 95: 235-251 (2000).

[21] McNamara, G.R. and Zanetti, G., Use of the Boltzmann Equation to Simulate Lattice-Gas Automata, Phys. Rev. Lett., 61, Number 20, (1988), 2332.

[22] Chen, S., and Doolen, G. D., Lattice Boltzmann Method for Fluid Flows, Annu. Rev. Fluid Mech., 30 (4), 329-364.

[23] Rothman, D.H., and Zaleski, S., Lattice-Gas Cellular Automata: Simple models of complex hydrodynamics, Cambridge University Press, 1997.

[24] D. Jacqmin. Calculation of two-phase Navier-Stokes flows using phase-field modeling. J. Comput. Phys. 155(1): 96-127 (1999).

[25] D.M. Anderson, G.B. McFadden, and A.A. Wheeler. Diffusive-interface methods in fluid mechanics. Ann. Rev. Fluid Mech. 30:193-165 (1998).

[26] J.W. Cahn and J.E. Hilliard. Free energy of a non-uniform system I. Interfacial energy. J. Chem. Phys. 28:258-267 (1958).

[27]M.R. Swift, E. Orlandini, W.R. Osborne and J.M. Yeomans. Lattice Boltz-mann simulations of liquid-gas and binary fluid mixtures. Phys. Rev. E 54: 5041-5052 (1996).

[28]G. Gompper and M. Schick. Self-assembing amphiphilic systems. Academic Press, London (1994).

[29]H. Diamant, G. Ariel, and D. Andelman. Kinetics of surfactant adsorption: the free energy approach. Colloids Surf. A 183-185: 259-276 (2001).

[30]M.R. Swift, E. Orlandini, W.R. Osborne and J.M. Yeomans. Lattice Boltz-mann simulations of liquid-gas and binary fluid mixtures. Phys. Rev. E 54: 5041-5052 (1996).

[31]Lamura, G. Gonella, and J.M. Yeomans. A Lattice Boltzmann model of ternary fluid mixtures. Europhys. Lett. 45(3):314-320 (1999).

[32]A.J. Abrahamse, R. van Lierop, R.G.M. vander Sman, A. van der Padt and R.M. Boom, Ánalysis of droplet formation and interaction during cross-flow membrane emulsification, J. Membr. Sc. 5247 (2002) 1-13.

[33] D. J. Burt, J. W. J. Ferguson and H. Pordal, Numerical Computation of Surface Tension Effects, Proc. ASME 3 (1996) 439.

[34] C.J.M. van Rijn et al., to be submitted to J. of Membrane Science.

[35] S.J. Peng, R.A. Williams, Controlled production of emulsions using a cross flow membrane, Part I: Droplet formation from a single pore, Trans IchemE, 76, (1998) 894-901.

[36] G.Rubin, Widerstands- und Auftriebsbeiwerte von ruhenden, kugel-förmigen Partikeln in stationären, wandnahen laminaren Grenz-schichten, Dissertation Universität Karlsruhe, 1977.

[37] P.G. de Gennes, Scaling Concepts in Polymer Physics, 1979, Cornell University, ISBN 0180141203X.

[38]V.Schröder and H. Schubert, Production of Emulsions Using Microporous Ceramic Membranes, Colloids Surf. A: Physicochem. Eng. Aspects 152 (1999) 103.

Relevant Author's Publications

Journals

Rijn, C.J.M. van., Wekken, M. van der., Nijdam, W., & Elwenspoek, M.(1997), 'Deflection and maximum load of microfiltration membrane sieve made with silicon micromachining.' Journal of microelectromechanical systems, 6 (nr: 1), (pp. 48-54). ISSN 1057-7157.

Rijn, C.J.M. van., Nijdam, W., Kuiper, S., Veldhuis, G.J., Wolferen, H.A.G.M. van., & Elwenspoek, M., (1999), 'Microsieves made with laser interference lithography for micro-filtration applications', Journal of micromechanics and microengineering, 9 (nr: 2), (pp. 170-172). ISSN 0960-1317.

Rijn, C.J.M. van., & Nijdam, W.(1995). Microfiltratie. Mikroniek, (nr: 2), (pp. 43-46). ISSN 0026-3699.

Rijn, C.J.M. van., Veldhuis, G.J., & Kuiper, S.(1998), 'Nanosieves with microsystem technology for microfiltration applications', Nanotechnology, (nr: 9), (pp. 343-345). ISSN 0957-4484.

Rijn, C.J.M. van, Raspe, O.J.A. en Nijdam, W (2000) Voedingsmiddelentechnologie, 'Snelle microbiologische detectie met microzeven', (nr. 21), pp.67 - 68

Kuiper, S., Rijn, C.J.M. van., Nijdam, W., & Elwenspoek, M., (1998), 'Development and applications of very high flux microfiltration membranes', Journal of membrane science, (nr: 150), (pp. 1-8). ISSN 0376-7388.

Exter, M.J. den., Bekkum, H. van., Rijn, C.J.M. van., Kapteijn, F., Moulijn, J.A., Schellevis, H., & Beenakker, C.I.N.(1997). 'Stability of oriented silicalite-1 films in view of zeolite membrane preparation', Zeolites, (nr: 19), (pp. 13-20). ISSN 0144-2449.

Conference contributions

Rijn, C.J.M. van., Nijdam, W., & Elwenspoek, M.(1996, October 31), 'A microsieve for leukocyte depletion of erythrocyte concentrates', [Proceedings IEEE Engineering in medicine and biology society], 2 pp, Amsterdam, ISBN CD-ROM 90-9010005-9.

Rijn, C.J.M. van., Nijdam, W., & Elwenspoek, M.(1995, November 15), 'High flowrate microsieve for biomedical applications', [1995 ASME International mechanical engineering congress and exposition], 2 pp. San Fransisco.

Rijn, C.J.M. van., Nijdam, W., Stappen, L.A.V.G. van der., Raspe, O.J.A., Broens, L., & Hoof, S.(1997, May 23), 'Innovation in yeast cell filtration: cost saving technology with high flux membranes', [Proceedings of European Brewery Conference 1997], (501-507), Maastricht.

Rijn, C.J.M. van., & Elwenspoek, M.(1995, January 29), 'Micro filtration membrane sieve with silicon micro machining for industrial and biomedical applications', [Micro Electro Mechanical Systems (MEMS) '95], (pp. 83-87), Amsterdam.

Rijn, C.J.M. van., Nijdam, W., Kuiper, S., Veldhuis, G.J., Wolferen, H.A.G.M. van., & Elwenspoek, M.(1998, June 03), 'Microsieves made with laser interference lithography for micro filtration applications', [Proceedings of the Micromechanics Europe 1998 Conference (MME)],(pp. 100-103), Ulvik, Norway.

Rijn, C.J.M. van., Kuiper, S., Nijdam, W., & Elwenspoek, M.C.(1997, June 23), 'Microsieves: development and applications of new high flux micro filtration membranes', [Proceedings of Euromembrane '97 Symposium], (pp. 437-440), Enschede. ISBN 90-356-09726.

Wissink, J.M., Nijdam, W., Heskamp, I., Rijn, C.J.M. van , 'Smart microengineered nozzles for monodispers aerosol generation with Rayleigh break-up', Respiratory Drug Delivery VIII, Tuscon, Arizona 12-16 may 2002.

Co-inventor and author of the following patent applications:

Rijn, C.J.M. van, WO9513860, Publication date: 26 Mai 1995, 'Membrane filter as well as a method of manufacturing the same'.

Elders, J. & Rijn, C.J.M. van., NL 1001220, Publication date: 03 June 1996, 'Mould, as well as a method to produce such a mould'.

Rijn, C.J.M. van, & Nijdam, W, WO0218058 , Publication date: 7 March 2002, 'Nozzle device for atomisation and methods for using the same'.

Wessling, M., Rijn, C.J.M. van, Barsema, J.N., & Nijdam, W., NL1016779, 2 december 2000, 'Mould, method for fabrication of precision products using a mould, as well as precision products, in particularly microsieves and filtration membranes, made with such mould'.

Rijn, C.J.M. van, Vogelaar, L., Nijdam, W., Barsema, J.N., & Wessling, M., WO0243937 , Publication date: 6 June 2002, 'Method of making a product with a micro or nano size structure'.

Co-inventor of the following patent applications

Rijn, C.J.M. van, Raspe, O., Hoof, S.C.J.van, EP19981125, Publication date: 25 November1998, Koninklijke Grolsch N.V. , 'Device for filtering a fermented liquid'

Kruithof, C.J., Knol, H., Rijn, C.J.M. van, & Nijdam, W. , WO9740213, Publication date: 30 November 1997, Aquamarijn & Stork Veco, 'Electroforming method, electroforming mandrel, electroformed product'.

Kromkamp J., & Rijn, C.J.M. van, WO0209527, Publication date: 7 July 2002, Friesland Brands BV, 'Method for filtering milk'.

N